THE ROLE OF STATISTICS IN BUSINESS AND INDUSTRY

THE ROLE OF STATISTICS IN BUSINESS AND INDUSTRY

GERALD J. HAHN
Retiree, GE Company
Schenectady, New York

NECIP DOGANAKSOY
GE Company
Schenectady, New York

With the assistance of William P. Alexander, Tim K. Keyes, William Q. Meeker, and Robert R. Starbuck

WILEY

A JOHN WILEY & SONS, INC., PUBLICATION

Published by John Wiley & Sons, Inc., Hoboken, New Jersey.
Published simultaneously in Canada.

Library of Congress Cataloging-in-Publication Data:
Hahn, Gerald J.
The role of statistics in business and industry / Gerald J. Hahn, Necip Doganaksoy.
p. cm.
Includes bibliographical references and index.
ISBN 978-0-471-21874-6 (pbk.)
1. Mathematical statistics. 2. Commercial statistics. I. Doganaksoy, Necip, 1960– II. Title.
QA276.H2195 2008
519.5–dc22

2008007357

Printed in the United States of America.

10 9 8 7 6 5 4 3 2 1

To Bea and Reyhan for their love and patience and to our children and grandchildren (current and future).

CONTENTS

MAJOR APPLICATIONS

PREFACE

OUR GOAL

We asked a recent Ph.D. in statistics what was the biggest surprise in her first year working as an applied statistician in business and industry. "I was amazed at how little I really knew, even about subjects about which I felt I knew a lot" she answered unhesitatingly. Her response motivated us to write this book.

Most texts on statistics describe the major technical tools (from summary statistics and statistical inference to analysis of variance and multiple regression) upon which the subject is based and then try to show how these are employed. We do the reverse. We describe key problems in business and industry from product design to field support and then show how statistics is used to address them. In doing so, we leverage the insights gained from our combined 70 years in the trenches as applied statisticians. Our major aim is to help readers worldwide become valuable contributors on the job by providing them with insights into the types of problems they are likely to encounter and how statistics is used to address them.

This book should be especially useful to:

- Students in statistics and related areas who want to learn how statistics is used in business and industry.
- Practicing statisticians who wish to increase their effectiveness.
- Those in fields other than statistics, who are, or wish to be, local statistical experts; for example, technically oriented Six Sigma black belts and master black belts.
- MBAs and other specialists who wish to use data for better business decision making.

In addition, this book should be of interest to:

- Students who have been intrigued by an introductory college, or possibly high school, statistics course and are trying to decide whether to go for a career in applied statistics.
- Academicians who want to add more practical applications to courses they teach in statistics.
- Managers who would like to get a better understanding of how their operations can benefit from statistics.
- Teachers of high school Advanced Placement courses in statistics in the U.S. who want to make their courses more meaningful and better guide students.

We address readers with a wide range of backgrounds – only a one-semester introductory course in statistics is assumed. Some technical concepts that we regard as important, but are not typically taught in an introductory course are reviewed, e.g., product life distributions (as an Appendix to Chapter 5) and the design of experiments and simulation (in Chapter 3).

The book can serve as a supplemental text and instructor's guide for introductory courses on applied statistics and statistical consulting. It is also well suited as the prime text for a second course in applied statistics and for self-study by early-career statisticians and practitioners.

TOPICS AND EMPHASIS

This book has three parts:

- Setting the stage: Overview of the role of statistics in general, and in business and industry, in particular.
- Manufactured product applications. Chapters on the role of statistics in product design (concepts and example), reliability assurance, manufacturing quality improvement (concepts and example), further manufacturing applications and product field support.
- Other applications: Chapters on the role of statistics for pharmaceutical products, financial services, business processes and further applications (including the food, beverage and related industries, semiconductor products, telecommunications and statistical imaging).

We place heavy emphasis on getting the right data – a fundamental subject that is often brushed aside in favor of discussions of statistical methods. Also we urge that those applying statistics play a proactive role in projects as team members. This calls, for example, for early involvement during the design of a product and continued participation throughout the product's life cycle.

SCOPE

Our comments reflect our many years as company-wide statistical resources for a global conglomerate (GE). This has involved us in a rich diversity of problems dealing with the design and manufacture of a wide variety of products (e.g., household appliances, locomotives, plastics, turbines, semiconductors). In addition, we have been engaged in service applications both with businesses whose major function is to provide a service (e.g., commercial and consumer finance, television networks) and with service or business applications (e.g., scheduling, marketing, field servicing) for operations that manufacture products.

To broaden the scope further, we have invited four contributing authors to prepare two chapters and critically review the others. They are:

- Robert R. Starbuck (Assistant Vice President, Special Projects, Wyeth Research) who wrote the chapter on pharmaceutical products.

- William P. Alexander (Director of Quantitative Research, Wachovia Securities) and Tim K. Keyes (Manager, Strategic Portfolio Decisioning, Corporate Financial Services, GE) who wrote the chapter on financial services applications.
- Professor William Q. Meeker (Professor of Statistics and Distinguished Professor of Liberal Arts and Sciences at Iowa State University) who added further perspective based on his academic and broad consulting experience.

Also, Colin McCulloch (Statistician, GE Global Research Center) wrote the section on statistical image analysis in the chapter on Further Applications.

Our original intent was to include in this volume a detailed discussion about the environment in which statisticians work today and some of the nontechnical challenges they face, and to provide some added important technical pointers. We have, in fact, drafted much of this material. Due to our (and our editors') desire to keep this book reasonably short, we have decided to leave these topics for future books.

STATISTICIANS AND PRACTITIONERS

In addition to discussing the role of statistics, we talk about what statisticians do. Our concept of a statistician, however, is not limited to those with formal degrees in the subject. It also encompasses those with significant statistical knowledge and understanding, often acquired by a combination of experience and self-study. Such statistically oriented individuals might not regard themselves as statisticians, but are still often considered to be local "statistical experts" by their peers.

We employ the term "practitioners" to refer to those who use statistics, but whose knowledge and interest in the subject is more limited, such as engineers who use point-and-click statistical software to analyze data. The number of such practitioners is increasing rapidly with the improved accessibility to databases and the availability of easy-to-use statistical software.

OTHER FEATURES

We rely heavily on specific examples and use these to illustrate key points. Some of these examples are presented in great detail and, on occasion, precede the discussion of the principles that they illustrate and run throughout an entire chapter. Most are adapted from actual problems that we, or the contributing authors, have encountered.

We conclude most chapters with a summary of "major takeaways" and two types of discussion questions. General questions are directed at readers with limited statistical training. Technical questions are meant for those who have knowledge of statistics beyond the introductory course. They also provide us the opportunity to raise added, and sometimes more complex, issues.

We have also created an ftp site for this book that can be easily located from Wiley's home page. This provides tabulations of data for some of the examples and, in some cases, technical details of their analyses. We also plan to post on this site a, hopefully very short, list of errors, thoughts that occur to us after submittal of the manuscript, and reader comments.

AN INVITATION

We hope that you will enjoy reading this book as much as we enjoyed writing it. In our continuing search for improvement (in possibly future editions), we invite and welcome your comments and suggestions. We can be reached at gerryhahn@yahoo.com and necipdoganaksoy@yahoo.com.

ACKNOWLEDGMENTS

In addition to contributing authors William Alexander, Tim Keyes, William Meeker and Robert Starbuck, and section author Colin McCulloch, we have asked various other knowledgeable people to comment critically on drafts of this book – and to set us straight, where needed. We were most fortunate to have received detailed and highly useful inputs on large parts of this manuscript from Ted Hardwicke, Roger Hoerl, Ron Kenett, Antonio Possolo and Gipsie Ranney. We also received very useful comments on individual chapters from Srinivas Bollapragada, Robert Bress, John Chakmakas, Veronica Czitrom, Ron Frank, Judy Hahn, Lynne Hare, Diane Lambert, Alson Look, Wayne Nelson, Paul Tobias, Scott Vander Wiel, Mark VanDeven and Emmanuel Yashchin. We also thank Kunter Akbay, Mark Bailey, James Cawse, Bill Cheetham, Christy Chuang-Stein, Michael Clayton, George Domingo, Harold Dyck, Don Ellington, Frank Forbes, Al Greene, Adrienne Hahn, Bea Hahn, Canan Hardwicke, Brent Harrington, Thomas Hochkirchen, Kati Illouz, Fred Immermann, Bob Keating, Mary Lanzerotti, Yan Liu, Eric Maass, Bill Makuch, Gary McDonald, Tom Menten, Ron Menton, Rob Mischler, Solon Morgan, Norm Moskowitz, Fanni Natanegara, Walter Offen, Don Olsson, Pete Parker, Frank Rockhold, Charles Sampson, Rasik Shah, Jeff Sheaffer, Vladimir Shper, Muzaffer Sutcu, Roy Tamura, Ben Verschueren, Greg Watson, Gene Wiggs, Tom Willemain, J.D. Williams, Will Yancey and Rui Zhang for their helpful comments and support.

We greatly appreciate the unwavering support and encouragement by Necip's managers at GE, Roger Hoerl and Rusty Irving. We also would like to thank Heather Haselkorn, Jackie Palmieri, Steve Quigley, and Susanne Steitz-Filler – our editors and their associates at Wiley – for their patience, encouragement and helpful comments.

We begin many sections with a quote. Some of these are taken from Gaither and Cavazos-Gaither (1996).

Finally, we thank our wives, Bea Hahn and Reyhan Doganaksoy, for their encouragement, strong support and remarkable perseverance over the years that we spent writing this book.

PART I

SETTING THE STAGE

IN THIS part of the book we provide a brief review of the field and its many applications with introductory chapters on:

- Statistics: An Overview
- Statistics in Business and Industry

In these chapters, as well as the book as a whole, we aim to impart our excitement about statistics and its role in business and industry.

CHAPTER 1

STATISTICS: AN OVERVIEW

1.1 ABOUT THIS CHAPTER

We begin this general overview chapter with a brief discussion on "What is Statistics?" and briefly describe some applications. We then indicate the importance of statistical literacy to our daily lives, discuss why better decisions require better data, and introduce the concept of statistical thinking. We conclude with a brief consideration of the relationship of statistics to mathematics, computer science, and operations research/management science (OR/MS).

1.2 WHAT IS STATISTICS?

> **There are never in nature two beings which are exactly alike.**
> —G.W. Leibniz

The Merriam-Webster Online Collegiate Dictionary defines statistics as "a branch of mathematics dealing with the collection, analysis, interpretation, and presentation of masses of numerical data." Our preference is to exclude the "masses of" from the definition. Even today, many situations still require us to deal with small data sets. We might also quibble about the word "numerical." Statisticians sometimes deal with data on preferences, rankings, grades, and colors that are nonnumerical. Finally, we have strong reservations about statistics being a "branch of mathematics" – more on that shortly. Apart from that, we agree with the dictionary definition.

Jon Kettenring, 1997 President of the American Statistical Association (ASA), has more simply referred to statistics as "the science of learning from data" that "is essential for the proper running of government, central to decision making in industry, and a core component of modern educational curricula at all levels." Sallie McNulty, the 2006 ASA President, describes statistics as "the quintessential interdisciplinary science." We agree.

We live in an uncertain world. We can predict the time of the rising and setting of the sun for a specified location with great precision. But the exact outcomes of most other things in life are far less known. Two cars built on the same manufacturing line, even on the same day, will still differ in performance, reliability, appearance, and so forth. One of our major goals, in fact, is to minimize such variability.

The Role of Statistics in Business and Industry, by Gerald J. Hahn & Necip Doganaksoy

Statistics allows us to quantify uncertainty, and, as a consequence, to express the likelihood of future happenings probabilistically – and, when possible, act to impact such outcomes favorably. This typically involves the gathering of appropriate data and their proper analysis.

1.3 AREAS OF APPLICATION

Statisticians are employed in academia, business and industry, government, and by social and research agencies. Statistics is, in fact, used in almost all fields of human endeavor, ranging from actuarial science to zoology. At the time of this writing, the ASA Internet site lists 45 application areas. And we recently learned of one more with the publication of a book on statistics in musicology (see Beran 2003).

So on what kinds of problems do applied statisticians work?

1.3.1 Automobile Gasoline Consumption

An automobile manufacturer wishes to compare the gasoline consumption of one of its models against that of competitors. The results are to be used as the basis of a possible advertising claim. Samples of different model cars are to be test-driven to obtain a statistically valid comparison. How many cars of each model need to be evaluated? How should these be selected? Under what conditions should they be driven? For how long? Who should do the driving? How can one factor in variability in gasoline? How are the resulting data to be analyzed? And how should an advertising claim be worded so that it is both clear to the audience at which it is directed and valid? And more fundamentally, what is the impact of fuel economy on customer purchasing behavior and satisfaction? Statistical concepts and methods, often as part of a multidisciplinary effort, help answer such questions and many more.

This example suggests a common thread in addressing problems statistically: define the question of interest, obtain appropriate data, analyze the data, and then help make decisions from the findings.

1.3.2 Some Other Examples

Peck, Casella, Cobb, Hoerl, Nolan, Starbuck, and Stern (2005) aim to show a general audience "how statistics can provide an organized way of learning from data and how the ensuing knowledge can be used to address important, social, environmental and economic problems." Table 1.1 shows the topics of the 25 essays in this volume.

Also, Peck, Haugh, and Goodman (1998) provide case studies from 20 different application areas, including biology/environment, medical and health care, pharmaceutical, marketing and survey research, and manufacturing.

Chance and *Stats*, two ASA magazines, give further examples. And the 2,000 or so papers presented at the yearly national Joint Statistical Meetings, listed on ASA's Internet site, provide both examples of current hot topics and overwhelming evidence of the rich diversity of subjects addressed by statistics.

TABLE 1.1 Essays in Peck, Casella, Cobb, Hoerl, Nolan, Starbuck, and Stern (2005)

Public Policy and Social Science
• Statistics in the Courtroom
• The Anatomy of a Pre-election Poll
• Counting and Apportionment: Foundations of America's Democracy
• Evaluating School Choice Programs
• Designing National Health Care Surveys to Inform Health Policy
Science and Technology
• Monitoring Tiger Prey Abundance in the Russian Far East
• Predicting the Africanized Bee Invasion
• Statistics and the War on Spam
• Should You Measure the Radon Concentration in Your Home?
• Statistical Weather Forecasting
• Space Debris: Yet Another Environmental Problem
Biology and Medicine
• Modeling an Outbreak of Anthrax
• The Last Frontier: Understanding the Human Mind
• Leveraging Chance in HIV Research
• Statistical Genetics: Associating Genotypic Differences with Measurable Outcomes
• DNA Fingerprinting
• How many Genes? Mapping Mouse Traits
Business and Industry
• To Catch a Thief: Detecting Cell Phone Fraud
• Reducing Junk Mail Using Data Mining Techniques
• Improving the Accuracy of a Newspaper
• Assessing Product Reliability and Safety
• Randomness in the Stock Market
• Advertising as an Engineering Science
Hobbies and Recreation
• Baseball Decision Making
• Predicting the Quality and Price of Bordeaux Wines

And to add a personal element, we describe briefly in Sidebar 1.1 three interesting, and somewhat unusual, problems in which we have been involved.

SIDEBAR 1.1: ALL IN A DAY'S WORK: SOME MEMORABLE APPLICATIONS

Birds, Mating Calls, and Airplane Crashes

Crashes of jet airliners due to the plane's engine ingesting birds upon landing or takeoff are of obvious concern to the manufacturers of airplane engines. An article in the *New York Times* suggested that birds might mistake the sound emanating from certain model jet engines to be mating calls. As a result, instead of flying away, they fly toward the engine and are trapped in it, resulting in malfunction.

This assertion needed to be tested scientifically. So an experiment was to be conducted, at a carefully selected site to evaluate the claim. This involved playing recordings of mating calls and jet engine noise to groups of birds and observing their behavior. Our job was to help plan an experiment that would yield statistically valid results and to analyze the resulting data. It demonstrated that birds are no fools; they are smart enough to differentiate jet engines from mating calls.

Which Brand Food Do Dogs Prefer?

We have been engaged in various consumer preference studies. One dealt with determining preferences among dogs between different brands of food products. The results were to be used to substantiate a proposed advertising claim. Dogs were offered food from the competing brands by selecting from among various bowls placed in front of them over a period of time. The amount of food consumed from each bowl was subsequently measured. We were asked to assess the statistical validity of the experiment used to make this comparison and of the analysis of the resulting data.

Combating Barnacles

Ships and submarines that spend time in harbors are susceptible to marine growth in the form of barnacles, mussels, algae, and other tiny organisms on their hulls. These limit the vessel's maneuverability and speed. Constant plugging up of water inlets by invading organisms is a particular challenge for nuclear-powered vessels. Various coatings to inhibit permanent attachment of marine organisms to submarines had been developed. We were asked to plan a test to compare the effectiveness of alternative coatings and analyze the results. This involved submerging specimens with different types of coating in water and allowing marine growth to occur. The primary measured response was the force required to remove the organisms from the specimens.

1.3.3 Examples from Business and Industry

> **Statistics, as applied to products and services, is not an end in itself, but exists principally to illuminate the way to progress in other fields that require the proper handling of uncertainty.**
> —*Antonio Possolo*

As a prologue to the discussion in subsequent chapters, we present a few examples of how business and industry use statistics:

- Pharmaceutical companies request the appropriate regulating agency to license a new drug – based upon a statistical analysis of the results of carefully planned studies of the drug's performance in treating a specified condition and of its side effects. The request also specifies what type of patients should (and should not) take the drug, how frequently, and at what dosages.
- Providers of consumer credit use statistical models to decide whether or not to approve new loan requests (and establish maximum allowable credit limits) to provide the best possible balance between business gained from authorizing loans and losses suffered due to payment default. Such models are developed

by relating applicant and loan characteristics to payment performance, based upon the analyses of data from past loans.

- A manufacturer of a chemical product uses recent process information to determine how much catalyst to add to the product mix during manufacture so as to come as close as possible to the desired target performance. The algorithm is based upon past data used to characterize the relationship between the process and performance variables.
- Another manufacturer monitors engine measurements (e.g., oil pressure and temperature, coolant temperature, and fuel consumption) of a locomotive in operation on an ongoing basis so as to trigger preventive maintenance, and, if necessary, shut down the engine to avoid an impending field failure. The statistical algorithm used for this purpose is based upon an analysis of past data; it aims to detect the maximum number of impending failures, while minimizing the false alarm rate.
- Broadcasters of television commercials need decide between selling advertising timeslots in a pre-season auction versus on the "spot market" during the year. Spot market prices are determined by recent ratings of the show during which the commercial is to be aired. The preseason auction price is determined based on the expected audience and provides early cash flow and minimal risk. However, it can result in loss of revenue if a show delivers higher ratings than expected. Statistical studies help broadcasters arrive at the best decision by providing algorithms to predict future show performance.

1.4 STATISTICAL LITERACY

> **Statistical thinking will one day be as necessary for efficient citizenship as the ability to read and write.**
> —H.G. Wells

The need for an understanding of basic statistical concepts, or statistical literacy, is not limited to the workplace, but pervades our daily lives. It is an important part of being an educated citizen.

1.4.1 Some Examples

We need only pick up a newspaper or watch a television newscast to learn about statistical studies that have important social and national implications. For example:

- The almost daily opinion polls that provide the "pulse of the nation" (with associated statistical error bands) on key issues such as consumer confidence, same-sex marriage, or fast food preferences – based upon a "scientifically selected" sample of, say, 2,000 respondents in the United States. Such polls become especially important prior to elections. But how good are the results from an appropriately selected sample of 2,000 out of a population of over 225 million citizens of voting age?

- Television ratings estimate the number of people watching different programs nationwide, again based upon a relatively tiny sample of the population of television viewers.
- Environmental studies to assess the impact of global warming rely heavily on statistical analyses of long-term trends in weather data.
- Studies of the harmful side effects of specified drugs or the benefits of a particular type of vitamin.
- The link between war service and various specified illnesses.

1.4.2 Looking at Studies Critically

It is the mark of a truly intelligent person to be moved by statistics.
—George Bernard Shaw

Training in statistics allows us to look critically at studies reported in the press. Such investigations are often committed to selling a point of view. For example, a widely reported study by Waite and Gallagher (2000) alleged "marriage improves longevity." A statistically oriented person would want to know such things as:

- How were the data obtained?
- How was the analysis conducted?
- How many days, weeks, months or years, on the average, does marriage, supposedly, add to our lives?
- Are the results the same for men and women (some claim they apply to men only) and different age groups, etc.?

We would question, moreover, the implication in the article that marriage is what improves longevity. Even if married people live longer, is marriage the reason? Could it simply be that those in good health are more likely to get (and stay) married?

It is well known (although, perhaps, sometimes ignored by politicians) that statistical association found in, often crude, analyses of historical data should not be confused with proof of cause and effect. That is why many scientific investigations, such as the evaluation of the effectiveness of a proposed new drug, employ controlled investigations. In such studies, some subjects are often randomly selected to receive a placebo, instead of the treatments (e.g., different dosages of the new drug) being evaluated, and the results for the various groups are compared. Controlled studies are contrasted with observational studies (often conducted when controlled studies are not feasible) in Sidebar 1.2.

SIDEBAR 1.2: CONTROLLED VERSUS OBSERVATIONAL STUDIES

Controlled studies typically involve a random selection of sample units from a specified "frame"[1] and a random assignment of these units to different "treatments." Controlled studies are highly desirable in many situations, especially when the goal is to establish cause-and-effect relationships. They presumably eliminate bias due to self-selection (by subjects selecting a preferred treatment) and "confounding" variables (discussed shortly). The statistical design of experiments is one form of controlled study that has gained much popularity (Section 2.2.7) in business and industry.

In many applications, it is, however, inappropriate, or even unethical, to conduct a controlled study. Obvious examples are assessing the effect on health of cigarette smoking or of alcohol abuse, or the impact of alternative forms of punishment of past crimes on the incidence of future crimes. In such cases, we are often restricted to conducting an observational study.

An observational study, in contrast to a controlled study, involves the analysis of existing or new data that are typically obtained on some process whose major purpose is *not* that of providing data for analysis or decision making. Instead of randomizing, we have little or no say in how the treatments or exposures are allocated. It is, therefore, often difficult to draw definitive conclusions about cause-and-effect relationships and conclusions based on such studies must be carefully scrutinized. Yet such studies, when carefully planned and conducted, can still provide useful insights and suggest areas for further exploration.

The Framingham Heart Study is one of the best-known observational studies. Starting in 1948, some 5,000 volunteers from Framingham, a Boston, Massachusetts suburb, were monitored for decades through periodic physical examinations, blood tests, and in-depth interviews about their lifestyles. This study provided the first important clues into the effects of a person's diet, exercise, and smoking on cardiovascular health. Some of these associations were subsequently supported by further investigations (see Levy and Brink 2005). The overall study is continuing and now includes some of the original volunteers' children and grandchildren.

For further discussion contrasting these two types of studies, see Flanagan-Hyde (2006), Katz (2006, Chapter 1), and Utts (2005, Chapter 5).

It is not feasible to conduct a controlled study in the "marriage improves longevity" example. So we try, by statistical analyses of data on married and unmarried individuals, to adjust for "confounding" variables that also impact longevity and that may well be correlated with marital status, such as an individual's health or wealth. Adjusting for these variables is usually better than ignoring them, but it is unlikely to provide an unambiguous answer to the question posed. This is because it is usually not possible to identify, measure, and appropriately adjust for *all* confounding variables – and we don't know whether we have.

Campbell (1998), Gelman and Nolan (2002), Gigerenzer (2002), and Spirer, Spirer, and Jaffe (1998) use general interest examples, such as breast cancer

1. A sampling frame is a list or map or other means of identifying and accessing members or sampling units of the target population.

screening, AIDS counseling, domestic violence, experts on trial, and DNA fingerprinting to illustrate questionable analyses of historical data and common misuses of statistics.

1.4.3 Enumerative versus Analytic Studies

Statistical studies typically deal with data obtained in the past, or possibly the present, but the real interest is often about the future. In recognition of this, the renowned statistician, W. Edwards Deming urged us to distinguish between enumerative and analytic studies (see Deming 1975).[2]

Enumerative Studies In an enumerative study, one has a sampling frame that presumably provides a reasonable representation of the target population. A random sample is then taken from that frame. Examples of enumerative studies are:

- Estimating the proportion of defective units in a production lot, based on a random sample drawn from that lot.
- Estimating how many people in the United States are watching a TV show, using a sample of 2,000 homes randomly selected from among all U.S. homes.
- Using an "exit poll" of randomly selected voters on election day to estimate the proportion of voters who assert that they had voted for a particular candidate.

Analytic Studies In many actual situations in business and industry, one is, however, dealing, not with a static population, but with a dynamically changing one or with a future process. In such "analytic studies," one typically wishes to use the results from a current population or process and environment to draw inferences about what will happen in the future for a likely changing population or process and environment. For example:

- In assessing demand for a proposed product, one may use the results of a past survey of consumers to draw conclusions about future purchasing behavior. However, this behavior might change due, for example, to changes in the economy or new competitive offerings.
- In designing a product, one wishes to use the results from a sample of individually built prototype units of a new product built today to extrapolate performance under high-volume production some time in the future.
- In developing a model for credit scoring for loan approval, the payment performance of a past sample of credit applicants is used to decide to whom to give credit in the future. But the payment patterns of future applicants may differ from that of past applicants, especially if there has been a change in economic conditions.

2. Deming used the terminology "analytic problems" rather than "analytic studies."

Thus, the results of an analytic study on an existing process – and the statistical inferences we draw from these – apply to a future process only to the degree to which the two processes are the same.[3]

An understanding of the limitations of analytic studies strongly argues for making investigations as broad as possible. For example, in a study to predict the performance of a future production process, one should try to introduce, as much as possible, the type of variability that is expected to occur in future production – such as day-to-day differences in the environment, differences between operators, and raw material lot differences. This does not eliminate the unknown uncertainty of projecting into the future, but it should help reduce it.

1.4.4 The Role of Statistics in Addressing Life's Problems

People who don't count won't count.
—Anatole France

Statistics helps us address our daily life concerns, such as in the following examples (also see Sidebar 1.3):

- Proper interpretation of medical test results is essential for decision making by patients and caregivers. A series of tests may indicate that an 80-year-old patient has a 1-in-20 and a 1-in-3 chance of dying from a cancer within a year and within 10 years, respectively. Does this warrant a difficult operation, which carries its own estimated risks, given the patient's overall health condition?
- An understanding of how credit scores are constructed, using statistical models and past payment performance, can help us improve our own scores.
- Our utility bills, especially when there is a meter failure, may be estimated statistically from past usage. Such estimates could be slanted in favor of the utility. Our knowledge of statistics can help us achieve an equitable resolution.

SIDEBAR 1.3: STATISTICS IN EVERYDAY LIFE: ANOTHER EXAMPLE

One of us had an elderly mother living in the borough of Queens in New York City. About once a month, he took the train from Schenectady to New York to visit her. When he got there, he needed to make a decision – go to her house from the station by cab or by subway. From past experience, he knew that (adjusting for time of day and day of week) the median time for the cab ride is about 45 minutes. There is, however, a 1-in-10 chance that there are long traffic delays and the ride will take an hour and a half or more. The subway time is an hour, on the average, and 9 times out of 10, the ride does not exceed an hour and 10 minutes. So, forgetting about cost and comfort, he made the following decision. On the way down, he took a cab. This allowed him to increase the time that, in the long run, he

3. Statistical confidence intervals, for example, apply only to enumerative studies. If used in an analytic study, they must be considered as lower bounds reflecting only the *statistical* uncertainty.

could spend with mom. On the way back, he took the subway. This way he reduced the chances of missing the train home. Statistically speaking, he minimized the average time of the trip on the way down and reduced variability (so as to reduce the chances of missing the train) on the way back.

Also, we find it puzzling to have online weather forecasting software tell us that the probability of rain *some time* on a future day is 50%, and then have the same software assert that the probability of rain for *each* of 8 hours during that day is also 50%.

An appreciation of statistics also creates a passion for procuring as much relevant data as possible *before* making a decision. In purchasing a car, we avidly seek data on repair frequency to help us select a model. Or in deciding whether or not to see a movie, we not only seek out the opinions of others who have seen the movie, but also decide how much weight to give to each such assessment – based upon how well the assessor's past judgments correlate with our own.

1.5 BETTER DECISIONS REQUIRE BETTER DATA

> **The value of statistics in industry lies in its ability to influence decision making on a strategic level.**
> —Lynne Hare

Frequently, the ultimate goal in using statistics is to help make informed, and thereby better, decisions in the face of uncertainty. This involves procuring appropriate data and converting such data into useful information with the help of well-selected statistical and/or graphical methods.

Having "appropriate data" is key to the process. Such data are generally not just sitting there for us to analyze, but must be obtained, often by well-targeted and carefully planned investigations, Thus, helping ensure good data is a highly important (and not often fully appreciated) part of our job and is often more important than the statistical analysis itself. A study with good data can, after all, frequently be well analyzed by simple graphical methods. But the world's most sophisticated statistical analysis is unlikely to save a flawed investigation.

Frequently, a series of iterations is required before there is enough information to make a decision. After defining the problem, the available data are evaluated to determine what relevant information may be gleaned therefrom. This often leads to the need for gathering more and better data, which in turn are then analyzed. The process may be repeated as part of what Box, Hunter and Hunter (2005) call an "iterative learning process."

1.6 STATISTICAL THINKING

In contrast to deterministic thinking, statistical thinking recognizes the importance of variability. Its basic concepts are:

- All work occurs in a system of interconnected processes.
- Variation exists in all processes.
- Using data to understand and reduce variation is key to success.

Statistical thinking has broad applicability in situations where there is an opportunity to reduce variation, rework and waste, and for improving quality and productivity. Therefore, a disciplined and databased approach toward identifying the root causes of variability is required. At the higher levels of an organization, this involves clearly communicating the vision and the strategy, adapting a disciplined project management and review system, and taking process variation into consideration in setting goals.

The key ideas of statistical thinking were developed under the sponsorship of the Statistics Division of the American Society for Quality (ASQ) (see Britz, Emerling, Hare, Hoerl, Janis and Shade 2000). Also, Hoerl and Snee (2001) provide further information and examples of business process improvement based on statistical thinking. We illustrate this concept in various places and especially in discussing business processes (Section 12.3.9). And we note that, although statistical thinking was developed principally for business and industry applications, it has much more general applicability.

1.7 RELATED DISCIPLINES

1.7.1 Statistics and Mathematics

The Relationship Statistical methods are derived from mathematical theory; probability theory, in particular, is deeply rooted in mathematics.

At the same time, however, statistics – and especially applied statistics – is much more than a branch of mathematics. Like physics, which has much of its foundation in mathematics, statistics is a discipline of its own. There is much within the realm of statistics that has little to do with mathematics. The planning of studies to get good data and graphical data analysis are two examples.

So How Much Mathematics Do I Need? Practitioners may have little or no training in the theory underlying the tools that they are using. This may not be a serious problem for users of packaged software – as long as they are fully aware of the assumptions underlying these methods (generally a consequence of the theory) and know when to call for expert guidance.

Applied statisticians, in contrast, are likely to be working on more complex problems and frequently need to adapt a particular method to fit the problem and data at hand. This cannot be done without an intimate understanding of the theoretical foundations of statistics. Applied statisticians need to have, as a minimum, the basic knowledge provided by a one-year introductory course in mathematical statistics.[4]

4. Statisticians, mostly in academia, who are developing new methods, need to know additional theory. This, in turn, may require added course work in mathematics.

The application domain, in addition, often involves some inherent mathematical complexity. For example, optimization of a chemical process requires understanding of the differential equations that characterize the system.

1.7.2 Statistics and Computers

Statisticians have been closely associated with computers from the beginning of the "computer age." The renowned statistician John Tukey is credited with coining the computer terms "bit" (for binary digit) in 1946 and "software" (in a computing context) in 1958. We shall be discussing the role of computers and computations in the application of statistics in business and industry throughout this book, starting with Section 2.4.1.

Those who apply statistics need be proficient computer analysts. In that sense, they are hardly different from the rest of today's scientific community. In addition, they also need to be familiar with statistical software and have a strong interest in data storage, representation, and manipulation.

1.7.3 Statistics and OR/MS (Operations Research/Management Science)

OR/MS, also sometimes referred to as decision science or decision technology, is a field that is closely related to statistics. The Institute for Operations Research and the Management Sciences (INFORMS) states that "Members of the OR/MS profession apply scientific tools and methods to improve systems and operations and to assist in managerial decision making." OR/MS has been described as a scientific approach to analyzing problems and making decisions. This field emerged in an effort to leverage for nonmilitary applications – such as factory layout planning, manufacturing planning, transportation, and capacity planning – some of the highly successful logistics work done during World War II.

The INFORMS Internet site states that OR/MS professionals typically concern themselves with such problems as:

- What new sanitary facilities will be needed to serve the population of Sun Valley, Idaho in some designated future year?
- How can a dress manufacturer lay out patterns to minimize wasted material?
- How often should the sales force of a frozen yogurt company call on its customers?
- How many elevators should be installed in a new office building?

These questions sound very similar to ones that statisticians are likely to encounter. In fact, statistics is an important part of OR/MS and OR/MS professionals generally receive training in statistics. In turn, statisticians employ OR/MS methods extensively and need to be knowledgeable in them. It is often difficult to differentiate whether one is engaged in statistics or in OR/MS work – and quite unnecessary to do so.

We provide an introductory discussion of OR/MS methods in Section 12.7.

1.7.4 Other Domains

Problems that are addressed by statistics can also often be usefully analyzed by recently developed tools from artificial intelligence (AI), knowledge discovery in databases (KDD) and data mining, and soft computing. These methods include case-based reasoning, fuzzy logic, genetic algorithms, machine learning, and neural networks. Those who apply statistics, therefore, need, as a minimum, to be familiar with these tools (and, hopefully, those engaged in such work are similarly knowledgeable in statistics).

1.7.5 Some Closely Related Specialized Areas

In some areas of application, statistical and related quantitative approaches have led to knowledge areas of their own. Notable among these is actuarial science, focusing on the uncertainties associated with granting insurance and setting rates and terms for insurance policies (Section 11.7.6). In addition, there are such hybrid areas as biometrics, chemometrics (Section 13.6.7), econometrics, and environmetrics. These focus on the application of statistical, mathematical, OR/MS, and other quantitative approaches to problems in biology, chemistry, economics, and the environment, respectively. They require a high degree of knowledge both in the subject area and in statistics, and often also rely heavily on computer technology.

1.8 MAJOR TAKEAWAYS

- Statistics has been broadly defined as "the science of learning from data." It has important application in most fields of human endeavor from actuarial science to zoology – and notably in business and industry.
- Statistical literacy is an important part of being an educated citizen; it helps us look at studies reported in the press critically and make better decisions in addressing life's problems.
- Many studies are analytic (one is interested in what will happen in a future process) rather than enumerative (one wants to draw inferences about an existing population). A future process is likely to differ from the current process under investigation. We, therefore, need to make analytic studies as broad as possible and recognize their limitations.
- Better business decisions require better data. Planning to get good data is a key part of most statistical studies and may often be more important than the subsequent statistical analysis.
- Statistical thinking emphasizes that all work occurs in a system of interconnected processes, that variability is present in every process, and that such variability needs to be addressed by databased studies.
- Statistical methods are derived from mathematical theory. However, statistics is a discipline in its own right.

- Statisticians have been closely associated with computers from the beginning of the computer age.
- Statistics is closely related to OR/MS.
- Problems that are addressed by statistics can also often be usefully analyzed by recently developed tools from AI, KDD and data mining, and soft computing.
- Knowledge areas closely akin to statistics, such as actuarial science (and biometrics, chemometrics, econometrics, and environmetrics), have also evolved.

DISCUSSION QUESTIONS

General

1. Identify an application of statistics currently in the news. How did statistics contribute?
2. Identify a decision-making situation from your own experience for which a statistically based approach led, or might have led, to a better decision.
3. What, in your favorite sport, are some tactical or strategic decisions that can benefit from a statistical approach? How would you go about using such an approach?
4. We state in Section 1.4.1 that "environmental studies to assess the impact of global warming rely heavily on statistical analyses of long-term trends in weather data." Research this comment and elaborate.
5. Give some further examples of analytic studies.
6. We argue in Section 1.4.3 for making analytic "investigations as broad as possible." How would you go about doing this in the three cited examples of such studies?
7. In the description of Statistical Thinking in Section 1.6, it is asserted that "all work occurs in a system of interconnected processes." Elaborate on this statement and give some examples.

Technical

1. Consider again an application of statistics in the news. What do you think was the technical approach used? Critique, from a statistical perspective, the discussion of this application by the media.
2. What are some of the issues that needed to be considered in planning the study (Section 1.3.2) to obtain a valid assessment of whether birds can differentiate between mating calls and the sound of jet engines? How would you address these?
3. In discussing public opinion polls in Section 1.4.1, we raise the question "how good are the results from an appropriately selected sample of 2,000 out of a population of over 225 million citizens of voting age?" Respond to this question and indicate the degree of precision about a population that a sample of 2,000 can provide. What assumptions are being made and how might these be violated in practice?
4. Elaborate on how you would go about identifying potential confounding variables in a particular observational study and adjusting for them.

5. In each of the three examples of enumerative studies in Section 1.4.3, what is the target population and the sampling frame? How well does the frame represent the population? How might you select a random sample from the frame?

6. We assert in Section 1.4.4 that it is "puzzling to have online weather forecasting software tell us that the probability of rain *some time* on a future day is 50%, and then have the same software assert that the probability of rain for *each* of 8 hours during that day is also 50%." Explain.

CHAPTER 2

STATISTICS IN BUSINESS AND INDUSTRY

2.1 ABOUT THIS CHAPTER

In this chapter we turn to a general discussion of how statistics is used in business and industry today. We briefly consider:

- The evolution of statistics for manufactured product applications (the topic of Part II of this book).
- Applications in business and industry beyond manufactured products (the topic of Part III).
- Forces impacting the use of statistics today.

2.2 EVOLUTION OF STATISTICS FOR MANUFACTURED PRODUCT APPLICATIONS

2.2.1 A Good Student

William Sealy Gosset, a pioneer in the use of statistical methods in industry, is best known for the *t*-distribution, which is taught in most introductory statistics courses. He was employed by the Guinness Brewery in Dublin, Ireland in the early 1900s and published his work under the name Student. Gosset advanced to become head brewer at Guinness – a true inspiration to generations of future statisticians. See Sidebar 2.1 and Box (1981) and Boland (1984) for more on Gosset's career.

SIDEBAR 2.1: GOSSET'S PROBLEM

Gosset joined the Guinness brewing company in 1899 as a scientist trained in chemistry and mathematics. He soon became involved with the collection and analysis of data for quality assessment. Gosset was a self-learner and quickly developed an appreciation of how statistics could help his work. This led him to take a year of absence to study under Karl Pearson at the University College in London. At the time, most statistical methods were based on large sample theory. The resulting approximations were not adequate for

The Role of Statistics in Business and Industry, by Gerald J. Hahn & Necip Doganaksoy

the small samples that Gosset often encountered. His curiosity led him to use a form of Monte Carlo simulation (Section 3.5.4) – long before modern computers – to show that the distribution of the sample mean differed markedly from a normal distribution in small samples. He then tabulated the "true" distribution based upon his empirical studies and numerical approximations. The mathematical foundation for this distribution (which became known as Student's *t*-distribution) was subsequently developed by the highly influential statistician of the early twentieth century, Sir Ronald A. Fisher.

2.2.2 Shewhart and Control Charts

During the 1920s and 1930s, Walter Shewhart of the Western Electric Company (a manufacturer of telephone hardware), and subsequently Bell Telephone Laboratories, introduced the concept of statistical control and the control charts that bear his name. Ed Deming, then a young statistician working at the U.S. Department of Agriculture, helped document and expand Shewhart's work. These simple tools of what later became known as statistical process control (SPC) provided the foundation for the early use of statistics in business and industry. They and their extensions are still used widely today (Section 6.7).

2.2.3 World War II, Acceptance Sampling and Beyond

The demands of fault-free mass production of munitions and complex systems during wartime helped bring statistics in industry prominently on the scene. The Statistical Research Group at Columbia University became an important center for research on applied statistics and involved many of the future leaders of the field. Much of the work was centered on assuring quality by rejecting incoming materials or outgoing product through acceptance sampling[1] (Section 8.5.2).

These and other developments gave rise to the quality engineering profession, with its own professional organizations, certifications, periodicals, and courses. The American Society for Quality Control was formed in 1946. Quality engineering has broadened appreciably over the years to include tools and methods from many disciplines outside statistics. Nevertheless, statistical concepts and tools continue to play an important role.

1. The popular World War II doodle "Kilroy was here," featuring the character shown at the beginning of this section, was said to have its origins with J.J. Kilroy, an inspector of rivets manufactured at a U.S. shipyard, who left his mark on the product he inspected.

2.2.4 Formation of Central Statistical Groups

The preceding and other developments led to the hiring of statisticians in business and industry in the United States in the 1950s and 1960s. In addition, statistics groups were established at various government or government-sponsored scientific agencies (e.g., Statistical Engineering Division at the National Institute of Science and Technology (NIST) and Los Alamos and Sandia National Laboratories) and at various large corporations (e.g., Bell Labs, DuPont, Ford, GE, and GM). In addition, individual statisticians were hired by smaller companies or in local plants of larger companies.

2.2.5 Deming and Developments in Japan

Deming to the Rescue Deming was enticed to come to Japan after World War II. He was soon providing guidance in using statistical methods to build high-quality products.

Japan was eminently successful – a reputation for shoddy merchandise was rapidly replaced by one for world-class quality – and became a leader in world markets, ranging from electronics to automobiles. Much of this success was attributed to Deming's guidance. The emperor of Japan conferred the Second Order Medal of the Sacred Treasure on him and he became a national hero.

The teachings of various Japanese quality experts, such as Kaoru Ishikawa and Shigeo Shingo, also received worldwide attention.

The Plan-Do-Study-Act (PDSA) Cycle Deming, expanding upon work by Shewhart, proposed the PDSA cycle (see Figure 2.1) as a mechanism for methodically and continuously developing knowledge and making improvements using the following steps:

- Plan – Develop a plan for improvement (including data gathering)
- Do – Implement the plan (initially on a small or pilot scale)
- Study – Assess the outcome and adjust the plan, if needed
- Act – If the outcome is successful, make the improvement part of the process

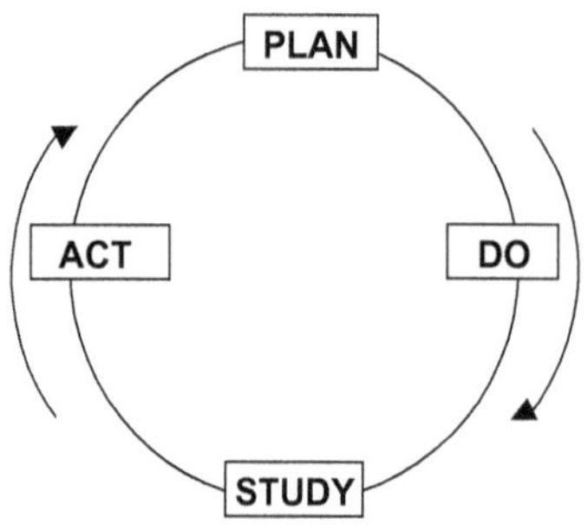

Figure 2.1 The Plan-Do-Study-Act cycle.

TABLE 2.1 Basic Quality Tools

Check sheet	A predefined form to collect and display data; e.g., Table 7.1 in Chapter 7.
Control chart	Graphical display of data on a product or process characteristic over time with statistically based alarm limits to identify process changes; e.g., Figure 6.7 in Chapter 6.
Fishbone diagram	A tool for displaying and categorizing problem causes (also called cause-and-effect or Ishikawa diagram); e.g., Figure 7.6 in Chapter 7.
Histogram	Bar chart display of a frequency distribution; e.g., Figure 12.5 in Chapter 12.
Pareto chart	Bar chart of frequency of defects by type ordered by frequency of occurrence; e.g., Figure 7.2 in Chapter 7.
Process flow diagram	Graphical display of inputs, processing steps, decisions and outputs of a process; e.g., Figure 12.4 in Chapter 12.
Scatter plot	Graphical display of two numerical variables plotted against each other to show their association; e.g., Figure 12.2 in Chapter 12.
Segmentation	Analysis of data by sub-groups (e.g., machine, supplier, time period) to localize problems or find similarities (sometimes also referred to as stratification); e.g., Figure 9.4 in Chapter 9.

See Langley, Nolan, Nolan, Norman, and Provost (1996) for further discussion and examples of applications.

Some Simple Tools Japanese companies successfully used elementary statistical and other tools to seek the root causes of problems. This led to the basic quality tools shown in Table 2.1.

This list of tools is an outgrowth (with the addition of the process flow diagram) of what was once referred to as "The Magnificent Seven." Others have added to the list various further tools, such as box plots, multi-vari charts, star plots, and pie plots (see Hoerl and Snee 2001; Tague 2004; and the American Society for Quality (ASQ) Internet site for further discussion and examples).

2.2.6 North America Recognizes the Significance of Quality

> **We must remember that measures were made for man and not man for measures.**
> —Isaac Asimov

Deming Returns Japan's success, much at the expense of the United States and other Western nations, could not go unnoticed. A key event was the 1980 U.S. NBC-TV documentary, starring Deming, with the provocative title "If Japan Can, Why Can't We?". Deming considered understanding variability a core competence for management and claimed that management's lack of understanding variability often led to unnecessary tampering with processes. He strongly criticized the overemphasis placed on "the most recent numbers," such as last month's sales, in making management decisions.

This helped trigger the so-called Quality Revolution in the United States. Deming was rapidly transformed from an obscure statistician to an eminent one by the media,[2] who typically referred to him as "Dr. Deming." His 4-day short courses attracted about 100,000 participants, many from top management, and his books, *Out of the Crisis* (Deming 1986) and *The New Economics for Industry, Government, Education* (Deming 1993), became highly popular. He influenced American CEOs and other leaders (then U.S. House Speaker Newt Gingrich spoke of him frequently). In 1987, President Reagan awarded Deming the National Medal of Technology. (See Hahn 2002; Kilian 1992; and Mann 1985 for further discussion and references on Deming and his legacy.)

The Baldrige Award and Consequences The United States in 1987 established the Baldrige Award to recognize organizations for their achievements in quality and performance and to raise awareness of the importance of these in gaining a competitive edge. Although not its centerpiece, statistical concepts and tools were an integral element.

A consequence of these developments was that quality improvement, and the concomitant application of statistical methods, was no longer regarded as the exclusive job of a centralized quality group, but became everybody's responsibility. The reliability of products in the hands of consumers also received extensive recognition.

The broad acceptance of Six Sigma (Section 2.4.2) was a further important driver for quality improvement.

The Age of Gurus The quality improvement movement was significantly impacted by a number of "gurus." In the United States, in addition to Deming, these included Phil Crosby, Armand Feigenbaum, Bill Golomski, Joe Juran, and Dorian Shainin. Like Deming's, many of their teachings were initially promulgated in Japan, but soon became widely adapted in North America. Their works were, to varying degrees, less statistically oriented than Deming's.

Other Developments The concept of Total Quality Management (TQM) gained much popularity in the 1980s. Its main principles are focus on the customer, total involvement of the organization and continuous improvement. TQM's emphasis on quality improvement represented a major change from the earlier focus on inspection (see Bisgaard and DeMast 2006).

Another key development was the evolution of international quality standards, culminating with the ISO 9000 series by the International Organization for Standardization. These were mostly oriented to process standardization, documentation, and management.

Process capability indices (Section 6.6.1), a basically statistical concept, also received widespread attention subsequent to their introduction in the 1980s.

2. Deming was already eminent within the statistical profession for his work with Shewhart, and at the U.S. Department of Agriculture and Bureau of the Census.

Global Aspects The Japanese economy has gone through some difficult times – even though Japanese products are still, by and large, known for their high quality. However, the emergence of other countries, and especially South Korea, China and potentially India, as important manufacturers of consumer and industrial products is providing additional impetus to the quality improvement movement as manufacturers strive to compete – and survive – in a global economy.

2.2.7 The Design of Experiments

Over the years, the statistical design of experiments (DOE) came to play an important role. Examples are provided throughout this book. In Sidebar 2.2, we briefly sketch the evolution of the use of DOE in business and industry.

SIDEBAR 2.2: THE DESIGN OF EXPERIMENTS IN BUSINESS AND INDUSTRY

Important developments in statistics have been in direct response to real-life problems. The use of DOE in business and industry provides a good illustration.

Early Work

Fisher pioneered DOE while working at the Rothamsted Experimental Station in the 1920s and 1930s. Much of the early work was motivated by applications in agriculture. A typical example is that of evaluating the effect of different fertilizers, watering regimes, and growing techniques on the growth, taste, quality, and usable life of fruits and vegetables.

Adaptation to Business and Industrial Problems

In the 1940s and 1950s, researchers in the chemical and textile industries, and subsequently in industry in general began to recognize the usefulness of experimental design in their work. In industrial applications, unlike agriculture, it is often possible to complete experiments in a relatively short elapsed time.[3] This allows one to conduct experiments in sequence, building upon one another. Moreover, many modern industrial applications deal with a large number of variables, the importance and impact of each of which needs to be evaluated. Pioneers, such as George Box, Cuthbert Daniel, Stuart Hunter, and Jack Youden worked toward developing designs that were responsive to industry's needs, resulting, for example, in the development of response surface designs (Section 3.5).

Training Programs

The heightened interest and continually wider applications of DOE led to various training programs aimed at practitioners. Today, many short courses on DOE are offered to industry, often centered on computer software that facilitates design and data analysis.

Robust Design

In the 1950s and 1960s, Genichi Taguchi, a Japanese engineer, recognized the need for designing products that operate well over a wide spectrum of manufacturing and usage conditions – that is, products that are robust to variability in manufacturing conditions and to customer use and misuse (Section 3.5.3). He also emphasized the use of planned

3. A notable exception is many applications involving product life assessment.

experimentation (see Taguchi, Chowdhuri and Yuin 2004). This work generated much interest in experimental design among engineers, scientists, and managers, especially those involved with electronics products.

Today's Challenges

Advances in instrumentation and rapid analytic testing have led, especially in the chemical and pharmaceutical industries, to the conduct of massive experiments with a large number of variables over short time periods (Section 3.5.1). In addition, DOE is being employed increasingly in conjunction with engineering modeling (Section 3.5.2). This is creating the need for new experimental designs and analysis methods.

2.2.8 Further Reading

See Montgomery (2004) and Bisgaard (2005) for a more detailed discussion of the evolution of statistics for manufactured product applications.

2.3 APPLICATIONS BEYOND MANUFACTURED PRODUCTS

The successful application of statistics to the design, reliability assurance, production, and servicing of manufactured products led to the realization that statistical methods can also play an important role in other application areas in business and industry, as discussed in Part III of this book. The high human stakes involved in areas, such as pharmaceuticals and health care, and the large sums of money that enter into many financial transactions have prompted particular interest in these areas. Advances in information technology (Section 2.4.1) have been an important force leading to the more general use of statistical methods, especially in such "data-rich" areas as consumer finance, telecommunications, and medical imaging.

The need for rigorous methods to protect consumers is yet another reason for the broader use of statistics by business and industry. A statistical approach is mandated by some government agencies. For example, in the United States, a new drug is required by the Food and Drug Administration to go through rigorous statistically based testing (see Chapter 10). Enforcement of environmental and energy requirements, such as limiting undesirable emissions or gasoline consumption rates of automobiles likewise call for careful statistical assessments.

2.4 FORCES IMPACTING THE USE OF STATISTICS TODAY

2.4.1 Information Technology

> **Polygor existed to give numbers to the Universe. ... Each day millions and billions and zillions of numbers were distributed to every world that needed them.**
>
> —Cat Stevens

Computers today provide the backbone of much statistical activity in business and industry. In particular,

- Massive data sets are increasingly encountered, especially in such areas as financial services and in measurement intensive manufacturing environments, as in the process industries. For example, in one typical chemical process, 20 on-line measurements (such as feed rates, torque, and temperature) from 24 parallel manufacturing lines are captured and stored every two seconds.
- User-friendly statistical software is readily accessible to practitioners (Section 2.4.3).
- Many popular statistical methods are highly computer-intensive (Section 13.7.1).
- Data exploration, especially through the use of graphical displays, is becoming increasingly commonplace. These might be dynamic, involving rotations in multidimensional space.

This has led to numerous exciting new applications in areas ranging from remote monitoring and diagnostics of equipment to credit scoring, as discussed throughout this book.

The special challenges and needs that arise in the analysis of large data sets helped prompt the emergence of KDD and data mining methods (Section 1.7.4). These aim to detect unsuspected patterns and associations in large databases and build heavily on statistical concepts and tools.

2.4.2 The Six Sigma Initiative

Background Over the years, various strategies for deploying quality improvement – with varying degrees of emphasis on statistics – have evolved. Such methods have much in common in their basic approach. Statistical thinking has been an important element of some of them (see Hare 2005).

The Six Sigma initiative is one of the most far-reaching and influential of such approaches. Introduced at Motorola in the mid-1980s, it may have helped the company win one of the first Baldrige Awards in 1988. Six Sigma achieved national prominence about a decade later when Honeywell, GE, and various other large companies publicly embraced it. The Six Sigma effort is perhaps best known for its laudable – but not always attainable or even practical – goal to reduce the level of defective products and services to "3.4 defects per million opportunities."

The Six Sigma Approach More importantly, Six Sigma provides a highly disciplined approach to quality improvement involving a series of steps, discussed in Sections 3.2.2 and 6.5; these bear much similarity to the PDSA cycle.

Six Sigma also places heavy emphasis on basing decisions on data and the use of statistical techniques; one of its key slogans is "In God we trust, all others bring data." As a result, Six Sigma has led to increased focus on statistical concepts and methods in the companies in which it has been adapted (see Hahn, Hill, Hoerl, and Zinkgraf 1999).

Six Sigma efforts are typically led and implemented by a hierarchy of so-called:

- Six Sigma champions, providing the needed resources and management leadership;
- Master black belts, integrating the work and providing the training;
- Black belts, leading the projects; and
- Green belts, implementing the work.[4]

Training and Certification Most employees of companies that have adopted Six Sigma get a crash course that includes statistical methods and other tools. This typically involves up to four weeks of training, generally over a 3-month period during which Six Sigma concepts are applied on the job. Particular emphasis is given to such topics as DOE, simple graphical displays, measurement accuracy and precision, and process stability. Participants obtain and get to use a statistical software package. (See Hoerl 2001; Hahn, Doganaksoy, and Stanard 2001 for further discussion.)

Certification of green belts and black belts by a master black belt typically also calls for the successful completion of three Six Sigma projects. In addition, master black belts may be required by management to mentor 10 to 20 Six Sigma projects for certification.

As a result of this training, all those involved with Six Sigma attain some familiarity and proficiency in the use of selected statistical tools. Master black belts and many black belts are, in addition, expected to have further knowledge and experience beyond the standard training and may use this in guiding green belts.

Evolution In most companies that use Six Sigma, the *initial* focus tends to be in manufacturing (Section 6.5). Based on successful results, companies quickly extend its use to other operations, such as product design (Section 3.2.2), information systems, business processes (Section 12.2.7), and customer support (see also Hahn, Doganaksoy, and Hoerl 2000).

Six Sigma has been closely allied lately with another initiative known as lean manufacturing (see Section 8.6.2).

2.4.3 The Democratization of Statistics

Earlier in our careers, the use of statistical methods was left principally to professional statisticians. Today, easy-to-use statistical software with online tutoring is readily accessible to practitioners in business and industry. Introductory courses in statistics, inside and outside of universities, routinely teach the use of such software. On-the-job training programs, such as Six Sigma, provide additional instruction. In

4. The adaptation of these terms from karate training has been ascribed to the need for mental discipline, systematic training, speed and decisiveness that both karate and Six Sigma are claimed to require.

addition, technical questions may be addressed in response to web-based queries. As a result, many nonstatisticians now feel they have the tools, motivation, and self-confidence to conduct their own statistical analyses.[5]

These developments have been characterized as "the democratization of statistics." They have resulted in a change of roles for many professional statisticians. Instead of being heavily involved in conducting statistical analyses themselves, statisticians' focus is likely to be on those problems that are most important to the business and most complex technically.

2.4.4 The Need to be Proactive

Statistics is not a spectator sport.
—E.B. Wilson

The Reactive Mode Statistics used to be typically employed reactively, especially when dealing with quality assurance. There was heavy reliance on end-of-line product testing and fixing problems in the field after they occurred. Much effort was spent responding to crises and fixing problems after they had already created some damage. Typical questions addressed by statistics were "Is a recall of product in the field (or for sale on the shelf) needed?" and if so, "What segment of the product population should be recalled?" Some manufacturing lots may, for example, be more susceptible to failure than others. Or failures might be more likely to occur in locations with high temperature and high humidity or under certain usage modes. This often required quantifying the magnitude of a problem and evaluating alternative ways of minimizing the damage.

A Proactive Mode Forward-looking leaders in business and industry have come to realize that achieving quality by reactive measures is unacceptably expensive and potentially disastrous to retaining customer confidence. There is now general agreement that quality needs to be built into the design of products and processes *proactively*. This can have significant payoff, both in customer satisfaction and cost reduction; problems discovered in design, though often more difficult to identify, are usually less costly and much easier to fix.

In contrast, problems found after the design has been frozen, and especially after significant quantities of product have been built, although easy to identify, are often difficult and expensive to fix. With time, "build it right the first time" has become the true goal of companies, rather than just a slogan. The American Society for Quality Control in 1997 took out "Control" from its name to remove the reactive connotation associated with this word.

The need for a proactive approach is not limited to manufacturing applications or business and industry. It has applicability to the world around us (see Sidebar 2.3).

5. Pharmaceutical applications appear to be an exception (see Chapter 10).

SIDEBAR 2.3: A MORE PROACTIVE WORLD

When the impact of not addressing potential concerns can be disastrous, the use of a proactive approach is evident. Nobody questions the need for air controllers to avert potential crashes or for periodic physical checkups to identify and address emerging health issues.

Proactive protection against disease and injury is becoming increasingly prevalent and long-range focused. For example:

- Flu shots are developed yearly to immunize the most susceptible – and, increasingly, those less susceptible.
- Dentists used to focus on filling cavities and removing bad teeth. Today, they emphasize steps to avoid tooth decay.
- Injury prevention programs are mandated in the workplace to identify and remove potential hazards, rather than treating their consequences.
- Sports have become safer with the use, for example, of padded goalposts in soccer and protective helmets in baseball and cricket.

The Proactive Statistician To be maximally effective in today's environment statisticians, and statistically oriented individuals, take on increasingly proactive roles and actively seek and leverage opportunities for improvement in all endeavors. Table 2.2 contrasts the proactive with the reactive statistician.

In summary, a proactive statistician strives to address problems before they happen so as to avoid them completely, or, to the degree this is not possible, mitigate their adverse impact. Deming was an early advocate of this tactic (see Sidebar 2.4); also Box et al. (2006) focus on using statistics for improving processes and products proactively.

TABLE 2.2 Comparison of the Reactive and the Proactive Statistician

The Reactive Statistician	The Proactive Statistician
Responds to requests.	Seeks and leverages opportunities for improvement.
Typically serves as consultant.	Often is team member (from outset) and, sometimes, team leader.
Engages in firefighting: Quantifies problems and helps to contain them (e.g., determine whether a product recall is needed and what and when to recall.)	Helps avoid problems before they happen (e.g., design a new product for long-term high reliability.)
Addresses problems as presented.	Considers key strategic needs of business.
Principally concerned with a single (often manufacturing) process.	Addresses entire system (recognizes assertion that 85% of quality problems are in the system).
Focuses on analyzing existing data.	Focuses on getting good data.
Finds it relatively easy to quantify impact.	Often finds it difficult to quantify impact (because of its long-term nature).

SIDEBAR 2.4: DEMING'S CASE FOR BEING PROACTIVE

Although Deming might not have used the term "proactive," this concept was clearly part of his message. In *Out of the Crisis*, Deming (1986) states "Every appalling example in this book turned up because I was there on the line, on the job, trying to be helpful by looking for sources of improvement and strong practices. If I had waited for them to come for help, I'd still be waiting." And in *The New Economics for Industry, Government, Education*, Deming (1993) says "It is necessary to innovate, to predict the needs of the consumer, give him more. ... The secret for reduction in time of development is to put more effort into early stages."

As a consequence of becoming more proactive, statistics, rather than being associated mainly with manufacturing or possibly field-service operations, has moved upstream and is being used to support engineering and new product development and introduction activities; this is illustrated in subsequent chapters and by the example in Sidebar 2.5.

SIDEBAR 2.5: PROACTIVE IMPROVEMENT: AN EXAMPLE

Customers routinely order plastic pellets from a manufacturer for molding into parts with specified characteristics – such as viscosity, tensile strength, and flammability – to meet their unique needs. In response, product developers traditionally conducted small "tweaking" (typically one-factor-at-a-time) experiments to determine the formulation that best addressed each order's requirements. This resulted in much repetitive work and often substandard solutions. Statisticians lobbied for DOE to replace *ad hoc* experimentation and provided across-the-business training in DOE and the associated software. The benefits became rapidly evident through faster design cycles, better product designs, and more effective utilization of resources. Product developers quickly gained competence in designing and analyzing such experiments. (See the example in Chapter 4 for an illustration.)

This allowed statisticians to look at the bigger picture. They identified the need to leverage the large amounts of data generated from the many DOEs conducted throughout the business. They then proposed and helped lead the development of a web-based system that provides product developers access to the results of all past DOEs, lets them use such data to find a region in design space to encompass the desired processing conditions for each product, and then generates a test plan to search for an optimum within this region. The new data are fed back into the system, thus continuously improving it. As a result, expensive, individualized DOEs were replaced by a comprehensive business-wide product development system.

2.4.5 Addressing Obstacles

> **Some of the most important figures for management are unknown and unknowable.**
> —W.E. Deming

Despite the compelling arguments in favor of a proactive approach, it has traditionally been easier to gain management support for responding to a tangible, existing problem than for avoiding, possibly undefined, future problems.

The reluctance in going proactive may stem from the fact that the benefits from reactive projects – typically aimed at reducing scrap, rework, and warranty costs, or increasing market penetration – are often readily quantifiable in advance and easily demonstrated shortly after implementation. In contrast, for most proactive projects, there is often a large upfront cost associated with proactive measures – generally incurred long before any benefit is achieved. The biggest gains, moreover, come from cost *avoidance*, such as averted field repairs and recalls, and from greater customer satisfaction – often resulting in increased business in the long run. The savings from proactive work are, therefore, more speculative, accrue over time, and, sometimes, are unknown or unknowable – especially for problems that have been avoided. However, such efforts are critical to the long-run health of the business. Thus, both costs and long-term gains need be considered in arriving at a reasonable trade-off between a reactive and a proactive approach.

Proactive statisticians leverage the contacts and goodwill gained from their reactive work to lobby for more proactive efforts in support of the long-range goals of the business. An ideal time to suggest the implementation of processes to avoid future problems is, in fact, immediately after successfully helping identify and remove a serious existing problem.

2.5 MAJOR TAKEAWAYS

- The roots of industrial statistics are found in the early work of Gosset, the development of control charts by Shewhart, and the need for acceptance sampling in mass production during World War II.
- After World War II, hiring statisticians in business and industry became common and some companies formed their own statistical groups.
- Japan, with guidance from Deming and others, pioneered the use of statistical methods in building high-quality products. The basic quality tools of Table 2.1 played an important role.
- North America in the 1980s came to recognize the significance of quality and the associated contributions of statistics. The Baldrige Award was established and quality gurus had much influence.
- The role of the design of experiments in business and industry became recognized with Box, Daniel, Hunter, Youden, and others building upon earlier work, principally in agriculture, of Fisher and his colleagues. Taguchi helped popularize DOE and expanded its focus to robust designs.
- Statistics is being used extensively today in important applications beyond manufactured products.
- Computers provide the backbone of much statistical activity in business and industry.
- Six Sigma, a disciplined approach for quality improvement, focuses on reducing defects and basing decisions on data. Originally emphasizing manufacturing operations, it has applicability throughout a business.
- The availability of easy-to-use statistical software has led to the "democratization of statistics," allowing practitioners to do their own statistical analyses.

- Forward-looking management has come to recognize the importance of a proactive approach. In this mode, statisticians seek opportunities for improvement and strive to avert problems before they happen.

DISCUSSION QUESTIONS

General

1. Put yourself into Gosset's shoes. What were some obstacles to his work that we take for granted today?
2. Describe how you might apply the PDSA cycle for improvement to one or more of the following:
   a. Improving the study habits of a high-school student.
   b. Addressing chronic unpunctuality.
   c. Reducing personal credit card debt.
3. Research three recent Baldrige Awards and describe how statistics likely contributed to the work recognized by the award.
4. Various quality gurus were identified in Section 2.2.6. What were the major contributions of each of these?
5. It has been asserted that there are not many prominent quality gurus around today. Do you agree? If so, what do you think is the reason?
6. How is Six Sigma being used today? Is it being overshadowed by new approaches? If so, what are they and how, if at all, does Six Sigma fit in?
7. Identify some specific situations from the news or your own life (e.g., car maintenance, health, and flood protection) in which there was a choice between applying a reactive and a proactive approach, and compare these approaches.

Technical

1. In discussing Gosset's career, we state (Section 2.2.1) that "at the time, most statistical methods were based on large sample theory. The resulting approximations were not adequate for the small samples that Gosset often encountered." What does large sample theory mean in this context? How does it relate to the central limit theorem?
2. Elaborate further on the differences between using designed experiments in most industrial and business settings versus their use in agriculture.
3. The Six Sigma initiative may be best known for its laudable – but not always attainable or even practical – goal to reduce the level of defective products and services to 3.4 defects per million opportunities. By reviewing the Six Sigma literature, explain how 3.4 defects per million opportunities translate into Six Sigma and the associated assumptions. What is meant by an "opportunity?" What are some possible deficiencies of this approach?
4. Consider the "bigger picture" system to help the plastic pellet manufacturer determine formulations to meet customer needs in the proactive improvement example stated in Sidebar 2.5. How might you go about developing such a system?
5. We suggest (in Table 2.2) that the proactive statistician "sometimes finds it difficult to quantify impact (because of its long-term nature)." Elaborate and suggest some ways of addressing this in a particular situation.

PART **II**

MANUFACTURED PRODUCT APPLICATIONS

IN THIS part of the book we describe applications of statistics aimed directly at designing, building, and improving manufactured products.[1]

The following sketch provides an overview of *some* of the key steps in the life cycle of a manufactured product.

Product Design, Reliability Assurance and Scale-Up Chapters 3, 4, 5	Manufacturing Chapters 6, 7, 8	Field Support Chapter 9
• Determine what customers want (and will pay for) • Identify nature, purpose and scope of product • Define product performance characteristics • Develop alternative product concepts • Identify "best" design • Establish design specifications • Ensure measurement capability • Select materials and parts • Validate and test components, materials, assemblies, prototypes • Conduct reliability testing and evaluation • Scale up design to manufacturing	• Translate design specifications to manufacturing tolerances for parts, processes, materials • Evaluate materials and parts received from suppliers • Stabilize processes and assess capability • Build product • Monitor to ensure product quality • Strive to reduce costs • Improve product quality and reliability	• Ensure smooth product start up • Speedily identify and fix (hopefully few) remaining problems • Feedback information for product improvement to design and manufacturing • Service product

Companies differ greatly in the way they organize, staff, and execute these activities. But all strive for faster product development cycles, higher quality and reliability, and more satisfied customers – with the overall goal

1. The discussion in this part of the book has benefited greatly from inputs provided by Gipsie Ranney.

of increasing sales and profits, and reducing costs. Statistics plays an important role in meeting these objectives.

The chapters in this part of the book deal with:

- Product design: Concepts and example (two chapters).
- Product reliability assurance.
- Manufacturing quality improvement: Concepts and example (two chapters).
- Product manufacturing: Further applications.
- Product field support.

CHAPTER 3

PRODUCT DESIGN: CONCEPTS

3.1 ABOUT THIS CHAPTER

The use of statistics in product design used to be limited mainly to validation studies long after the creative work of design was complete. Over the years, statistical concepts and tools have come to play an increasing role in quantifying the impact of key variables on product performance and identifying and measuring sources of variation. Today, the use of statistical methods is an important element of a proactively oriented product design process.

In this and the following chapter, we describe some of the ways statistics is used during product design and preparation for manufacture. We begin with a general discussion of the design process. We then describe the role of statistics in:

- Setting the design goals: Defining what the product aims to achieve.
- Ensuring measurement capability.
- Design and development: Designing a product that meets or exceeds the design goals.
- Design validation: Helping ensure that the design has succeeded.
- Transition to manufacturing.

We illustrate these concepts in the next chapter by a case study.

3.2 THE PRODUCT DESIGN PROCESS

3.2.1 The Key Role of Rigorous Product Design

The highly competitive market dynamics in many industries has placed businesses under pressure to develop and introduce new and better products in shorter time cycles. Failure to come up with the right product at the right time has had devastating results. Similarly, the release of a poorly designed product, or the premature release of a potentially good product, can lead to fatal "show-stoppers." Some classical examples are:

- The Pinto was meant to be Ford's response to increased competition from Japanese car manufacturers in the compact car market in the early 1970s. The

The Role of Statistics in Business and Industry, by Gerald J. Hahn & Necip Doganaksoy

car was designed and moved to production on an accelerated schedule; this is often cited as a major contributor to subsequent problems. Most notably, the fuel system on some cars ruptured and caught fire in even mild rear-end collisions, resulting in numerous fatalities. Ford had to recall 1.4 million Pintos in 1978 to make safety modifications (see Schlager 1994).

- A Concorde plane caught fire and crashed in July 2000 during takeoff at Charles de Gaulle Airport in Paris. An investigation revealed that debris on the runway caused a tire to burst. Its fragments were thrown against the wing structure, puncturing a fuel tank, resulting in a fire under the left wing. Upon investigation, it was determined that the tanks provided insufficient protection for such situations.
- In 2007, DaimlerChrysler had to recall over 270,000 minivans in regions in the U.S. where roads are salted heavily during the winter to avert icing; salting tended to corrode airbag sensors thus preventing the airbag from deploying when needed.

A key lesson learned from these examples is the importance of gaining an understanding in product design of the environments in which the product might be required to operate and how these might impact its performance and reliability.

Each of the examples deals with designs that were inadequate in addressing some special, but not highly unusual, circumstance. Additional examples dealing with situations in which the design did not provide the desired product reliability are given in Section 5.2.3. Taken together, these illustrate some of the performance, quality and reliability problems that need to be addressed proactively in the design of a product.

3.2.2 Design for Six Sigma

Following the initial success of Six Sigma in cutting defect rates in manufacturing, it was soon realized that there was a point beyond which most processes could not be improved further. This is because the quality of a product is basically limited by its design. Thus, to achieve the desired product performance and quality, one needs to start with the design process, and apply Six Sigma concepts there.

The Six Sigma approach that had been developed for improving manufacturing operations was, however, not completely suitable for design because it assumed an existing product or process. Thus, a methodology specifically tailored to new product design – Design for Six Sigma (DFSS) – was proposed (see Chowdhury 2002).

DFSS is a disciplined approach that integrates engineering design methods with statistical and other techniques to develop new products and processes (or redesign existing ones) with high levels of quality and reliability. The goal is to eliminate the unpleasant surprises – both in manufacturing and field operation – that had frequently been associated with the introduction of new or redesigned products, services, and processes. The process used to implement DFSS, called DMADV (define, measure, analyze, design, and verify), is described in Sidebar 3.1.

SIDEBAR 3.1: THE DMADV PROCESS

DMADV involves the following steps:

Define (D): Define the new product, service or process to be designed, and develop a team charter, including goals, scope, business justification and milestones; also determine the needed resources and develop a project plan.

Measure (M): Determine customer needs (or wishes) and the associated product requirements; translate these into measurable critical-to-quality (CTQ) characteristics for the product or process, and ensure that these can be accurately measured.

Analyze (A): Develop conceptual and alternative design concepts; select the concept chosen for development and predict the capability of the design to meet requirements.

Design (D): Develop the detailed design; evaluate the capability of the proposed design and develop plans to pilot test the new product or process.

Verify (V): Build and pilot a full-function, but perhaps limited scale, version of the new product or process. At this stage, the production process is developed, start-up and validation activities are performed, and the new or redesigned product is transitioned to the process owners.

These stages of DFSS relate to the discussion in this chapter as follows:

- The Define and Measure stages apply principally to setting design goals and ensuring measurement capability.
- The Analyze and Design stages apply principally to design and development.
- The Verify stage applies principally to design validation and transition to manufacturing.

3.2.3 Interconnectivity of the Design Process

It has long been known that the process of learning is a motivated iteration between theory and practice.
—George Box

Product design is an iterative and intertwined process. Decisions made during the early phases of design have consequences downstream on material specifications, manufacturing tolerances, and ultimately manufacturing costs and reliability. Companies are increasingly embracing practices that reflect the interconnected nature of the design, manufacturing and use of their products. Advances in Computer-Aided Design and Computer-Aided Manufacturing (CAD/CAM) systems, concurrent engineering, and design for manufacturability are some examples (see Benhabib 2003).

This new environment allows everybody, including statisticians, to contribute more broadly and effectively. In order to facilitate this complex process, most design organizations utilize project tracking systems and employ management tools for designing and introducing new products. (See Ulrich and Eppinger 1995 for a comprehensive overview.)

3.3 SETTING THE DESIGN GOALS

Customers do not always know what they need but they will treasure it if you give it to them.
—W.E. Deming

Businesses rely on data from many sources, both internal and external, in making new product introduction decisions. Trends in the general economy, market, competitive environment, pricing, and technological innovations are major considerations. The marketing team, in collaboration with design engineering, defines new product opportunities, based on its knowledge of the marketplace and product technology. Much of this involves obtaining and properly interpreting appropriate data.

3.3.1 Determining Customer Needs

Design teams develop products that satisfy customer needs and wishes subject to meeting business goals. Once these have been identified through market research, they need to be prioritized. Competitive, operating environment, technological and economic considerations weigh heavily at this phase.

Market Research A basic goal is to determine precisely what features customers want in a product and how much they are willing to pay for each. In addition, businesses desire information on such matters as likely consumer response to advertising campaigns, brand loyalty, and customer retention.

There are a variety of ways to get information from past, current, and potential customers. These include unsolicited customer feedback, interviews and focus groups, and sample surveys. The specific means of gathering information depends on the industry and the type of product. Manufacturers of consumer goods often have a large base of customers or potential customers whose needs they have to assess. Producers of specialty or industrial products, ranging from compressors for turbines to machinery for assembling engine parts, usually have fewer customers and face somewhat different data acquisition challenges.

Statistical concepts help ensure that the information that is obtained is as relevant as possible and that meaningful analyses are conducted on the resulting data. Churchill and Iacobucci (2004) and Gryna (1998) provide further details of the use of statistics in market research.

Interviews and Focus Groups Consumer views might be solicited by direct interviews or by assembling a so-called focus group of current and potential customers who are questioned interactively. These methods are directed at exploring attitudes and feelings, and defining and probing into, mainly qualitative, issues. They generally involve less statistical planning than surveys, but may provide useful information for subsequent more formal studies.

Sample Surveys Sample surveys are frequently conducted to assess what customers like most and least about an existing or proposed product. In addition, current and past customers, purchasers of competitive products, and potential customers are surveyed to determine what new product features would be of greatest interest to them and how highly these are valued. Such studies may be performed in person, by telephone, on the Internet or by mail. The results are often broken down or segmented by customer demographics, such as age, gender, income range, and lifestyle, for use in developing marketing and advertising strategies.

Specialized methods that address specific situations and aim to get information most efficiently, such as cluster sampling and stratified sampling, are often used. There are also many pitfalls associated with conducting sample surveys, ranging from poor questionnaire construction to appropriate sample selection. Nonresponse and inaccurate responses are of particular concern. Statistical concepts help market researchers develop sampling schemes that aim to mitigate such problems and to assess the possible bias that remains. As in other applications, this requires more than an appreciation of statistics – product and marketing experience is essential; psychological expertise is also useful.

Various books discuss survey sampling. These include Kish (1995), Lohr (1999), Groves, Fowler, Couper, Lepkowski, Singer, and Tourangeau (2004), and Scheaffer, Mendenhall, and Ott (2005). We also return to this topic in Section 13.2.3 in discussing the food, beverage, and related industries.

Analysis Methods Methods for analyzing the data from sample surveys range from elementary descriptive statistics to specialized multivariate techniques that explore relationships between product features and customer preferences. Conjoint analysis (see Sidebar 3.2) is often used to assess consumers' preferences for product features.

SIDEBAR 3.2: CONJOINT ANALYSIS FOR EVALUATING CUSTOMER PREFERENCES

Conjoint analysis uses statistically planned experimentation to help identify the most important product characteristics and trade-off considerations (e.g., cost versus performance). For example, in designing a new bar of soap, it was desired to assess customer reactions to product weight, size, shape, color, surface feel, type of fragrance, and intensity of fragrance. Participants in a survey were asked to use samples, involving different combinations of these variables, selected by a statistical plan, and to indicate what they considered the most appropriate purpose for each (e.g., moisturizing, deep cleaning, use for oily skin, deodorant). Analysis of the resulting data provided important inputs for designing the final product and for advertising.

Green and Wind (1975) give a detailed description of conjoint analysis and further details of the preceding example. Also Green and Srivasan (1990) discuss some experiences and developments and provide numerous further references.

3.3.2 Tailoring Product Requirements to Customer Needs

Customer needs result in product requirements that must be translated into measurable product characteristics, which, in turn, provide actionable goals for the design team (see Table 3.1 for three examples).

TABLE 3.1 From Customer Needs to Product Characteristics

Customer Need	Product Requirement(s)	Measurable Product Characteristic(s)
Convenient, readily usable, easily transportable, dependable laptop computer	• Light weight • Long battery life • Easy to carry • Ease of use in a variety of environments • Reliable	• Total weight • Battery life • Physical dimensions • Ergonomic design and feedback from user panel • MTBF (Section 5.5.3)
Quiet operation of dishwasher	• Vibration insulation • Acoustic dampening	• Noise levels of selected washes
Dent resistant, reliable car bumper	• Energy absorption • Paint surface durability	• Elastic yield strength • Ultimate tensile strength • Paint peel resistance

Identifying Product Characteristics Product characteristics reflect different attributes of a product such as it features, performance, aesthetics, or reliability. Companies aim to gain a competitive edge by providing new and imaginative product characteristics to serve stated and perceived customer needs. The introduction of new features in cars at the beginning of each model season provides a good example.

Langley, Nolan, Nolan, Norman, and Provost (1996, Chapter 12) elaborate on the characterization of product features relative to customer expectations, originally proposed by Noriaki Kano:

- Must-have characteristics: Customers expect the product to have these features and may not even state this expectation. For example, drivers expect a car to start when the key is turned in the ignition.
- More-the-better characteristics: Gas mileage of a new car is a typical example.
- Delighter characteristics: Innovative features that excite the customer. DVD players and GPS systems were delighters when first introduced in cars.

These definitions often change rapidly over time. What is considered a delighter in an earlier generation product may become a must-have for a future generation.

Identifying product characteristics, once customer needs have been identified, is principally a product domain expertise and marketing and engineering task. Statistics contributes initially by helping establish and assess meaningful and reproducible numerical measurements for such characteristics.

Quality Function Deployment (QFD) QFD is a tool that facilitates translating multiple customer needs into product characteristics and, subsequently, product characteristics into design requirements, and design requirements into manufacturing requirements. It may lead to prioritizing competing product characteristics. QFD is described through a case study in the next chapter. (See Akao 1990 and Tague 2004 for further description and Wheeler 2005 for a discussion of some limitations.)

3.3.3 Setting Targets and Specification Limits on Product Characteristics

To help assure that customer and business needs are met (and to evaluate the design as it evolves), quantifiable goals need to be established for each important product (or performance) characteristic. These are generally expressed in terms of target values and specification limits.[1]

Target-Is-Best Product Characteristics For some product characteristics, it is desirable to be as close as possible to a target value. Taguchi called these "target-is-best" product characteristics. For such characteristics, the goal is to have the statistical distribution of the characteristic be centered on its target value with minimum variability. A few examples are:

- The dimensions of parts that need to be fitted together.
- The actual sizes of shoes that have the same nominal size.
- The temperature maintained by a thermostat.
- The quantity of milk in a labeled "one-gallon" container.

Two-sided specification limits around the target value are typically established for target-is-best product characteristics. (The target value is usually the midpoint of the resulting specification range.) These define the "permissible" variability of individual units. For example, the specification limits on the length of a part with target length of 15 cm might be set as 14.99 cm to 15.01 cm.

Higher-Is-Better or Smaller-Is-Better Product Characteristics For other product characteristics, "higher-is-better" or "smaller-is-better." For such characteristics, we strive to attain consistently high or consistently low average values. Some examples are:

- The image resolution of a photograph.
- The degree to which a washing machine removes stains from soiled clothing.
- The life of most products.
- The gasoline consumption and carbon monoxide emission of an automobile.

Product characteristics for which the goal is "higher-is-better" or "smaller-is-better" typically have one-sided specification limits.[2] To be acceptable, a product needs to exceed its lower specification limit or fall below its upper specification limit. For example, gasoline mileage for an automobile needs to exceed a specified lower limit and gas emissions need to fall below a specified upper limit.

1. Product specification limits need to be clearly differentiated from control chart limits (Section 6.7) and from statistical tolerance limits (Hahn and Meeker 1991).
2. To drive for consistency it is useful to define a target value also.

Meeting Specs Is Not Enough All product that meets specification limits is *not* the same, as illustrated by the example in Sidebar 3.3. There is often little difference between a product that just meets specifications versus one that just falls outside.

SIDEBAR 3.3: A TALE OF TWO MANUFACTURERS

Taguchi and Clausing (1990) report that transmissions for a particular model automobile were built, according to the same parts' specifications, by both Ford and Mazda. However, the transmissions manufactured by Ford generated greater scrap and rework, and also higher warranty costs and customer complaints about noise. Study revealed that while the transmission parts were within specification limits for both manufacturers, those built by Mazda, unlike those from Ford, exhibited very little variability around their target values. As these parts were assembled into transmissions, the deviations from target added up, resulting in the Ford-built transmissions having greater friction and more vibration than those assembled by Mazda.

The Need for Low Variability Instead of focusing solely on meeting specifications, we need to also recognize that customers demand consistent products and services. Variability may arise in *repetitive use* of a single unit, such as a washing machine, a camera, or a mobile phone. Or it may involve variability *from one unit to the next* for a frequently purchased product. The consumer of a soft drink, for example, expects the same taste for each purchase of a particular brand.

Other products, such as a specific item of clothing or a particular toy or brand bed, are typically bought only once by a particular consumer. Thus, homogeneity among units may not be of equal concern. However, there is still a need to minimize variability in, say, the actual size of different shirts of ostensibly the same size and the differences in appearance of a left and right shoe of the same pair or of chairs in the same set.

Achieving low variability is also a key internal business goal. Variations in manufacturing processes increase variability in product performance and costs and hurt productivity. For example, variation in part sizes can cause delays in assembly, as well as problems in the performance of the final product.

Reduction of variability can also result in less obvious benefits. Consider a tire whose life is guaranteed to exceed 30,000 miles. A manufacturer is producing tires with a mean life of 45,000 miles and a standard deviation of 5,000 miles (Figure 3.1a). Essentially no tires have a life less than 30,000 miles and, at first glance, reducing variability does not seem needed. However, if the standard deviation were reduced to say 1,500 miles (Figure 3.1b), the life guarantee can be raised to 40,000 miles, resulting in greater sales or permitting a higher price. Alternately, such reduction in variability may let the manufacturer cut costs by using less raw materials, reducing mean life to 35,000 miles.

Achieving low variability must, of course, be combined with attaining satisfactory average performance. Golfers do not want to get consistently poor scores and manufacturers do not want to build uniformly inferior product. There are, moreover, some applications for which a certain degree of variability in product performance is actually desirable (see Sidebar 3.4).

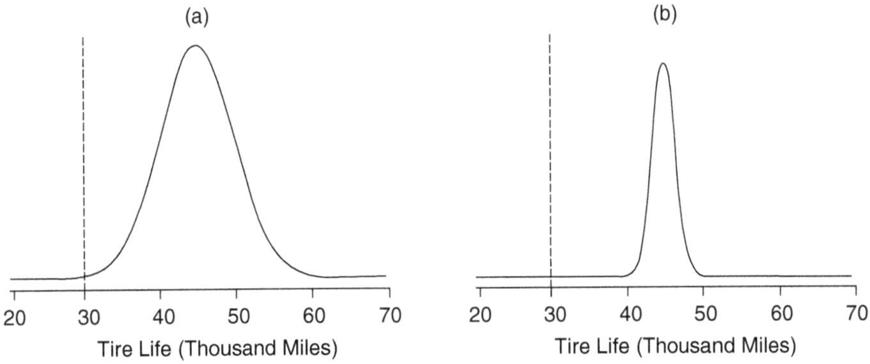

Figure 3.1 Tire lifetime distributions before (a) and after (b) variability reduction (dashed line shows manufacturer's warranty).

SIDEBAR 3.4: VARIABILITY IS NOT *ALWAYS* EVIL

Variety is the spice of life.
—Popular saying

Some examples of *desirable variability* are[3]:

- Candies of many different colors in a bag of candy make them more attractive.
- Variety in the size and shape of vegetables in frozen food entrees tend to be preferred by consumers and diminishes their machine-made appearance.
- Varied popping times of kernels in a bag of popcorn make the popping experience more interesting.
- Financial portfolio managers minimize the uncertainty in total return by increasing the diversity of their individual holdings (thereby averaging out losses).

Even in these cases, however, it is important to have some *consistency* in the variability, as, for example, among batches of manufactured products.

Quantifying the Cost of Variability Taguchi proposed formal methods for quantifying the cost of variability (see Phadke 1989 and Taguchi, Chowdhuri, and Yuin 2004). For target-is-best characteristics, he proposed the quadratic loss function around the target value, illustrated in Figure 3.2, to quantify the cost of deviating from target. To minimize cost, designers must seek design solutions that favor on-target performance with minimum variability (see Creveling 1997 for examples).

Multiple Product Characteristics There are frequently several, often conflicting, product characteristics that may compete with one another. Some examples are provided in Sidebar 3.5. For simplicity, these examples deal with just two product characteristics. In practice, there are frequently more than two, and sometimes a large number of, such characteristics. For a product to be acceptable, all critical characteristics must meet their specification limits.

3. These examples extend beyond product design.

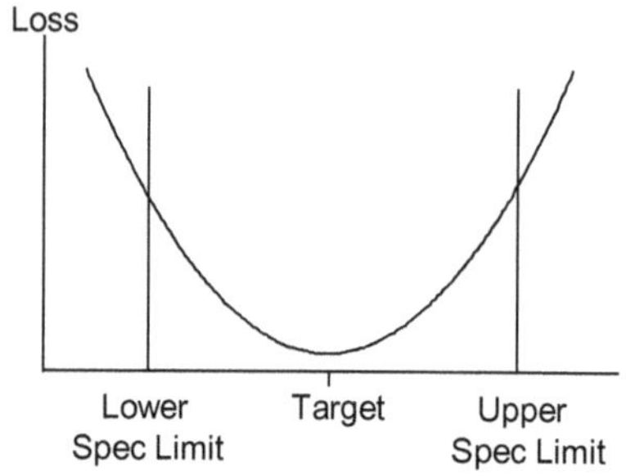

Figure 3.2 Taguchi's quadratic loss function.

SIDEBAR 3.5: A BALANCING ACT: SOME COMPETING PRODUCT CHARACTERISTICS

- Laptop computers: Customers desire both increased functionality and reduced weight.
- Aircraft engines: Manufacturers desire higher engine thrust and reduced emissions.
- Fruit: Customers desire a ripe product, but suppliers wish to minimize spoilage.
- Adhesive sealants: Decreased curing times are desirable to improve manufacturing efficiency, but these might compromise adhesion strength.
- Automobiles: Customers want roomy cars, but still get good gas mileage.

As the examples in the sidebar illustrate, designers often need to make trade-off decisions to meet conflicting requirements. A specific method for trading off competing product characteristics, using desirability functions, is illustrated by the example in Section 4.6.2.

The Art of Setting Targets and Specification Limits Often the stated needs of the customer are insufficient to define targets and specification limits. Statements such as "as strong as possible," "aesthetically appealing," or "better than the existing product" need to be converted into measurable and attainable goals. For example:

- What is sufficiently high resolution in digital photography to achieve high image quality?
- What defines cleanliness for laundry from a washing machine?
- What is "long life" for a tire?

Targets and specification limits are based on customer wishes and also on competitive, technical, and financial considerations. A key to making such determinations is an understanding of how a product is to be used, for example, the intended operating conditions and use environment. Businesses often try to gain competitive advantage by broadening the ranges set on operating conditions relative to current market product offerings; this sometimes necessitates a major redesign. It is, therefore, important to understand what can be reasonably achieved in practice and the cost associated with alternative product requirements.

In addition, an appreciation of the characteristics of current, and likely future, competitive products is required. Sometimes, one may utilize data obtained from past comparable products built by the manufacturer, and from competitive products, to provide benchmarks for the new design requirements. Frequently, however, the needed data, especially from competitors, is not readily available. In that case, such information might have to be obtained, often by appropriate sampling studies of competitive products (Section 3.3.1).

3.4 ENSURING MEASUREMENT CAPABILITY

Measure twice, cut once.
—Anonymous

To design a product successfully, we need to measure important product properties, as well as those of materials, components, and subsystems. Also, we require accurate measurements of the manufacturing processing conditions, such as temperatures, pressures, and flow rates. Engineers rely on such measurements to evaluate design alternatives, to assess the impact of design variables, and to identify appropriate adjustments in product design and processing.

Statistical methods help assess and ensure the integrity of the measurement system. This requires:

- Evaluating and removing possible bias in measurements.
- Ensuring that the measurement variability is appropriately small.
- Surveillance of the measurement system to ensure that it continues to perform well.

We briefly consider each of these elements (see also Snee 2006 and the "Measure" chapter of the online *Engineering Statistics Handbook* developed by NIST and Sematech[4]).

3.4.1 Evaluating and Removing Bias

This step deals with evaluating whether repeated measurements of the same object are on target *on the average*. A scale, for example, needs to target in on the true weight of the object it weighs.

Gold Standards In assessing whether a measurement system is biased, one would ideally like to know the true value of what is being measured, or something close to it. Sometimes there is a so-called "gold standard" available for such calibration, obtained in the United States, for example, from NIST or a professional group, such as the American Society for Testing and Materials (ASTM). Bias is then estimated as the difference between the average of a large number of measurements on the gold standard and its "true" value.

4. A consortium of semiconductor manufacturers.

Follow-Up If a bias or other consistent difference is found, the next step is to take appropriate steps to remove it – or, if that is not possible, correcting measurements accordingly, taking care not to overcorrect. An important element of this is determining that the bias is constant over time.

Other Methods In many situations, especially when the measurement is subjective (e.g., evaluating softness of a facial tissue, taste of a chocolate chip cookie, or appeal of a mobile phone cover) a gold standard may not be available. In such cases, we strive to ensure, as a minimum, that our measurements are in agreement with those taken by our customers and/or suppliers, who may be using different measurement systems. This frequently calls for conducting a well-planned comparative assessment of the system. Such evaluations are, in fact, useful even if there is a gold standard.

3.4.2 Reducing Measurement Variability

This step is concerned with *precision* of the measurement system. Variability is present in all measurement systems and may arise, for example, due to the use of multiple measuring devices, multiple test operators, differences in preparation of samples for testing, and many other factors. It is desirable to reduce measurement variability – and necessary to do so when measurement error is unacceptably high. This typically requires breaking down total measurement error into its constituent components and estimating each.

Repeatability and Reproducibility In assessing measurement error, we need to differentiate between repeatability and reproducibility. Repeatability is the measurement variability that arises when the same item is measured repeatedly under identical conditions – in, for example, the same setup by the same operator and under essentially the same ambient conditions. Reproducibility is the variability resulting from introducing the type of changes that are normally encountered in taking measurements at different times, due, for example, to differences between operators and setups, and to fluctuations in ambient conditions.

Documented Measurement Variability Manufacturers often provide statements about an instrument's measurement capability. There are, however, no universal standards on how to report such information. A digital thermocouple manufacturer might claim device accuracy of ±0.5%. This is typically interpreted to mean that "most" measurements are within ±0.5% of their true values, where "most" remains undefined. Such statements, moreover, generally deal with repeatability only and do not encompass reproducibility.

Gauge Repeatability and Reproducibility (GRR) Study The set of data that is typically available, from say past manufacturing, often does not lend itself to estimating measurement error, much less its constituent components. A special investigation, often referred to as a GRR study, may thus be required.

TABLE 3.2 Typical Layout for a GRR Study

	Operator	A						B						C					
	Setup	1		2		3		1		2		3		1		2		3	
	Measurement	1	2	1	2	1	2	1	2	1	2	1	2	1	2	1	2	1	2
Part Number	1																		
	2																		
	3																		
	4																		
	5																		

Table 3.2 shows the layout of a typical GRR study. This example involves two repeat measurements on each of three different setups by each of three operators on each of five parts.

A GRR study needs to be carefully planned to suit the specific circumstances (see Sidebar 3.6).

SIDEBAR 3.6: PLANNING A GRR STUDY

A GRR study needs to be designed under sufficiently broad conditions to incorporate all major sources of measurement variability encountered during normal operations. The standard GRR study shown in Table 3.2 is thus often extended to include other pertinent sources of measurement variation. For example, if multiple instruments or test facilities are used, these need be incorporated into the study. In many applications, parts go through a special process (e.g., drying, moisture removal, and molding test bars) prior to testing. Such preparatory steps also need to be included in the GRR study since they contribute to measurement variability as well.

A poorly designed GRR study is likely to underestimate measurement variability. For example, when the study calls for replicate measurements on the same part by the same operator, operators – in the absence of clear instructions to the contrary – may find it convenient to perform these measurements back-to-back without changing the setup.[5] This, however, may not permit estimation of variability between setups.

The nature of the measurement need also be taken into consideration. For example:

- Measurement systems that yield attribute (e.g., pass/fail or go/no-go) responses. These usually require larger samples to achieve a desired level of precision than studies with continuous measurements.
- Destructive testing, permitting only a single measurement on any given part.

5. A setup may involve sample preparation, alignment of test equipment, running calibration samples, assay preparation, etc.

GRR Data Analysis The results of a GRR study are frequently presented graphically and evaluated statistically. Analysis of variance of a well-designed study permits breaking down the estimated total variability into individual components of variation (in the form of standard deviations) associated, for example, with operator, equipment and setup variability (quantifying reproducibility), and residual (or "pure") measurement error (quantifying repeatability). A further key outcome of such evaluations is the estimate $\hat{\sigma}_{MEASUREMENT}$ of the overall measurement error standard deviation, $\sigma_{MEASUREMENT}$.

GRR studies are considered further in Section 6.6.2 and a specific example dealing with assessing measurement error is provided in Section 7.4.2. Also see Ott, Schilling, and Neubauer (2005, Chapter 16) and Burdick, Borror and Montgomery (2005) for further details.

GRR Index The GRR index (also known as the precision-to-tolerance ratio) is a widely used metric to assess the overall capability of a measurement system relative to specification limits. This is usually expressed for measurements with two-sided specification limits as the percentage

$$\text{GRR Index} = 100\frac{5.15\sigma_{MEASUREMENT}}{(\text{Upper Specification Limit} - \text{Lower Specification Limit})}.$$

The basis for the constant 5.15 is that for normally distributed and centered measurement errors, 99% of the measured values are within $\pm 2.575\ \sigma_{MEASUREMENT}$ of their true values. Thus, the GRR index quantifies the proportion of "permissible" variability defined by the specification limits taken up by measurement error.

As a general (but not always appropriate) guideline, a GRR index greater than 30% is deemed unacceptable, 10% to 30% is considered marginal, and 10% or less is desirable (see Automotive Industry Action Group 1990).

In situations involving a one sided specification limit, it is customary to use

$$\text{GRR Index} = 100\frac{5.15\sigma_{MEASUREMENT}}{2(\text{Specification Limit} - \text{Target})}.$$

Improving the Measurement System The preceding analyses provide important information to help reduce measurement variability. Possible corrective actions to reduce individual components of measurement variability might include:

- Providing improved training to operators.
- Obtaining more precise measurement equipment.
- Improving sample preparation procedures.
- Taking (added) repeat measurements and averaging the results (often the least desirable course).

After the appropriate actions are taken, the previous studies might be repeated, perhaps in reduced form, to verify that the desired results have been obtained.

3.4.3 Continued Surveillance

Evaluation of measurement error is not a one-time undertaking; changes are likely to occur over time. Appropriate monitoring is required to check that measurement error continues to be at a reasonable level.

This may call for repeat measurements on a standard part or calibration sample, or on the gold standard if there is one, over time. The GRR study can provide benchmark information for such ongoing investigations. The resulting data may be plotted on control charts (Section 6.7) to signal measurement bias or changes in the magnitude of measurement error.

3.5 DESIGN AND DEVELOPMENT

We need statistical techniques that are designed to catalyze rather than to replace engineering inventiveness.
—George Box

The design of a new product is driven by subject domain science and engineering knowledge. As part of the design process, organizations rely heavily on analytic approaches, models, hard data, and industry-specific practices and regulations. Statistically planned investigations, that is, the *design of experiments*, can play an important role throughout the product design and development cycle to assess new materials, compare alternative designs, evaluate design changes, make competitive evaluations, test prototypes, optimize manufacturing conditions, and so forth (see Sidebar 3.7). The resulting data might then be analyzed by methods ranging from simple plots to advanced statistical methods.

SIDEBAR 3.7: DESIGN OF EXPERIMENTS: A BRIEF OVERVIEW

There is no such thing as a failed experiment, only experiments with unexpected outcomes.
—R.B. Fuller

An experiment is typically conducted under controlled conditions to assess the impact of one or more process variables (*X*s) upon one or more response variables (*Y*s). The (statistical) design of experiments (DOE) deals with the development of test plans to obtain the maximum amount of unambiguous information from a statistical (and/or graphical) analysis of the resulting data. In a DOE, multiple experimental variables are changed *together*, as opposed to the traditional approach of varying variables one at a time. This permits evaluation of interactions among designated process variables.[6]

Popular Test Plans

Types of plans that have been found especially useful in business and industry include:

6. Two variables are said to interact if the impact of one of the variables upon the response variable is dependent upon the associated setting of the second variable.

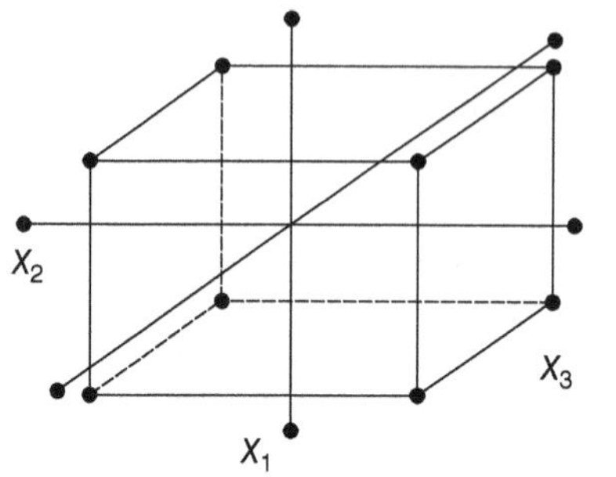

Figure 3.3 A central composite design with three experimental variables.

- Full factorial designs: These require running all combinations of all conditions of the experimental variables. For example a 2^6 full factorial involves running all 64 combinations of six variables at each of two levels or conditions. Full factorials are especially useful when there are a relatively small number of variables and levels.
- Fractional factorial designs: These involve running a selected fraction of experimental runs from a full factorial plan. Fractional factorials are especially useful for screening purposes to identify the most important variables.
- Response surface designs, such as a central composite plan (see Figure 3.3): These are especially useful in dealing with quantitative variables whose impact on one or more response variables is to be estimated by regression analyses.

Specific applications of these plans are shown in Section 4.6.2.

Sequential Experimentation

DOE can be considered as a learning tool in a scientific process, as stressed by Box, Hunter, and Hunter (2005) and Box et al. (2006). This results in an iterative cycle of producing a hypothesis, gathering data through experimentation, revising the hypothesis based on such data, gathering more data, and so on.

Where to Go to Learn More

There are numerous texts on DOE. Box, Hunter, and Hunter (2005) is a popular easy-to-read introduction. Myers and Montgomery (2002) is at a more advanced level and emphasizes designs for response surface estimation, optimization and robust product design (see below). Coleman and Montgomery (1993) and Hahn (1977) focus on the steps involved in DOE.

3.5.1 Conceptual Product Design

A key to successful product design is to develop and evaluate alternative concepts early on. The ultimate selection of a design concept is based on a multitude of criteria, including cost, time to market, technology risk, existing manufacturing process, and competition, in addition to customer wishes. Design teams utilize various engineering and project management tools, such as Failure Mode and Effects Analysis (FMEA, see Section 5.7.4); Quality Function Deployment (QFD, see Section 4.4.2);

and Pugh concept evaluation (see Pugh 1990 to facilitate such assessments; see also Tague 2004). The evaluations at this stage – though at a high level since all the details have not yet been worked out – are important in identifying the potential promise and weaknesses of alternative design concepts.

Redesign In some situations, requirements for a new product can be met through a modification of an existing product, often referred to as a redesigned, tweaked or "derivative" product.

Say, for example, that we need to substitute a type of material used in the product with a different type because of economic, environmental, logistical or cost considerations. Scientists and engineers with fundamental physical knowledge of the product identify suitable alternative materials, develop the redesign, and appropriately modify the manufacturing process. DOEs may be employed to compare alternative materials, identify interactions, estimate sensitivity to variability in material properties, and so forth. In addition, empirical evidence is usually required to ensure that the change does not have an adverse impact on reliability.

New Design In other cases, existing technology cannot be stretched further to meet customer needs and expectations in an economical manner. Conventional automobile engines could, for example, be enhanced up to a point to meet the demand for better fuel efficiency and lower emissions. Beyond that point, new technologies, such as the hybrid engine, were required to provide the desired performance.

Some important discoveries have occurred by a combination of accident and keen observation. We can, however, not leave things to chance in today's environment in which companies compete ferociously to achieve valuable patent rights and to be the first on the market. After a preliminary design has been developed, many such efforts require the disciplined empirical approach provided by DOE – closely coupled with physical understanding and assessments – to test models and theories, provide proof of concept, and sometimes even aid in the discovery itself. Sidebar 3.8 gives an example from the chemical industry.

SIDEBAR 3.8: COMBINATORIAL CHEMISTRY FOR NEW PRODUCT DISCOVERY AND IDENTIFICATION

Basic Concept

Combinatorial chemistry[7] provides a major, and to date still somewhat untapped, opportunity area for the statistical planning of experiments. The goal is to establish a library of performance responses and approximate functional relationships for different combinations of materials that make up a chemical compound. These are then searched to tailor-make a product to address the specialized needs of an important customer and to discover new materials with desired properties.

7. Combinatorial chemistry evolved from an approach, often referred to as "high throughput screening," used by pharmaceutical companies to identify potentially promising drugs.

Combinatorial chemistry has become "practical because of the convergence of low-cost computer systems, reliable robotic systems, sophisticated molecular modeling, statistical experimental strategies, and database software tools" (see Cawse 2002). It is now, for example, possible to make up and assess 100 or more vials a day of different combinations of product.

In one typical application (see Cawse 2000), there were:

- Three qualitative formulation variables with 20, 20 and 10 conditions, respectively.
- Three quantitative formulation variables, each with three conditions.
- Three quantitative process variables, each with three conditions.

This resulted in 2,916,000 possible combinations; even with 100 vials per day, one could explore only a small fraction of these in the time available. Thus, it made sense to turn to DOE and, in particular, fractional factorial plans for such "finding a needle in a haystack" experimentation.

3.5.2 Identifying and Quantifying the Impact of Critical Variables

Among the factors to be considered there will usually be the vital few and the trivial many.
—J.M. Juran

In moving from a basic design concept to a practical product, the system is typically broken down into subsystems or assemblies, and individual components (see Table 3.3). The individual parts are developed, usually working from the bottom up, by individual designers or small teams. Their integration into the overall design, however, requires ongoing collaboration among all involved.

Designers need to determine the critical process variables (*X*s) that impact product performance (*Y*s). For complex products, the number of *X*s can be quite large. Table 3.4 provides some examples.

TABLE 3.3 System Breakdown Examples

System	Subsystems	Components
Automobile	Engine, transmission, body and chassis, fuel and exhaust, electrical system, suspension and steering, brakes, and climate controls.	Transmission subsystem: Housing, input shaft, output shaft, gears, roller bearings, oil pump, clutches, seals and gaskets, activation linkages, and drive axles.
Air conditioner	Condenser, evaporator, compressor, heat exchanger, and electrical controls.	Compressor subsystem: Filters, valves, and piston rod.
Personal computer	Input devices, storage, processor(s), communication, and output devices.	Storage subsystem: RAM, hard drive, and tape reader.

TABLE 3.4 Examples of Product Characteristics and Impacting Variables

Type of Product	Product Characteristics (*Y*s)	Some Impacting Variables (*X*s)
Concrete mixture	Strength, setting time, and elasticity	Type of cement, amount of water, mixing time, and curing conditions
Machine tool	Quality of cut and tool life	Cutting speed, tool geometry, tool material, cutting angle, and age
Chemical reaction	Yield, purity, and viscosity	Concentration, temperature, pressure, agitation rate, mixing method, reaction time, and amount of catalyst

The goal is to quantify the underlying relationships between the *X*s and *Y*s, based upon assumed engineering and/or empirical models. These relationships are then used to find optimal settings of the design variables that do the best job of meeting the product targets and specifications. This often requires trade-offs among the product characteristics.

Engineering Models Much of the understanding at this stage comes from physical models, domain knowledge, and experience. Some examples are:

- kinetic models for a chemical reaction,
- circuit model for an electronic system, and
- differential equations describing heat flow in a mechanical system.

Such models enable designers to vary the settings of the impacting variables in order to evaluate their effects on product characteristics. Computer models are widely utilized in the early stages of design before any prototypes are built.

Example: Finite Element Analysis (FEA) FEA (see Benhabib 2003) is a computer-modeling technique used by designers to assess, for example, the impact of stresses and vibrations in a gas turbine assembly, cooling of a car engine, and metal fatigue in an aircraft engine. As these examples suggest, FEA typically involves stresses and strains, fracture and crack propagation, and heat and fluid flow.

FEA for a Gas Turbine Component The transition piece of a gas turbine helps hot combustion gases move from the combustor to the turbine. The piece squeezes gases, thereby increasing their flow and exit velocity. A major goal in designing this component is to maximize exit temperature and gas velocity, subject to ensuring that the component functions satisfactorily between, as a minimum, periodic inspections typically conducted every two years. The inner surface of the transition piece is coated with a thermal barrier coating (TBC) to protect the underlying metal from oxidation and creep deformation.

The designer needs to understand the effect of alternative design configurations on the exit temperature and velocity of the gases. For such complex problems, a closed-form analytic model is often unavailable. Instead, FEA is used to convert

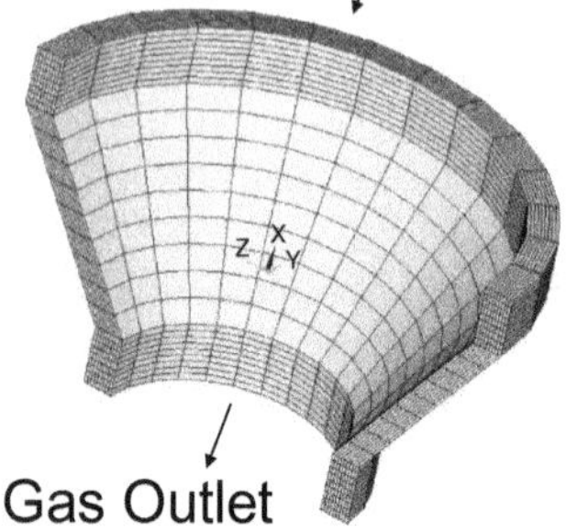

Figure 3.4 Finite element model mesh of a gas turbine transition piece.

the problem into many smaller ones, each of which can be evaluated by intensive computer calculations.

The starting point for a FEA is a solid model – essentially a three-dimensional graphical representation of the component. The model represents the part as a collection of individual finite elements that are connected through nodes, commonly referred to as a mesh, and created by CAD software for geometric modeling. Figure 3.4 illustrates the mesh for a cross section of the transition piece.

Underlying the finite element model are the mechanical and thermal relationships that govern the behavior of each element in the mesh. These are expressed by a set of equations based on fundamental domain knowledge built into the FEA software. The software evaluates the equations at each node and aggregates the results to determine the stresses, thermal displacements, and velocities. These are used, in turn, to determine the exit temperature, velocity, and expected life of the component (the *Y*s).

FEA and DOE So where does statistics fit in?

The outcomes of the analyses in the example depend on the settings of numerous impacting variables shown in Table 3.5. In this example, each analysis for a specified set of *X*s requires intensive computations, taking possibly hours of computer time to yield the resulting *Y*s. Multiple runs, at different combinations of the *X*s are then required to search for the best combination of these impacting variables. This can become a daunting task; in Table 3.5 there are 12 *X*s whose impacts on temperature, velocity and life (the *Y*s) needs to be evaluated.

Statistical DOE and analyses provide a framework for addressing this situation. The resulting plan designates the combinations of the *X*s for which computer runs are to be performed, probably in a sequential or multistep manner. These are then used to develop models to approximate the relationships between the *X*s and the *Y*s.

Computer Experiments The preceding example calls for a special use of a DOE, sometimes referred to as a "computer experiment." The *Y*s are not

TABLE 3.5 Impacting Variables in a Finite Element Analysis of Gas Turbine Component

Impacting Variable	Comments
MATERIAL CHARACTERISTICS Thermal conductivity of TBC Thermal conductivity of metal Elastic modulus of metal Thermal expansion coefficient of metal Strain-to-fail value of TBC	*Once the type of materials is specified by the designer, the values of the variables are obtained from material databases or lab tests. These are also impacted by industry regulations, safety considerations, etc.*
PART CHARACTERISTICS Thickness of TBC Thickness of metallic wall Cone angle variation in axial direction Ratio of exit area to inlet area Overall length of transition piece	*These variables are related to the geometric dimensions of the part and are set by the designer.*
AMBIENT CONDITIONS Inlet velocity Inlet temperature	*The values of these variables are determined from past operational data and customer inputs on the planned use of the equipment. Ability to handle higher inlet velocity and temperature provides a competitive advantage to the manufacturer.*

random observations, but the result of a deterministic computer calculation; Kenett and Steinberg (2006) describe this approach and illustrate it with a piston design example.

Limited Domain Knowledge In the gas turbine component example, our evaluations were guided by finite element analysis. In other situations, the underlying theory and knowledge may be insufficient to provide such explicit models. In such cases, physical understanding might, at best, suggest *general model forms* that relate the appropriate *X*s to the *Y*s, without knowledge of some or all parameters contained in the models. Some simple examples are provided by the life acceleration models described in Section 5.7.8. In such cases, DOE can be used to get the needed data to estimate the unknown model parameters statistically.

In yet other cases, we do not know the general form of the underlying model and may not even know which among a set of potential *X*s are the important impacting variables. In such cases, we typically use DOEs to first identify the important *X*s and then to obtain the data to fit a statistical relationship between these *X*s and the *Y*s. This generally involves assuming some form of response surface model, as illustrated by the example in the next chapter.

It is generally preferable, when practical, to build models based on physical knowledge, as opposed to doing so empirically; see Sidebar 3.9.

SIDEBAR 3.9: GAMBLING MACHINE ANALYSIS

Say that you wish to assess the probability of coming out ahead after a specified number of trials in playing a slot machine. This requires knowing the probability of each possible outcome in a single use of the machine. These probabilities can be determined from knowledge of how the machine was designed to operate, that is, physical knowledge. Failing such knowledge, one can conduct an empirical study. This approach, however, is likely to be expensive and time-consuming and will not yield an exact result.

Of course, even if physical knowledge is available, one might still want to conduct an empirical study to assess whether the machine is operating as it should.

3.5.3 Achieving Design Robustness

> **No "optimal" scheme is ever *in practice* optimal. What we must aim for are schemes which are *robust* and *good* over a wide range of circumstances.**
> —George Box

Concept of Product Robustness The major emphasis in traditional product design was on assessing the impact of process variables on the *mean* of a response variable. Consequently, statistical DOEs were used mainly to achieve *average* product performance. Building a product that performs satisfactorily, on the average, however, is, as we have seen, not enough. Its performance also needs to be satisfactory *consistently*. This requires striving for so-called "product robustness" during design.

Robustness to Manufacturing Variability Products need be designed to be robust (or forgiving) to variations that occur during manufacture, such as machine wear or differences between batches of incoming materials or variability in ambient conditions. Product robustness is especially important in the food and beverage industry (Section 13.2). An early application was said to have taken place at that hotbed of statistics, the Guinness Brewery, in seeking barley that, though grown in different regions, would still result in a consistently good-tasting (and good-looking) beer.

Robustness to Variability in Product Use Products also need to be designed to operate consistently, and safely under a wide variety of operating environments and customer applications (and, sometimes, misapplications). For example:

- A photocopy machine needs to operate smoothly and not jam when loaded with different weight paper.
- A cake mix should result in similar looking and tasting cakes regardless of differences in oven type, variability in oven temperature around the set value, and the sequence of mixing ingredients.
- A car battery needs to run for a specified lifetime (or longer) in different operating environments.

- A washing machine should perform well and not spill over water, irrespective of the frequency of use, load, and type and amount of water and detergent.

Robustness to variability in the way customers use a product is especially relevant when the product is used globally. TV viewers want good pictures despite differences in power supply, weather conditions, or the turning on and off of other electric appliances. Likewise, cultural factors may be important. For example, manufacturers of kitchen appliances need their designs to be robust to differences in family sizes, types of bowls used for food storage, preparation and dining, and popular types of foods.

Alternative Approaches One way to address diversity in manufacturing conditions and in customer use requirements is to build different product models for different markets; for example, sell different car batteries with different operating features and characteristics in different parts of a country or the world. This, however, is not always cost-effective or practical. It can also lead to potential customer dissatisfaction – for example, when a car bought in a hot-weather location is driven to a cold-weather location.

Historically, designers have strived to achieve design robustness by:

- tightening manufacturing tolerances and/or specifications on incoming materials;
- designing for the worst-case scenario; and
- using safety factors in design – especially when risk of injury or human life is involved. A fire escape ladder required to carry a 500-lb load may, for example, be designed to carry 2,000 lbs.

The adequacy of the resulting designs was then typically evaluated by testing at what were believed to be the harshest conditions that the product might encounter in the field.

The preceding approaches might result in overdesigning a product and making it too costly. Also, they could compromise reliability. Suppose, for example, that a computer chip is designed to operate below a junction temperature of 150 °C. To be safe, one may then design the chip so as to keep its junction temperature below 130 °C. This may require a highly sophisticated (and costly) thermal control system, which, in turn, could compromise the reliability of the overall system.

Control and Noise Variables It is important to distinguish between two types of variables impacting product performance: control (or design) variables and noise variables.

Control variables are ones whose settings are under the direct control of the designer. For example:

- Relative amounts of different raw materials in a chemical formulation.
- Physical dimensions of an assembly.
- Properties of materials used in a mechanical component.
- Transistor gain in an electric power supply.

In contrast, noise variables are not controllable by the designer. Their values are, instead, determined by the production and use environments. Differences in incoming materials received from suppliers and fluctuations in ambient conditions beyond those being controlled are typical sources of variability in manufacturing. Similarly, differences in the operating environment, such as in temperature, humidity, dust, visibility, and so forth, create variability during product use.

In practice, a particular variable is often *both* a control and a noise variable. A manufacturer might, for example, control the properties of incoming material by setting and enforcing specification limits. The fluctuations within these limits can, however, not be controlled and, thus, create variability in product performance.

In the gas turbine transition piece example, the material and part characteristics are control variables since their nominal values are specified. In contrast, inlet temperature and velocity vary over a broad range, depending on the use profile of the customer, and, therefore, are noise variables. Moreover, some of the control variables also fluctuate around their nominal values in operation, and, therefore, also include a noise component.

Parameter Design: Concept Robust design aims to select the settings of the design variables at which the effects of the noise variables on performance are minimized. Technically, this involves exploiting interactions between the control and noise variables.

Taguchi advocated the use of statistically planned experimentation to help obtain the needed information to explore the impact of both the control variables and the noise variables, as well as their interactions, on the product characteristics. (Even though noise variables cannot be controlled in normal production and field use, it is often possible to do so in a controlled experiment.)

This approach differs from that used traditionally in designed experiments in that it considers the standard deviation as well as the mean of each product characteristic to be a response variable. The results are then used to design a robust product by setting the conditions of the control variables in a manner that takes into consideration their impact on the variability as well as the average of the product performance characteristics. This approach is often referred to as parameter design. Sidebar 3.10 describes some issues encountered in robust design applications.

SIDEBAR 3.10: AN ISSUE IN ROBUST DESIGN

The concept of robust design was originally introduced for applications dealing mainly with electronic products. In such applications, one can frequently control mean performance by appropriate adjustments, such as adding resistors, without much difficulty. Thus, defining the design conditions so as to control variability, and then setting the control variables to achieve the desired average performance, is often feasible.

In contrast, in many situations outside electronics, such as mechanical or chemical applications, it is often *not* possible to readily set the control variables so as to obtain the desired mean performance. This then results in a balancing act to optimize both mean performance and variability. All of this becomes even more complicated when there are multiple performance characteristics, some of which might conflict with one another, as in the example in the next chapter.

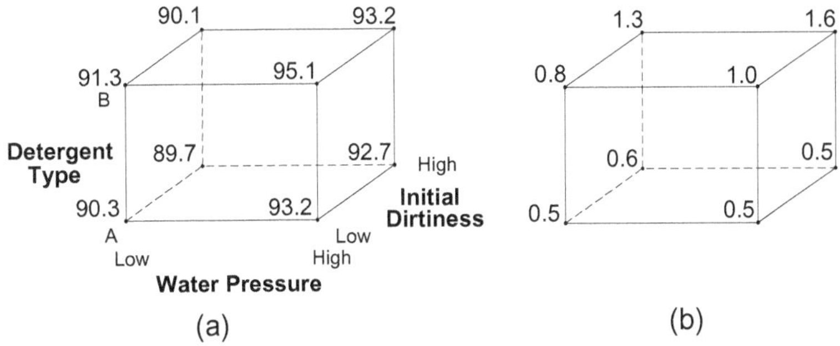

Figure 3.5 Cube plots for (a) mean cleanliness and (b) for standard deviation of cleanliness after wash.

Parameter Design: A Simple Illustration In a car wash operation, the key performance variable is cleanliness after wash, as measured by special instrumentation. We consider two control variables that impact after-wash cleanliness – detergent type (Type A versus Type B) and water pressure. There is also one key uncontrollable or noise variable: the initial dirtiness (low and high) of the car, which can, however, be controlled during experimentation.

It is desired to determine conditions for the two control variables that would result in *consistently high* after-wash cleanliness, irrespective of initial dirtiness; in particular, we wish to select a detergent type and water pressure so that 99% of all cars will have an after-wash cleanliness score of 90 or more.

To assess the impact of detergent type and water pressure on cleanliness for cars of low and high initial dirtiness, it was decided to conduct the two-level three-factor full factorial experiment, illustrated in Figure 3.5[8]. A total of 15 cars were washed at each of the eight experimental conditions. Figure 3.5 shows the mean and standard deviation of after-wash cleanliness attained at each condition.

It is clear from Figure 3.5 that in order to have a high proportion of cars achieve an after-wash cleanliness score exceeding 90, high water pressure is required. The decision with regard to detergent type is less clear-cut. We see from Figure 3.5a that detergent B tends to result in *higher average* cleanliness than detergent A. At the same time, we note from Figure 3.5b that there is appreciably *less variability* in cleanliness for detergent A than for detergent B, especially for cars with high initial dirtiness.

In the end, it was decided to go with detergent A – despite its lower average cleanliness – because it resulted in a larger percentage of cars, especially ones with initial high dirtiness, estimated to attain after-wash cleanliness scores above 90.[9]

Where to Go to Learn More Phadke (1989) and Taguchi (1986) provide detailed discussions of robust product design. Also see Taguchi and Clausing (1990)

8. This is a simplified example and ignores cost and throughput considerations. Details, such as how cleanliness is measured and randomization of testing, are also not discussed.
9. This conclusion was based on a statistical analysis that assumed cleanliness to be normally distributed.

and Taguchi, Chowdhuri, and Yuin (2004). Nair (1992) is a panel discussion and gives numerous further references. Grove and Davis (1992) and Myers and Montgomery (2002) blend classical DOE with experimental designs proposed for robust product design.

3.5.4 Quantifying System Variability from Component Variability

Final product (or system) performance depends directly upon numerous component variables, each of which generally exhibits variability. Part dimensions, raw material properties, and relative amounts of ingredients are just a few examples of the many variables that impact final product performance, but are themselves subject to variability.

To achieve the desired final product performance consistency, we need to understand how variability in the component variables impacts system performance. Such understanding allows us to perform trade-offs to attain the desired system performance consistency in an optimum manner. This leads to specifications on the degree to which the control variables may fluctuate during manufacturing.

Setting specification limits for product components and manufacturing processes, sometimes referred to as tolerance design, requires a disciplined up-front approach. Without this, serious problems are likely to arise when a design is transitioned to manufacturing (see Sidebar 3.11).

SIDEBAR 3.11: OVER-THE-WALL DESIGN

Historically, design engineering was limited principally to building a functional prototype of the product. The design team developed the product configuration and established component and process tolerances based on scientific and engineering principles, as well as judgment. This was often done with limited consideration of the variability of the parts used and of the manufacturing process, in general. It often resulted in tolerances on manufactured parts and processes being set too tightly or too loosely.

Manufacturing then had the task of building the product according to such "thrown over the wall" specifications. This often resulted in problems later, such as end-of-line rejects of the manufactured product. Typically, each unit would be inspected and those that did not meet specifications would be rejected and, possibly, reworked. Such methods were expensive and time-consuming. Recognition of these concerns has led to a close consideration of tolerances at the product design stage.

Simulation[10] and propagation of errors (POE) are two tools to relate component variability to product or system variability.

Example: Barrel Design We illustrate the methods through the design of a cylindrical barrel used for storage and transportation of a powder. A schematic of the

10. We will be principally concerned with simulation of randomly occurring events. This is often referred to as "Monte Carlo" simulation to differentiate it from simulation of deterministic systems. We do not make this distinction in wording.

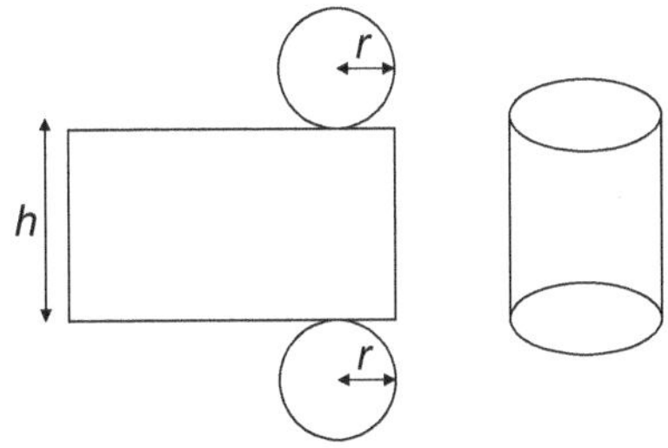

Figure 3.6 Schematic of cylindrical barrel.

barrel is shown in Figure 3.6. To build the barrel, a rectangular piece of material with height h is wrapped around two circular end pieces with radius r. The two end pieces are cut together and assumed to have essentially identical dimensions.

The barrel is required to enclose a volume (V) of 42 cubic inches (in^3). From elementary geometry:

$$V = \pi r^2 h. \tag{3.1}$$

Random samples of 163 rectangular pieces and 275 pairs of circular pieces were taken from inventory. These had been produced by processes similar to the ones to be used to manufacture the barrel. The data suggested that both r and h were approximately normally distributed and yielded the following estimates for their means and standard deviations:

$$\bar{x}_r = 1.811\,\text{inches},\ \bar{x}_h = 4.172\,\text{inches},\ s_r = 0.012\,\text{inches},\ s_h = 0.053\,\text{inches}.$$

We wish to determine the percentage of barrels that will meet the minimum volume requirement of $42\,\text{in}^3$ and make appropriate adjustments if this proportion is less than 99%.[11]

Simulation Simulation lets you evaluate the statistical distribution of system performance (volume in the example) from the assumed statistical distributions of the individual component variables (radius and height). It is a versatile and flexible tool that is applicable to a wide variety of problems.

Simulation involves building a model of a process or operation on a computer and then exercising that model repetitively, using randomly generated values of the input variables to approximate the distributions of the output variables. To use simulation, you need know the functional relationship between the input and output variables as well as the statistical distribution and associated parameters of each of the input variables.

Application to Example In the barrel design example, r and h were assumed to follow a normal distribution with the previously estimated means and standard deviations.[12] One then randomly generates components (values of r and h) from each

11. In practice, additional factors, such as cost, competitive offerings, and long-term strategic goals also need to be considered.

12. It was also assumed that r and h are uncorrelated. This was reasonable since the circular pieces and end pieces were built separately.

TABLE 3.6 First Four Simulated Values of Barrel Volume

Simulation Run	Radius r Randomly selected from $N(1.811, 0.012)$	Height h Randomly selected from $N(4.172, 0.053)$	Barrel Volume $V = \pi r^2 h$
1	1.812	4.171	43.024
2	1.806	4.277	43.825
3	1.799	4.172	42.419
4	1.801	4.227	43.073
...			

$N(\mu, \sigma)$ denotes normal distribution with mean μ and standard deviation σ.

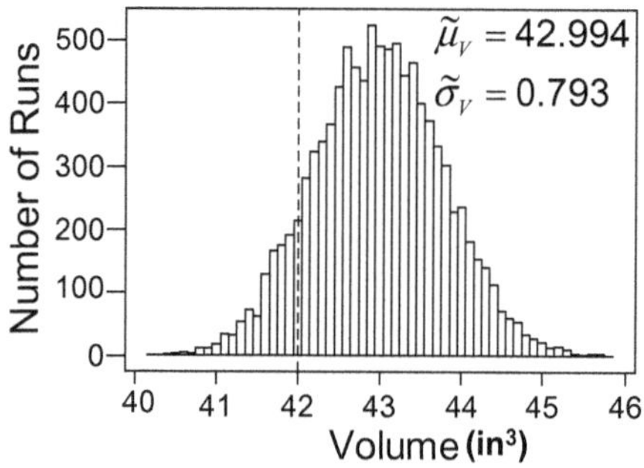

Figure 3.7 Simulated distribution of barrel volume.

of their distributions, and, using equation (3.1), determines the volume of each of the resulting computer-simulated barrels. This procedure is repeated 10,000 times to obtain a statistical distribution for barrel volume. (See Sidebar 3.12 for an alternative simulation approach.)

The results from the first four simulation runs are shown in Table 3.6. The values of V from the 10,000 simulations are displayed in a histogram in Figure 3.7. The resulting distribution looks approximately normal with the estimated mean $\tilde{\mu}_V = 42.994$ and standard deviation $\tilde{\sigma}_V = 0.793$, (the notation ~ is used to designate estimates based on simulation).

A total of 1,101 of the 10,000 simulated barrel volumes, or 11.0%, had values that fell below the minimum required V of 42 in^3. This was unacceptable.

Trade-off in Example The percentage of barrels with volume exceeding 42 in^3 can be increased to the desired 99% by appropriately reducing σ_r or σ_h, or by increasing μ_r or μ_h, or some combination of these. Reducing σ_r and/or σ_h was highly desirable and was set as a long-term goal. But, as an immediate step, measures were taken to increase μ_h to 4.256 – a value estimated to result in 99% of the barrels exceeding the specification limit.[13]

13. In light of the sampling error associated with the estimates of μ_r, μ_h, σ_r and σ_h, it would be prudent to select a value that will result in more than 99% of the barrels exceeding volume specifications. This concern can be addressed more precisely via a statistical tolerance interval, which takes the estimation uncertainty into consideration (see Hahn and Meeker 1991).

SIDEBAR 3.12: AN ALTERNATIVE SIMULATION APPROACH

The simulation in the example was based on an assumed normal distributions for r and h. An alternative approach is to randomly select samples from the 275 observed values of r and the 163 observed values of h directly and use these to estimate V. The remainder of the simulation analysis would then proceed as before.

An advantage of this approach is that it does not require a distribution to be assumed for r and h.

POE POE[14] is an analytic method that lets you approximate the mean and standard deviation of systems performance based on the means and standard deviations of the component variables and the functional relationship between the component variables and systems performance.

In particular, let $Y = g(X_1, X_2 \cdots X_k)$ denote a performance variable and $X_1, X_2 \cdots X_k$ the (assumed uncorrelated) component variables with means $\mu_1, \mu_2 \cdots \mu_k$ and standard deviations $\sigma_1, \sigma_2 \cdots \sigma_k$, respectively. The function g is the functional relationship between the Xs and Y.

The POE method lets you approximate μ_Y and σ_Y, the mean and standard deviation, respectively of Y, by the expressions

$$\mu_Y \cong g(\mu_1, \mu_2, \ldots \mu_k)$$

and

$$\sigma_Y \cong \sqrt{\left(\frac{\partial Y}{\partial X_1}\right)^2 \sigma_1^2 + \left(\frac{\partial Y}{\partial X_2}\right)^2 \sigma_2^2 + \cdots + \left(\frac{\partial Y}{\partial X_k}\right)^2 \sigma_k^2}\,.$$

where the derivatives are evaluated at the mean values of the Xs.[15]

See Cox (1986), Hahn and Shapiro (1967), or the online *Engineering Statistics Handbook* for further details.

Application to Example Applying POE to the example yields the approximate expressions:

$$\mu_V = \pi \mu_r^2 \mu_h \tag{3.2}$$

and

$$\sigma_V = \sqrt{(2\pi\mu_r\mu_h)^2\sigma_r^2 + (\pi\mu_r^2)^2\sigma_h^2}\,. \tag{3.3}$$

Evaluating equations (3.2) and (3.3) at the previously estimated values for μ_r, μ_h, σ_r and σ_h yields the estimates $\hat{\mu}_V = 42.986$ inches and $\hat{\sigma}_V = 0.789$ inches.

14. This method is also sometimes referred to as the delta method or generation of system moments.
15. Application of this formula to the special case of a linear relationship between Y and the Xs, that is, $Y = aX_1 + bX_2 + \cdots kX_k$ (where $a, b \cdots k$ are known constants), yields the well-known result (for uncorrelated Xs): $\sigma_Y = \sqrt{a^2\sigma_1^2 + b^2\sigma_2^2 \ldots + k^2\sigma_k^2}\,.$

These values are very close to those obtained by simulation. Also, if one assumes V to be normally distributed (as seems reasonable from simulation), the estimated fraction of barrels failing to meet the minimum volume is 10.6% – as compared to 11.0% obtained by simulation.

POE can also be employed to evaluate how much the mean height and/or radius need be increased and/or their standard deviations decreased to better meet the specified minimum volume requirements.

Multiple Product Characteristics The barrel volume problem was actually more complex than stated in that it also involved a second performance characteristic, surface area (A). Since reducing surface area reduces material costs, the goal was to minimize A while still ensuring that 99% of the barrels enclosed a volume of $42\,in^3$. This results in a so-called "multivariate constrained optimization problem," (see Everitt 1987 and Carlyle, Montgomery, and Runger 2000).

Assumed Model Both simulation and POE require a functional form for the relationship between the component variables and system performance, such as equation (3.1). This is often based on physical theory. In the absence of such theory, a general model form (such as a polynomial) is sometimes assumed and a designed experiment conducted to obtain the needed data to estimate unknown model parameters (see Bisgaard 1997).

3.6 DESIGN VALIDATION

In product design validation, the goal is to establish to the degree possible, prior to transition to manufacturing, that the final product or system meets its design goals. This frequently calls for a comprehensive program of testing to assess functionality, quality, manufacturability, usability and, especially, reliability (see Chapter 5). Such testing needs to be conducted as early as possible and preferably sequentially (on increasingly more "representative" product and in an increasingly more realistic environment) to allow ample time, prior to product release, to correct any problems that still remain.

3.6.1 Nature of Testing

During the early phases of design, it is often easier and less expensive to test components or subsystems, rather than the entire product (which might, in fact, not yet be finalized). The resulting information is then used for design decisions and to infer product or system performance. In design validation, the goal is to test the final product or system so as to, hopefully, confirm the earlier positive, but tentative, conclusions – and to speedily make necessary corrections if confirmation is not achieved.

Initial validation testing is typically performed on product made in the development lab on a small scale. Subsequent testing should, if possible, be conducted on prototype product, perhaps built, at least in small quantities, in a pilot plant.

Most initial validation testing is generally conducted in-house on special test equipment. It is desirable, however, that at least some of the later testing be performed directly in the field under use (or accelerated use) conditions. Particular emphasis in such testing needs to be on achieving the variety of environments that the product will experience in operation. Sometimes, it is in fact appropriate to skew the testing toward more, and especially the most, severe anticipated use environments.

3.6.2 Technical Considerations

Planning a comprehensive test program that will provide meaningful information is an important element of validation testing. Statistical concepts play an important role in helping ensure appropriate selection of test units and test conditions, allocation of test units to test conditions, and responding to the traditional question of how much testing is required – as well as in the analysis of the resulting data. Bisgaard and Steinberg (1997) present a case study illustrating the use of statistically designed experimentation on product prototypes.

The process that is sampled and the environment under which the product is tested usually differ markedly from that encountered in high-volume production and customer use. Therefore, it is important to ensure that such testing becomes increasingly broader over time to reflect the type of variability that can be expected in production and product use.

We continue this discussion in Section 5.8.

3.7 TRANSITION TO MANUFACTURING

3.7.1 Scale-Up Testing

The Need Validation testing has, presumably, demonstrated that it is *possible* to build a product with the desired characteristics. But designing a worthy product and building workable prototype units is one thing. Successfully transferring the design to mass production and assuring that a good product can be manufactured *consistently*, subject to the variability encountered in normal production, is quite another. The goal of scale-up testing is to ready a product for manufacturing. We want hard evidence to ensure, in a timely manner, that the high product quality, performance and reliability, presumably demonstrated prior to and during validation testing on prototypes, can be maintained in the ramp-up to high-volume production.

It is often possible to exert much tighter control in a lab environment than in high-volume production. Thus, production units tend to be more variable than prototype units.[16] We recall a situation in which temperature during processing was maintained within a much narrower range in the lab than was possible in manufacturing. Temperature turned out to be a critical variable impacting product performance. This resulted in unacceptable manufactured product and a scurry to pinpoint

16. There are exceptions. For example, production units built from a common tool or die might actually have less variation than prototypes made in a machine shop.

the problem to find a remedy. Moreover, the later a problem is discovered – especially as the project is nearing commercialization – the more costly it becomes to correct it. Scale-up testing, coupled with today's more disciplined approach to design, thus aims to identify and address potential manufacturing problems early on.

Nature of Scale-Up Testing Scale-up testing generally calls for manufacturing product directly in production facilities on a systems level and on a scale that is, at least approximately, comparable to what might be expected in routine manufacturing. For a chemical manufacturing process, for example, scale-up might involve going from 10-lb batches in a test laboratory to making thousands of pounds a day, under markedly different conditions, in a manufacturing plant.

Testing needs to be conducted over a sufficiently long time period to permit the variability that one would expect in manufacturing to be experienced. Production might, moreover, be spread out among sites around the world. These may be acquiring their raw materials from different suppliers, while employing somewhat different manufacturing practices and experiencing varying ambient conditions. There might also be differences in tools and manufacturing equipment and in the skill and dedication of the workforce. Scale-up testing at each site is, therefore, needed.

Finally, as in design validation, testing should be conducted under conditions that simulate customer use as closely as possible.

Addressing Practical Realities There are serious practical limitations associated with testing in a production environment. Such evaluations are generally very costly. As a result, it is often desirable to sell to customers the excess product built during scale-up testing. This can also yield added useful information about product performance. At the same time, care must be exercised to ensure that product shipped to customers is, indeed, of high quality so as not to damage a new product's reputation.

3.7.2 Quality Surveillance Planning

Once the product and process are qualified and successfully transitioned to manufacturing, and the ability to build product with the desired characteristics in routine manufacturing has been demonstrated, we need to ensure that we *continue* to make good product. Statistical concepts apply in establishing data collection and analysis methods for in-process and end-of-line quality checks and for taking appropriate actions thereon. This includes surveillance of:

- Incoming materials received from suppliers
- Tool wear
- Processing conditions, such as ambient humidity and temperature, speed of machinery, and material flow
- Parts dimensions
- Measurements at intermediate production steps
- Final product performance measurements

We shall return to this topic in Sections 6.7, 8.4, and 8.5. We have introduced it here to emphasize the importance of the design team working with manufacturing in defining and developing the appropriate systems for routine and timely monitoring – and to have these in place when needed.

3.8 MAJOR TAKEAWAYS

- Performance, quality, and reliability need to be addressed proactively in the design of a product. Statistics is an important element of proactive product design.
- DFSS provides a disciplined databased approach for product design.
- Product design is an iterative and intertwined process.
- Statistics contributes to setting design goals by helping:
  - Determine customer needs and wishes through market research.
  - Tailor product requirements to customer needs, identifying target-is-best, higher-is-better and smaller-is-better product characteristics and tying these to customer needs, using, for example, QFD.
  - Address the twin needs of improving average performance and reducing variability.
  - Set target values and specification limits on frequently multiple competing product characteristics, while recognizing the cost of variability.
- Statistics contributes to ensuring measurement capability by helping:
  - Evaluate and remove bias.
  - Reduce measurement variability, through, for example, GRR studies.
  - Establish systems for continued surveillance of the measurement process.
- Statistics contributes to product design and development by helping:
  - Assess conceptual issues in the design of new products and the redesign of existing ones.
  - Identify and quantify the impact of critical variables on product characteristics using designed experiments.
  - Achieve robustness of the design to fluctuations in manufacturing conditions and product use, based upon studies to quantify the impact of control and noise variables and their interactions.
  - Quantify system variability from component variability, using Monte Carlo simulation and POE, and use the results to achieve acceptable system variability.
- Statistics contributes to design validation by helping develop and analyze tests to confirm that the design has succeeded.
- Statistics contributes to the smooth transition of the design to manufacturing through the design and analysis of scale-up testing and quality surveillance systems.

DISCUSSION QUESTIONS

General

1. A cereal food manufacturer wishes to select a name and an advertising slogan for a new product from among a number suggested by marketing. Propose a consumer survey to help determine the most effective name and slogan. What are some likely obstacles that might bias the results and how might you try to overcome these?

2. Suggest likely desired customer product characteristics for:

   - An educational and entertaining toy for toddlers.
   - A plastic fork for use in fast-food restaurants.
   - A high-speed racing bicycle.

3. We suggested in Section 3.3.3 that "businesses often try to gain competitive advantage by broadening the ranges set on operating conditions relative to current market product offerings." Give some specific examples, indicating how such action might be helpful or detrimental to customers and/or the company.

4. Suggest some product characteristics for which you would expect the target value *not* to be the midpoint between the upper and lower specification limits (i.e., it is desirable to be closer to one of the limits).

5. Give some examples of situations for which it is necessary to balance the improvement of average performance and the reduction of variability.

6. Suggest some products that seem to lack robustness to customer use and some that appear to be robust. How might you improve robustness?

7. What characterizes a robust design for a DVD player? What are some likely noise and control variables? How might you study these and use the results to design a more robust product?

8. Propose a validation test for a newly designed high-speed racing bicycle.

9. Propose a production scale-up test for the newly designed high-speed racing bicycle.

Technical

1. Propose programs to assess current bias and to monitor potential future bias of a scale used to measure the weight of precious metals in a jewelry store, using one or more gold standards.

2. Evaluations of measurement error typically assume that the true value and the measurement error are uncorrelated. Under what conditions would this assumption be questionable and how would you address failure to meet these?

3. Research and report how you might conduct GRR studies for attributes and for situations that involve destructive testing.

4. How is the underlying model for a computer experiment different from one for which one is dealing with actual observations, and how does this impact the design and analysis of the experiment?

5. We state that the implementation of a robust design "involves exploiting interactions between the control and noise variables" (Section 3.5.3). Explain and give an example.

6. Suggest how the analysis that led to the selection of detergent A in the car wash example (Section 3.5.3) might have been conducted.

7. In the barrel volume example (Section 3.5.4), provide further details of how increasing μ_h to 4.256 will result in an estimated 99% of the product exceeding the specification limit of 42 in^3.

8. In the barrel volume example, it was stated that "reducing σ_r and/or σ_h was highly desirable and was set as a long-term goal." In so doing, it is useful to know the relative impact of reducing each of these two standard deviations. How would you make this determination?

9. Relate the "alternative simulation approach" suggested in the barrel volume example to statistical bootstrapping and resampling. How is it similar and how is it different?

10. Elaborate on the comment in the barrel volume example "Propagation of errors can also be employed to evaluate how much the mean height and/or radius need be increased and/or their standard deviations decreased to better meet the specified minimum volume requirements."

11. Elaborate on how you would use multivariate constrained optimization in the barrel volume problem to minimize surface area while still ensuring that 99% of the barrels enclosed a volume of 42 in^3.

12. We state (Section 3.6.1) that "Sometimes, it is ... appropriate to skew the testing toward more ... severe anticipated use environments." Suggest situations when this might be appropriate and how you might do this. Relate your approach to stratified sampling.

CHAPTER 4

PRODUCT DESIGN: EXAMPLE

4.1 ABOUT THIS CHAPTER

This chapter illustrates the concepts presented in the preceding chapter by applying them to a specific case study.

4.2 THERMOPLASTIC RESIN FOR CHILD CAR SEAT: INTRODUCTION

The ultimate product (or system) in this example is a child car seat that is attached to the backseat of a car to allow children to travel safely. Its manufacturer has decided to redesign the product to:

- simplify assembly by the customer;
- reduce the seat's weight and make it easier to move; and
- prevent cracking of parts, especially at the points where they are snap-fitted.

The example deals with the thermoplastic resin used in the manufacture of many of the parts that make up this product. The resin used in the current design does not meet the requirements of the new design and a new thermoplastic resin needs to be developed. The resin supplier must satisfy the requirements of two customers: the immediate customer (the car seat manufacturer) and the ultimate customer (purchasers of the car seat).

4.3 SOME BASIC ELEMENTS OF PLASTIC PART DESIGN AND MANUFACTURE

Numerous decisions are made in developing a design. These require both scientific and engineering knowledge, as well as a good understanding of the product. In this example, it calls for an appreciation of both the resin manufacturing process and of how the resin is to be processed by the car seat manufacturer in molding product parts.

Figure 4.1 provides some fundamentals of the design and manufacture of resin for molding car seat parts.

The Role of Statistics in Business and Industry, by Gerald J. Hahn & Necip Doganaksoy

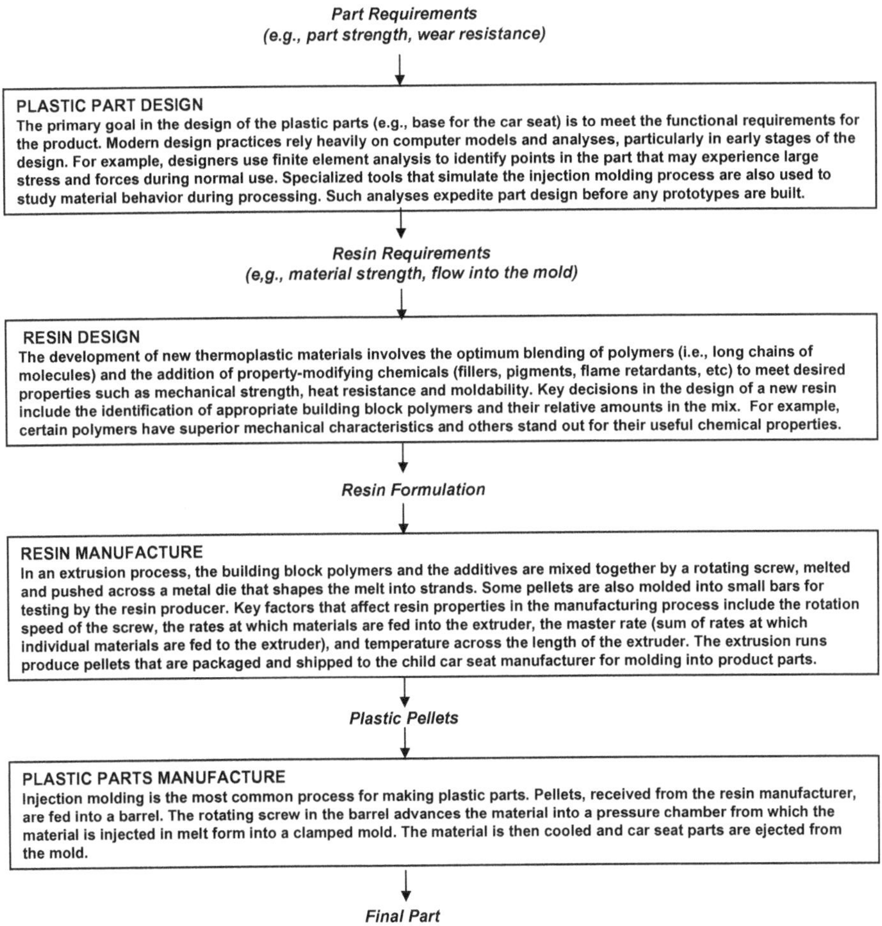

Figure 4.1 Basic elements of resin design.

4.4 SETTING THE DESIGN GOALS

4.4.1 Determining Customer Needs

Car Seat Customer The redesign of the child car seat was driven by both solicited and unsolicited feedback from customers and by recent improvements in competitive products. This effort was led by the car seat manufacturer's market research group and included careful evaluation of:

- customer comments on previous designs;
- features of recently introduced competitive products;
- regulatory agency requirements; and
- the environments in which the new product is to be used.

In collaboration with stores that sold car seats, a survey was also conducted to develop a customer "wish list" of desired product features and to assess the demand and the perceived dollar value of such features.

A frequent complaint dealt with cracking at various locations at which parts snap-fitted together; the new resin needed to provide higher endurance to eliminate this problem. Other final customer requirements included keeping down weight and having a highly fire-resistant product.

Car Seat Manufacturer In order to simplify assembly, the new child car seat was to be made up of fewer, but larger, molded parts. Engineering analyses revealed that the resin used in the previous design would not flow sufficiently fast during injection molding to fill the entire mold before cooling. Thus, a faster-flowing resin was needed. A further requirement was that the material be robust to variations in the processing conditions of the molding machines used during injection molding of the final parts. This is exacerbated by batch-to-batch variability in resin flow. The goal is to help ensure that molders do not have to adjust processing conditions from one resin batch to the next.

Summary The two key customer needs requiring important improvements over the current resin were resin flow during injection molding and part strength (i.e., ability to withstand impact). In addition, other customer requirements, such as fire resistance and weight, were to be maintained, as a minimum, at their existing levels.

4.4.2 Tailoring Product Requirements to Customer Needs

Identifying Product Characteristics The design team had to translate the customer requirements into measurable product characteristics for the new resin. They turned to standard tests proposed by the American Society for Testing and Materials (ASTM) and the International Organization for Standardization (ISO) for engineering thermoplastics. This led to the following product characteristics:

- Resin Flow: The need for a faster-flowing material to fill the larger molds required by the new part design is directly related to the viscosity of the material. A highly viscous material in melt form has higher resistance to flow; honey, for example, has higher viscosity than water. A *melt volume rate* (MVR) test conducted on molded test bars characterizes this property. It measures the volume (in cubic centimeters) of a polymer melt that is extruded for 10 minutes under a defined force and temperature through a standard die (and reported in units of cc/10 min). Increased MVR indicates increased resin flow (or decreased material viscosity).
- Part Strength: The part designers were concerned principally with loads imposed during the assembly of the car seat and when the car seat is snapped to its base. An ASTM *tensile strength* (TS) test on test bars is used for this purpose. This test measures the maximum pulling load (in mega Pascal or MPa) the bar can support before yielding or breaking.
- Weight: Final product weight is determined from measurements of the *specific gravity* (SG) of pellets.

- Fire resistance: Fire resistance can be measured directly by a *flammability test* (FT) on molded test bars. This involves subjecting the sample to flame ignition and measuring the rate and duration of burning.

Quality Function Deployment (QFD) QFD was used to relate customer needs to product characteristics based on the current product. This was accomplished by building the so-called House of Quality shown in Figure 4.2.

In this figure:

- The rows show the customer needs.
- The first numerical column provides a measure on a scale from 1 to 5 of the importance that customers assign to each of the needs (5 is most important).
- The next four columns show the measurable product characteristics. The numbers quantify the relationship between a customer need and a product characteristic. The scores 1, 4, and 9 designate weak, medium, and strong relationships, respectively.
- The next column provides an assessment on a scale of 1 to 5 (highest) of the degree to which each of the customer needs is being met by the current product.
- The final column provides an assessment on a scale of 1 to 5 (highest) of the degree to which each customer need is being met by competitive products.
- The bottom row (total score) provides a measure of the importance of each product characteristic. It is obtained for each column as the sum of the products

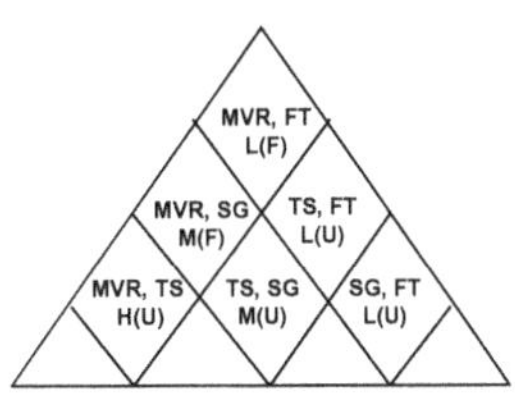

CUSTOMER NEEDS	IMPORTANCE	RESIN CHARACTERISTICS: Resin flow *Melt Volume Rate (MVR)*	Strength *Tensile Strength (TS)*	Weight *Specific Gravity (SG)*	Fire resistance *Flammability Test (FT)*	Met by Current Product	Met by Competitive Products
FINAL PRODUCT							
Withstand impact	5	4	9	4	1	2	4
Light weight	3	4	9	9	1	3	3
Fire resistant	5	4	1	1	9	5	4
PART MOLDING							
Ability to fill mold	5	9	4	4	1	2	3
Short processing time	3	9	9	1	1	3	3
Robustness to molding conditions	4	9	4	4	4	4	4
TOTAL SCORE		160	140	91	77		

Figure 4.2 House of quality for the thermoplastic resin used in the child car seat.

of each of the items in the column and their corresponding importance values (e.g., total score for Resin Flow is $5 \times 4 + 3 \times 4 + \ldots + 4 \times 9 = 160$.)

- The "roof" of the house of quality shows the relationships – high (H), medium (M), and low (L) – among product characteristics. A parenthetical F or U indicates whether the relationship is favorable (i.e., improvement in one characteristic is associated with improvement in the other characteristic) or unfavorable. MVR and TS, for example, are highly, but unfavorably, related.

Implications of QFD The QFD shows that:

- Ability to withstand impact is a critical need with which final customers are currently displeased; there is, in addition, an unfavorable difference between the perceived performance of the current product and that of competitors. TS, moreover, is key to withstanding impact.
- Ability to fill the mold, the other critical need for the new product, cannot be met satisfactorily by the current product and requires improvements in resin flow.
- Resin flow and TS are related in an unfavorable manner. Thus, actions to improve TS may further worsen resin flow – unless appropriate counteracting measures are taken.
- Robustness to molding conditions, another customer need, is also heavily impacted by resin flow.
- Processing time, a further customer need, also needs to be improved. It tends to be determined by both resin flow and strength – another reason for focusing on these.
- Light weight is another customer need. New designs must be limited to ones that tend to decrease, rather than increase, weight. This is likely to require a trade-off between weight and strength since the two tend to be inversely related.
- Fire resistance is being met satisfactorily by the current design. The new product needs to be evaluated to ensure that this continues to be the case.

QFD reaffirms that resin flow and strength are the two most important product characteristics (highest total scores). In developing the new resin, we need to give top priority to these, while ensuring that other product characteristics (weight and flammability) remain at least at the levels achieved by the current product.

4.4.3 Setting Targets and Specification Limits on Product Characteristics

Designers used finite element analysis to explore the impact of material properties on the product characteristics of parts. These evaluations led to an improved appreciation of strength requirements. Computational fluid dynamics[1] was used to understand better the material requirements needed to fill the mold. These engineering

1. A variety of computational techniques to solve the equations that govern fluid flow.

tools permitted the requirements on the final parts to be translated into the following specifications on test bars:

- **Resin Flow.** Melt Volume Rate (MVR), a target-is-best product characteristic, was required to be within the range of 20 to 30 cc/10 min with a target of 25 cc/10 min. The lower limit was set to avoid slow-flowing material. Two-sided limits are necessary to maintain material consistency.
- **Strength.** TS, a higher-the-better product characteristic, needs to exceed the lower specification limit of 120 MPa, with a target value of 130 Mpa.

Similar specifications were established on specific gravity and flammability.

4.5 ENSURING MEASUREMENT CAPABILITY

Both the MVR and TS tests had been used extensively by the resin manufacturer on previous products. These showed the measurement system to be well-centered around true values. Based on recent data, the measurement error standard deviations for the MVR and TS tests were estimated to be 0.12 cc/10 min and 0.19 MPa, respectively. These led to GRR index values of 6.2% for MVR and 4.9% for TS. Measurement errors of this magnitude were felt to be acceptable for design purposes.

Bars, made up from homogeneous pellet samples, were tested periodically in the resin test lab to monitor the continuing accuracy and precision of these measurements.

4.6 DESIGN AND DEVELOPMENT

4.6.1 Conceptual Product Design

A particular challenge for the design team was the previously noted fact that the requirements to increase resin flow and strength tended to be in conflict. To address this challenge, the design team looked at three options:

- Blending existing polymers
- Adding property-modifying chemicals to an existing polymer
- Inventing new polymers

The last option required a breakthrough innovation and was considered too risky.

Theory, moreover, suggested that:

- Resin flow is largely determined by the chain lengths of the constituent polymers (i.e., number of molecules on the string). Decreasing polymer chain length allows material to flow more easily. This, however, decreases the strength of the molded part.
- Certain impact modifiers (e.g., rubber) provide outstanding impact resistance and can increase strength. Increasing the impact modifier, however, is likely to slow material flow.

The design team addressed this challenge by using an impact modifier in appropriate amounts to increase strength, and blending two polymers, with long and short chain lengths, respectively, to control flow. This needed to be done in a manner that would best balance flow and strength, without compromising flammability and weight.

4.6.2 Identifying and Quantifying the Impact of Critical Variables and Achieving Design Robustness

The gold is in the interactions.
—Lynne Hare

Turning the concept into an actual design included identifying appropriate polymers for the blend, determining their relative amounts, and assessing the impact of other design and environmental variables on resin performance.

Identifying Key Design and Environmental Variables The design team identified the variables that were likely to impact the product characteristics in a fishbone diagram (see Figure 4.3) and assigned each to one of four categories: raw materials, resin manufacturing, injection molding (by the child car seat manufacturer), and use environment. Also each variable was identified as control (C) or noise (N) or both; extruder screw speed, for example, is controlled to meet a target value, but varies within its specification limits.

Development Lab Testing The relationships between the process variables and the product variables could not be quantified from theory. They were, therefore, assessed empirically through statistically planned experimentation.

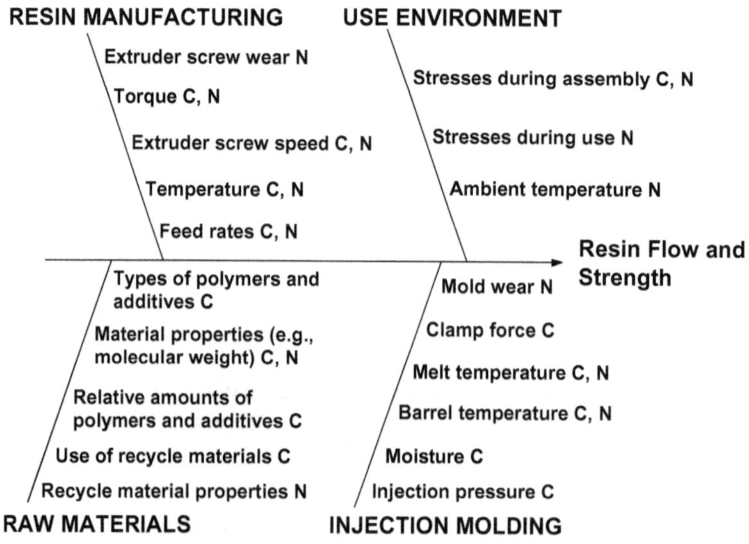

Figure 4.3 Fishbone diagram identifying noise (N) and control (C) variables that might impact infant car seat product characteristics.

It would have been most realistic to perform such testing on the actual production line. This, however, was expensive and impractical. Therefore, as much testing as possible was conducted in the development lab on a miniature version of the resin manufacturing process, making approximately 10 lbs. of pellets per run. The pellets were then molded by a small injection molding machine into test bars on which TS and MVR (as well as SG and FT) were measured. Thus, the four types of variables shown in the fishbone diagram could be addressed as follows:

- Raw materials: Meaningful results could be obtained by lab testing.
- Resin manufacturing: Screw speed could be evaluated in the lab. It was not possible to achieve a realistic representation of manufacturing conditions for the other variables because of the large scale difference between the lab and production extruders; their consideration was postponed to later scale-up studies.
- Injection molding: Barrel temperature (i.e., the temperature of the pellets in the barrel of the molding machine) and injection pressure could be reasonably assessed in the lab.
- Use environment: These variables were left for subsequent scale-up testing; they require testing on actual parts, rather than test bars.

A sequence of targeted experiments was conducted in the lab. Initial testing aimed to identify key polymers and additives. These were then studied further to quantify their impacts on the product characteristics and to assess interactions.

Materials Selection An initial experiment (not described here), involving six polymers and three impact modifiers, was conducted to assess the feasibility of different materials for this application. This led to the selection of two polymers (with relatively short and long chain lengths, respectively) and one impact modifier.

Screening Experiment Eight factors were now considered: Chain lengths of polymers 1 and 2, the proportion that polymer 1 and impact modifier constitute of the total polymer and formulation weights respectively, glass transition temperature of impact modifier (i.e., the temperature at which the material transitions from a glassy to a rubbery state), extruder screw speed, molding injection pressure, and molding barrel temperature.

A screening experiment was conducted to identify the most important of these variables. This consisted of a $2^{(8-4)}$ fractional factorial design with four center replicate points leading to 20 runs[2] (see Box, Hunter, Hunter 2005).

Variables for Response Surface and Robustness Exploration The screening experiment identified the following variables for further study:

- A: The proportion that polymer 1 constitutes of total polymer weight, that is:

2. This design, known as a Resolution 4 plan, permitted assessment of the main effects of each of the variables, independently of two factor interactions. Two factor interactions are, however, "confounded" with each other.

$$A = \frac{\text{Weight of Polymer 1}}{\text{Weight of Polymer 1} + \text{Weight of Polymer 2}}$$

- B: The proportion that impact modifier constitutes of total formulation weight[3], that is:

$$B = \frac{\text{Weight of Impact Modifier}}{\text{Weight of Polymer 1} + \text{Weight of Polymer 2} + \text{Weight of Impact Modifier}}$$

- Molding barrel temperature (T). This is a noise variable in production, varying between 510 °F to 530 °F – depending on how much material is kept in the barrel during the run, how long such material remains in the barrel, the configuration of the injection molding machine being used, and so forth. Barrel temperature can, however, be controlled in the experiment.

The goal of the next stage of experimentation was to quantify the impact on product performance of the key variables identified in the screening stage to determine conditions that will:

- meet (or exceed) the specifications on the product characteristics (principally resin flow and strength); and
- minimize the variability in these product characteristics due to the uncontrollable (in production) barrel temperature fluctuations.

Details of Experiment Based upon theoretical considerations and the results of the earlier screening experiments, it was decided to explore the relationships between A, B, and T and the product characteristics over the following ranges: A: 33% to 67%, B: 10% to 30% and T: 515 °F to 525 °F. A second order polynomial approximation was assumed to relate these variables to resin flow (measured by MVR) and strength (measured by TS).

To obtain the needed data for such response surface exploration, a central composite experiment requiring 20 runs was conducted. The center condition of the design (i.e., $A = 0.50$, $B = 0.20$, and $T = 520\,°\text{F}$) was run a total of six times. The test plan is shown in the first three columns of Table 4.1[4]. The sequence of running the 20 experimental runs was randomized.

Analysis of Results The measured values of MVR and TS obtained in the experiment are shown in the last two columns of Table 4.1.

The results were used to fit the assumed model using least squares regression analysis of the data with the following results (retaining statistically significant terms only):

$$\text{Resin flow: } \widehat{MVR} = -773.71 + 1267.39A + 273.93B + 1.58T - 585.18B^2 - 2.62AT. \tag{4.1}$$

3. The structuring of variables A and B was based on physical considerations. An alternative is to consider the weights of polymers 1 and 2 and of impact modifier as variables in a mixture experiment (Section 13.2.4) since their sum needs to add to 100%.
4. This plan provides an independent measure of repeatability at the center condition and also allows an evaluation of lack of model fit of the assumed (second order polynomial) model.

TABLE 4.1 Experimental Plan and Results for Child Car Seat Experiment

A	B	T	MVR	TS
0.33	0.1	515	33.4	121.2
0.67	0.1	515	6	125.2
0.33	0.3	515	41.2	130.1
0.67	0.3	515	14.2	135.4
0.33	0.1	525	40.4	116.1
0.67	0.1	525	4.3	120.6
0.33	0.3	525	48.3	125.6
0.67	0.3	525	12.2	130
0.22	0.2	520	57.4	121.2
0.78	0.2	520	4.3	129.7
0.5	0.03	520	7.2	117
0.5	0.37	520	20.8	133.7
0.5	0.2	512	28.6	129.8
0.5	0.2	528	33.1	121.8
0.5	0.2	520	30.7	126
0.5	0.2	520	30.8	125.1
0.5	0.2	520	30.9	125.5
0.5	0.2	520	30.9	125.6
0.5	0.2	520	31.2	125.9
0.5	0.2	520	31	125.5

$$\text{Strength: } \hat{TS} = 365.69 + 14.11A + 48.18B - 0.49T. \tag{4.2}$$

Similar models were also fitted for weight and fire resistance. In addition, the Finance operation helped quantify the costs for different relative amounts of each ingredient.

Plot of Results Figure 4.4 is a contour plot constructed from equations (4.1) and (4.2) for a barrel temperature of 520 °F. This shows:

- The values of A and B that result in $\hat{MVR}$ of 20 and 30 cc/10 min (curves).
- The values of A and B that result in $\hat{TS}$ of 120 MPa (straight line).

The following four regions are shown in Figure 4.4 for various combinations of A and B:

- Region I: The (unacceptable) region for which $\hat{MVR}$ is greater than 30 cc/10 min even though $\hat{TS}$ exceeds 120 MPa.
- Region II: The (unacceptable) region for which $\hat{MVR}$ is less than 20 cc/10 min even though $\hat{TS}$ exceeds 120 MPa.
- Region III: The (unacceptable) region for which $\hat{TS}$ is less than 120 MPa (consisting of subregions for which $\hat{MVR}$ is less than 20, between 20 and 30, and exceeds 30 cc/10 min).
- Region IV: The (acceptable) region, shown in white, for which $\hat{MVR}$ is between 20 and 30 cc/10 min and $\hat{TS}$ exceeds 120 MPa.

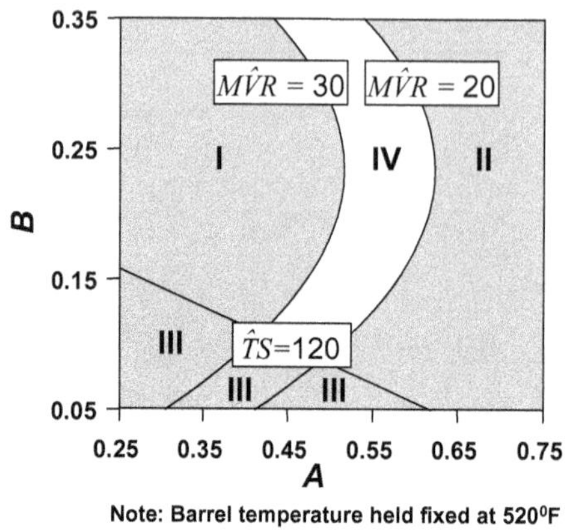

Figure 4.4 Contour plot for $\hat{MVR}$ (cc/10 min) and $\hat{TS}$ (MPa) as a function of A and B for barrel temperature of 520 °F, showing acceptable (in white) and unacceptable regions.

Robustness Evaluation In addition to meeting the specifications on $\hat{MVR}$ and $\hat{TS}$, we desire to find conditions of A and B for which both resin flow and strength are maximally robust to variations in barrel temperature. Thus, it was not enough to meet flow and strength requirements at a barrel temperature of 520 °F; we also needed to be able to do so over a range from 510 °F to 530 °F.

Equation (4.1) for $\hat{MVR}$ includes a term to represent the interaction between A and T. By applying propagation of errors to equation (4.1), we obtain the following equation for the estimated standard deviation of MVR:

$$\hat{\sigma}_{MVR} = |1.58 - 2.62A|\hat{\sigma}_T \tag{4.3}$$

where $\hat{\sigma}_T$ is the estimated standard deviation of barrel temperature during manufacture ($\hat{\sigma}_T = 5°$F was used to estimate this variability). This allows us to reduce the variability induced in flow due to the uncontrolled (in manufacture) variation of barrel temperature by appropriately setting A.

Unfortunately, we cannot similarly reduce the variability in strength induced by the variability in barrel temperature by appropriate setting of the control variables A and/or B. This is because equation (4.2) does not include an interaction term involving T.

Trade-Off Analysis Use of equations (4.1), (4.2), and (4.3) to find the best settings for the control variables A and B[5] requires a trade-off among the three criteria:

- MVR be as close as possible to its target value of 25 cc/10 min
- TS have a high value (exceeding 120, and preferably 130 MPa)
- Minimize σ_{MVR}[6]

5. Barrel temperature T was set at 520 °F.

6. Equation (4.3) indicates that the estimated effect of variability in barrel temperature on resin flow could be eliminated altogether by setting A = 0.60. This setting of A, however, does not meet the target MVR value of 25 cc/10 min, irrespective of how the control variable B is set.

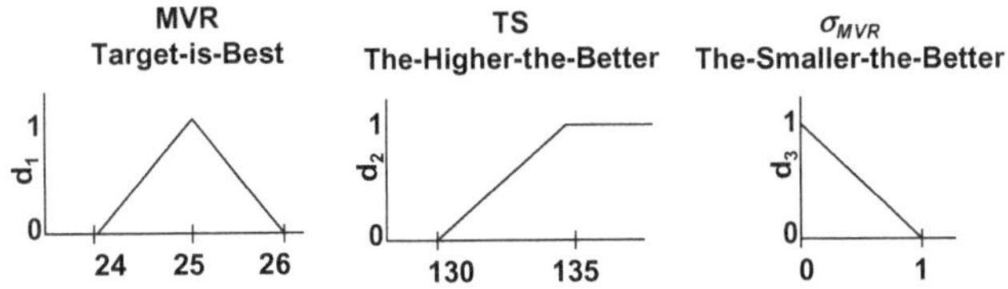

Figure 4.5 Desirability functions for three criteria in car seat example.

We achieve this trade-off by using an approach proposed by Derringer and Suich (1980), modifying earlier work by Harrington (1965; see also Myers and Montgomery 2002).

Individual Desirability Functions The "desirability" of each of the three criteria is expressed as a *desirability function*, shown in Figure 4.5:

- d_1, the desirability for MVR, increases linearly from 0 to 1 as MVR moves from 24 to the target value of 25 and decreases linearly from 1 to 0 as MVR moves from 25 to 26; it is set at 0 for MVR below 24 and above 26.
- d_2, the desirability for TS, increases linearly from 0 to 1 as TS moves from 130 to 135, and remains at 1 for values of TS above 135; it is set at 0 for TS less than 130.
- d_3, the desirability for σ_{MVR}, increases linearly from 0 to 1 as σ_{MVR} moves from 1 to 0; it is set to 0 for values of σ_{MVR} above 1.

Finding the Best Settings[7] The individual desirabilities are combined into a *composite desirability score D* by taking their geometric mean (i.e., $D = (d_1 d_2 d_3)^{1/3}$).

Numerical optimization using standard software for this purpose (see Myers and Montgomery 2002 or the (NIST/Sematech) online Engineering Statistics Handbook) is then used to search for the combination of control variables that maximize D. This leads to the solution $A = 0.564$ and $B = 0.283$, corresponding to optimum settings of 40.5%, 31.3%, and 28.2% for the relative amounts of polymer 1, polymer 2, and impact modifier, respectively.

These settings resulted in the following estimates:

- $\hat{MVR} = 24.96$ cc/10 min from equation (4.1) – desirability of $d_1 = 0.96$.
- $\hat{TS} = 132.48$ MPa from equation (4.2) – desirability of $d_2 = 0.48$.
- $\hat{\sigma}_{MVR} = 0.51$ cc/10 min from equation (4.3) – desirability of $d_3 = 0.49$.
- $\hat{\sigma}_{TS} = 2.45$ MPa by applying propagation of errors to equation (4.2).

4.6.3 Quantifying System Variability From Component Variability

The Problem In resin production, each of the three materials (i.e., polymer 1, polymer 2, and impact modifier) is fed through a mechanical feeder into the extruder.

7. These analyses ignored the statistical variability in equations (4.1) and (4.2). The impact of this variability on the results can be evaluated by simulation.

The feed rates remain essentially constant within a production batch. They fluctuate, however, around their target values between batches – due to such factors as machine wear and variation in material properties and in factory temperature. This, in turn, leads to added variability in resin flow and strength that needs to be assessed.

Analysis Methods We can use equations (4.1) and (4.2) and the definitions of A and B together with past data about feed rate variability to estimate the impact of such variability on resin flow and strength. This can be done using simulation (briefly described here) or propagation of errors.

Assumptions Say that the process is set to achieve the combinations of ingredients found to be optimum in the previous study and 100,000-lb. material batches are to be built. Each of the three feeders is assumed to operate independently around its target value with observed means and (batch-to-batch) standard deviations (estimated from past data on a similar product) shown in Table 4.2.

Also, barrel temperature is assumed to have a mean of 520°F and a standard deviation of 5°F and the amounts of material generated by each feeder and barrel temperature are both assumed to be normally distributed.

Results Using the preceding information and equations (4.1) and (4.2), 10,000 simulation runs were conducted for $\widehat{MVR}$ and $\widehat{TS}$, respectively. Figure 4.6 displays histograms of the results.

TABLE 4.2 Assumed Means and Standard Deviations for Three Raw Materials in Aimed-at 100,000-lb. Batch

	Mean (lbs.)	Standard Deviation (lbs.)
Amount of polymer 1	40,500	446
Amount of polymer 2	31,300	219
Amount of impact modifier	28,200	338

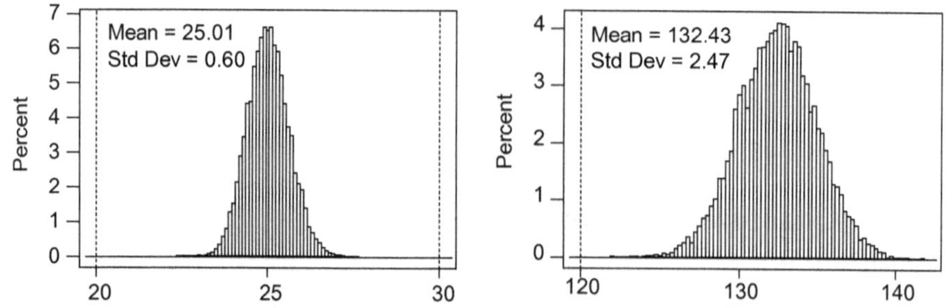

Figure 4.6 Histograms of $\widehat{MVR}$ (cc/10 min) and $\widehat{TS}$ (MPa), based on 10,000 simulation runs, showing combined variability induced by fluctuations in feed rates and barrel temperature (vertical lines represent specification limits).

The results show that the variability in feed rate:

- Increases $\hat{\sigma}_{MVR}$ from 0.51 to 0.60; the simulated values remain well within the specification limits of 20 to 30cc/10min for MVR.
- Increases $\hat{\sigma}_{TS}$ from 2.45 to 2.47; the simulated values continue to comfortably exceed the lower specification limit of 120MPa.

4.7 DESIGN VALIDATION

4.7.1 The Challenge

The evaluations on small test bars, molded from plastic pellets extruded in the development lab, yielded results that readily met the specifications on resin flow, strength, and other properties.

The next question was to validate that such material also met final product requirements. Will the pellets yield resin that will flow to fill the entire mold for car seat parts during injection molding? And will the molded parts, after assembly into child car seats, have the required strength and other desired properties?

4.7.2 Implementation

The manufacturer built sufficient pellets in multiple extrusion runs in the development lab to permit injection molding of one batch of car system parts that resembled those to be used in production. The injection molding runs were carefully monitored to measure the flow of the resin into the mold. The resulting parts were assembled into 20 child car seats that were then used to measure readily observable characteristics, such as car seat weight and flammability.

Some other properties, such as impact strength, required reliability testing to study durability over the life of the car seat. The car seat manufacturer's market studies had indicated that the car seat might be used as often as eight times a day for 10 years (for successive children). Snap-fitted parts were tested repeatedly under conditions that simulated this situation. The product was also tested under increased temperature, humidity, and light exposure to study degradation in strength and color over time. Such reliability testing is discussed further in the next chapter.

The resulting tests showed some minor flaws attributable to the resin (e.g., streaks on molded parts and color mismatches). This led to some modifications of the formulation that were then checked out successfully by further evaluations.

4.8 TRANSITION TO MANUFACTURING

4.8.1 Scale-Up Testing

Most of the evaluations so far, including the design validation, were based upon product built on a drastically reduced scale (around 10lbs. per run) in a principally laboratory environment. Lab testing can provide only a limited assessment of the performance and variability encountered in high-volume production. In addition, the impact of some key variables, such as screw speed and master feed rate, could not, to this point, be realistically assessed.

Evaluation of the scale-up to production was, therefore, a critical final step in the development of the new resin. This was done in close collaboration with the car seat manufacturer.

Initial Testing Scale-up required extruding multiple batches on the actual production line using the conditions optimized during design.

Three such runs were conducted on different days using different batches of raw material. Screw speed and master feed rate were held constant in these initial runs at 400 rpm and 2,025 lbs./hr, respectively – a condition that had been used for high volume manufacture for past resins, but might well not be optimum for the new design. Due to cost considerations, these initial runs were conducted for about 1 hour each, whereas the resin manufacturing process typically involves extrusion runs of 6 to 10 hours' duration.

Pellets were selected at random at approximately 15 equally spaced time intervals during each production run and molded into test bars for evaluation of resin flow, strength, and other properties. In addition, pellets were shipped to the car seat manufacturer, molded into car seat parts, assembled, and tested similar to design validation.

The production runs yielded more material than needed for testing. It was, therefore, decided that if the test results were favorable, the remaining material would be shipped to the car seat manufacturer to use in product to be sold to customers, subject to close scrutiny.

Results of Initial Testing The three production runs yielded satisfactory results with regard to strength, weight, and flame resistance. They also resulted in mean values of MVR of 28.7, 29.4, and 29.1, respectively. This suggested satisfactory batch-to-batch variability (within batch variability was also satisfactory); the mean MVR of 29.1 for the three batches, however, was higher than desired and required further action.

Search for Improved Processing Condition It was expected that flow could be improved by finding better combinations of extrusion screw speed and master feed rate – without impacting the other properties. To explore this, a cycle of five further 1-hour production runs was conducted at the combinations of screw speed and master feed rate shown in Table 4.3. These runs, performed in a randomized sequence, involved four perturbations around the current screw speed of 400 rpm and master feed rate of 2,025 lbs./hour, using a full factorial plan scheme and an additional run at the current condition.

Table 4.3 also shows the resulting mean MVR for each of the five production runs. (The other properties were not impacted.) The mean MVR of 24.7 attained at the screw speed of 375 rpm and master feed rate of 2,040 lbs./hr was felt to be acceptably close to the target value of 25 and was taken as the condition to be used in production, subject to further validation on normal-length production runs.

TABLE 4.3 Cycle of Production Runs to Search for Improved Mean MVR

Condition Number	Screw Speed (rpm)	Master Feed Rate (lbs/hr)	Mean MVR (cc/10 min)
1	375	2,010	23.1
2	375	2,040	24.7
3	425	2,010	26.4
4	425	2,040	27.0
5	400	2,025	29.0

If it were desired to get still closer to the MVR target, added assessments would be needed. Evolutionary operation (EVOP) is one possible approach for doing this. (See Box and Draper 1969 for a detailed discussion of EVOP; Box, Hunter, and Hunter 2005 and Myers and Montgomery 2002 for overviews; Spendley, Hext, and Himsworth 1962 for an alternative approach; and Hahn and Dershowitz 1974 for a survey of applications.)

Validation To complete the validation of the scale-up process, three additional extrusion runs were conducted at different times, days apart, at the optimized conditions, again using different batches of raw materials. Unlike the earlier 1-hour runs, these were of 6, 8, and 10 hours' duration, respectively.

Molded test bars were again evaluated from each of these batches by the resin manufacturer. The main assessments, however, were performed by the car seat manufacturer who conducted a series of molding production runs to check material flow and to make car seat parts. These were then assembled and evaluated in a manner similar to that for design validation. These added tests yielded satisfactory results.

Other Locations Similar scale-up testing was conducted at other plants as resin manufacturing was transitioned to them.

4.8.2 Quality Surveillance Planning

The design team continued to be involved even after the project was formally turned over to manufacturing. Team members helped devise control charts to monitor material feed rates; they also kept a watchful eye on other key manufacturing variables, such as screw speed. In addition, the team worked with manufacturing and suppliers to put in place procedures to monitor the quality of raw materials (e.g., molecular weight of incoming polymer batches).[8]

8. Within batch variability during the production of car seat parts was further reduced by thoroughly mixing the pellets from each manufacturing run (see Doganaksoy and Hahn 1996 for discussion of this approach).

4.9 MAJOR TAKEAWAYS

This chapter reinforces the discussion of the preceding chapter by applying the concepts to a specific example. It illustrates the use of

- QFD and the house of quality to relate customer needs to product characteristics.
- A fishbone diagram to display and categorize possible problem causes.
- The sequential use of DOE to quantify the impact of critical variable and to seek a robust product design.
- Desirability functions and a composite desirability score in trading off competing product characteristics to determine optimum settings for the control variables.
- Simulation to assess the impact of variability in the process variables on product characteristics.
- A stepwise approach to scale-up testing.

DISCUSSION QUESTIONS

General

1. We state in Section 3.3.2 that QFD can also be used to translate "product characteristics into design requirements, and then design requirements into manufacturing requirements." Check out this comment in the references and apply to the child car seat design, suggesting an appropriate house of quality.
2. You are part of a team responsible for the design of a new refrigerator aimed at the college dormitory market. Suggest some customer needs and how these might translate into quantifiable product characteristics; propose a possible house of quality.
3. Describe some of the criteria that could be used in comparing alternative design concepts for the new refrigerator in the preceding problem.
4. Suggest specific individual desirability functions for the following product characteristics of a laptop computer: weight, battery life, hard disk drive capacity, memory, and speed. Construct the resulting composite desirability score.
5. Suggest different ways in which the statistical design of experiments might be used in designing a new jet aircraft engine.

Technical

1. We note in a footnote in Section 4.6.2 that the selected fractional factorial design was "a Resolution 4 plan." How does a Resolution 4 plan contrast with a Resolution 3 and a Resolution 5 plan? How would you decide between such plans?
2. We state that equation (4.3) was obtained "by applying propagation of errors to equation (4.1)." Provide the details.
3. We state in a footnote in Section 4.6.2 that our use of desirability functions to establish the best settings ignored the statistical variability in equations (4.1) and (4.2), but that the impact of such variability can be evaluated by simulation. Elaborate on how to do this.

4. The optimum settings of the design variables in the example were based on a trade-off among MVR, TS, and σ_{MVT}. What would have been the design settings if resin strength had been the *sole* product characteristic considered? What would have been the resulting effect on $\widehat{MVR}$ and its variability?

5. Provide details on how simulation was presumably used to generate the histograms for $\widehat{MVR}$ and $\widehat{TS}$ in Figure 4.6.

6. Provide details on how you would use propagation of errors to evaluate the impact of variation in feed rates and barrel temperature on $\widehat{MVR}$ and $\widehat{TS}$.

CHAPTER 5

PRODUCT RELIABILITY ASSURANCE

5.1 ABOUT THIS CHAPTER

We devote a separate chapter to product reliability assurance because of its importance – even though it is an integral part of the design process that was discussed in Chapters 3 and 4.

We begin by discussing the key role of reliability and the importance of design for reliability. We comment on the reliability process and its participants, and then consider the following major steps in reliability assurance:

- Setting reliability goals.
- Reliability evaluation of a conceptual design.
- Product reliability development.
- Reliability validation.

We conclude with comments on product safety, reliability growth analysis, product burn-in, and software reliability. Some statistical distributions and concepts for product life analysis are reviewed in an Appendix.

Reliability needs to be considered in manufacturing and in the field, in addition to the design stage – if for no other reason than to learn how to design even more reliable products in the future.

5.2 THE KEY ROLE OF RELIABILITY

5.2.1 What Is Reliability?

The reliability of a product is the probability that the product (or equivalently a fraction or percentage of a product population) will satisfactorily perform its intended function under field conditions for a specified period of time (such as warranty or design life).[1] For example, if 1% of the units of a product population fail within the first 5 years of service, the product's 5-year reliability is 99%.

1. Due to the difficulty of characterizing and representing field conditions, especially during design, reliability is sometimes taken as the probability that a product will perform satisfactorily *under stated conditions* for a specified period of time.

Reliability has also been referred to as "quality over time" (see Condra 1993). Time might be expressed in terms of calendar time, operating cycles, number of start-ups, or whatever is most relevant for the product.

5.2.2 A Prime Customer Criterion

Customers experience reliability directly in terms of long-term trouble-free product operation. They expect a purchased product to meet or exceed life expectations and be safe. Thus, for many products, reliability is the key performance metric that, in the long run, stands out most vividly in customers' minds. This is especially so at the time of product replacement. The *New York Times* in 2003, for example, quoted a survey that suggested that consumers' top consideration in buying a car is its reputation for "long-term durability."[2]

And with continuingly increasing expectations and greater choices due to global competition, high reliability is becoming more important than ever. This is so despite the fact that information about reliability is often not readily accessible to customers. The purchaser of a new car knows its price, can judge its appearance by looking at it, and evaluate its performance by taking it on a test-drive. Assessing its reliability, however, is not that easy. Customers typically depend on a company's reputation, articles in consumer review magazines, and friends' recommendations. Further evidence is provided by product warranty terms – these are indicators of the confidence that producers have in the reliability of their products.

All of this sends a clear message about the need for high reliability to manufacturers who wish to delight customers, earn their repeat business, and have them recommend their product to friends – to say nothing of avoiding lawsuits.

5.2.3 Some Notable Reliability Failures

Reliability problems can lead to everything from minor inconveniences (e.g., no toast for breakfast) to potential human disasters (e.g., failed pacemaker or aircraft engine). We are all keenly aware of catastrophes that can result from poor reliability. Reliability problems that have received considerable visibility include[3]:

- The 1986 Challenger and 2003 Columbia space shuttle accidents resulting in the loss of all crew members.
- Rollovers in the late 1990s of Ford Explorer sport utility vehicles equipped with Bridgestone/Firestone tires, claiming nearly 90 lives and 250 injuries.
- The August 2003 power grid shutdown, resulting in loss of power to tens of millions of homes in the Eastern United States and Canada.
- Potential fire hazard of Sony laptop computer batteries, resulting in extensive recalls during 2006.

2. Durability is closely related to reliability. It is generally thought of in terms of time to replacement; reliability deals with time to failure.

3. See the book's ftp site for references to newspaper articles that describe most of these incidents.

- Minneapolis 2007 bridge collapse, asserted to be due to design flaw of gusset plates used to reinforce truss connections.

In each case, investigation suggested that the right data and analyses, coupled with prompt action, could have avoided, or at least appreciably, mitigated the severity of the problem. Many other reliability failures don't make the newspapers and don't have disastrous consequences, but still cause customers appreciable inconvenience or cost companies large amounts of money in warranty costs, as well as much goodwill.

5.3 DESIGN FOR RELIABILITY: THE NEED

The desire for reliability is not new. Throughout history, people learned from their successes and mistakes in building durable wheels, stronger ships, longer bridges, larger domes, and more competitive products. What is different today is the emphasis placed by many companies on assuring reliability of a product up-front during design and development, while still keeping the associated costs at a reasonable level.

In the past, reliability assurance was often an afterthought – even in organizations that emphasized quality. Companies rewarded immediate cost savings and profits, failing to recognize the benefits of long-term cost avoidance. The costs of reliability assurance during design are, after all, immediate – but the benefits come much later and may be in the form of unrecognized "non-events." Also, it takes time to design, test, and bring a reliable product to market. Management, moreover, is under intense pressure to beat competition to the market, and, perhaps, to release a product before its reliability has been fully validated.

If reliability problems arise after a product has been released for production and, especially, if units in the field need be recalled for retrofit, the cost can be severe and may rapidly dwarf a product's profit margin. Pennies saved per unit in selecting a supplier without adequate scrutiny can, for example, result in many millions of dollars in costs fixing a subsequent reliability problem. And because a few early dissatisfied customers can give a product a bad name, it is particularly important for the product to be reliable when first brought to market.

It is now abundantly clear that it is much less expensive to build reliability into products during design than to fix a flawed design – to say nothing of the impact on customer relations. Fortunately, the emergence of advanced material and design techniques and the increased access to information have made it easier to design a product for high reliability and design for reliability has become an important part of Design for Six Sigma.

5.4 THE PROCESS AND THE PARTICIPANTS

5.4.1 Reliability Engineering

Reliability engineering involves the application of engineering principles during design, development, validation, manufacture, and field use of a product to achieve reliability assurance and improvement (see O'Connor 2002). It has become a recognized profession with graduate degrees offered.

Reliability engineers need to be skilled in areas such as reliability project management, failure root cause analysis, and reliability testing and assessment. They also need to be knowledgeable in statistical tools for the planning and analysis of reliability studies.

5.4.2 Evolution of the Role of Statistics

As indicated in Section 2.4.4, the role of statistics in reliability was at one time limited to principally reactive "fire-fighting" – that is, to quantifying problems and helping minimize the damage.

As recognition grew of the importance of avoiding field failures, statistics was used to help plan reliability demonstration programs. This involved responding to such questions as "how large a sample is needed and for how long must one test to ensure with 90% confidence that 99% of the product will operate successfully for 10 years?" This question, however, was often asked at a time when design and development were essentially complete, and it was difficult and expensive to adequately address the problem if the test failed to provide the desired demonstration.

Focus on the design for reliability has led to using statistics proactively to help *improve* reliability during product design and development. This requires quantitative methods for predicting and assessing reliability and for providing early information on causes of failure, as well as for careful planning to ensure that the most meaningful information is obtained. (See Meeker and Hamada 1995 for further discussion of the use of statistical methods in developing highly reliable products.)

Moreover, in various applications requiring high reliability (e.g., aircraft, wind turbines, and power generators), regulatory and certification agencies now encourage, and sometimes require, up-front statistically based evidence of high reliability.

5.4.3 A Team Effort

Reliability engineers and statisticians, to be effective, cannot work in a vacuum or operate as isolated specialists. Instead, they need to have intimate knowledge about the products with which they are dealing, be immersed in the design process, and be sensitive to the business environment. Ensuring high reliability in the design of manufactured products, just as in developing the design itself, is, first and foremost, an engineering challenge – but statistics and other tools can help smooth the path.

5.5 SETTING RELIABILITY GOALS

The definition of reliability goals and requirements – like that of other product characteristics – starts with the identification of customer needs. These are often stated in general terms, such as that the product lasts for a long time, always starts up smoothly, operates well consistently, or does not require unscheduled shutdown or repair between routine servicing. The result is a broad range of interpretations of a product's reliability or lack thereof. For a dishwasher, this might range from a perception that the product is doing an increasingly less effective job in washing dishes to inability to start the machine.

To provide actionable goals for the design team, the reliability goals need be translated into measurable, quantifiable requirements along with a precise description of the stresses (e.g., current and voltage) and environmental conditions (e.g., temperature and humidity) under which the product is expected to be used. These goals are also impacted by a producer's desire to make product warranties as attractive as possible, without incurring high warranty costs.

The approach is somewhat different for nonrepairable and repairable products.[4]

5.5.1 Nonrepairable Products

Nonrepairable products are ones whose life comes to an end when the unit fails. Rather than being repaired, a failed unit may be replaced, presumably by a new one. Nonrepairable products might be stand-alones, such as a light bulb, or they may be components of a larger system, such as the various parts of a computer (e.g., memory, hard drive, power supply, and video card). For a nonrepairable product or component, we may, for example, require no more than 0.5% failures during the first year and no more than 3% failures during the first 5 years of life.

During design, one is typically concerned with the reliability of individual components and, therefore, much of the discussion in this chapter will be on non-repairable units.

Testing of nonrepairable units typically results in times to failure on the failed units and survival times on the unfailed ones (or so-called censored observations). The resulting data are often referred to as life data. Life data analysis typically requires concepts and tools different from those taught in most introductory courses on statistics. Some of these differences are discussed briefly in Sidebar 5.1. Further details are provided in the references cited in Section 5.13 (see also the Appendix to this chapter).

SIDEBAR 5.1: WHAT'S DIFFERENT ABOUT LIFFE DATA ANALYSIS

- As indicated, life data typically include unfailed units or censored observations for which only the running or survival times at the time of analysis are known; the lifetime for these units are known only to exceed their survival times. In analyzing such data, it is incorrect to treat the survival times as if they were failure times or to ignore the survived units altogether. Special graphical and analytic methods that take the censoring into consideration need to be used instead.
- One is often concerned with the tail of the life distribution and especially:

4. This distinction is not always clear-cut. Radios used to be repaired upon failure, but typically are not today; whether a failed tire is repaired or replaced may depend upon the nature of the failure, the age of the tire, and the wishes of the owner. In such cases, the analysis (i.e., repairable or nonrepairable) generally depends upon its purpose; sometimes both types of analysis are conducted.

  - The distribution percentile t_p – the time by which a (usually small) proportion p of units in the population fail, or
  - The proportion, p, of the product population failing (or the probability of failure) by a specified time.
- It is also often useful to study the product's hazard function (see Appendix).
- The normal distribution is generally *not* an appropriate model to represent lifetimes. Relatively short lifetimes may be quite likely, but negative ones are impossible; thus, a skewed distribution is suitable for many products. Various other distributions, most notably the Weibull, exponential and the lognormal, are used instead (see Appendix). Choice of an appropriate distribution (or distributions) is often an important part of product life data analysis.
- One often wishes to extrapolate beyond the range of the data, for example, to estimate the probability of survival to 5 years, based upon, say, only 2 years of life data.

5.5.2 Repairable Products

Most systems and some parts are repaired when a failure occurs. For systems, this may involve replacing a failed component or subsystem.[5] Repairable products typically generate a sequence of failure (and repair) times on the same unit. Repairable systems are also encountered extensively in the field and are considered in greater detail in Section 9.4.

For repairable systems, one is frequently concerned with the proportion of time that the system is available for operation, known as its availability. Thus, availability = (uptime) / (uptime + downtime), where uptime and downtime are the amount of time that the system is operational and nonoperational, respectively, over a specified period. We might, for example, require a repairable system to have 99% availability over a 10-year period.[6]

5.5.3 Mean Time *to* Failure (MTTF) and Mean Time *between* Failures (MTBF)

Traditionally, and especially for electronic parts, the reliability of *nonrepairable* products has been characterized by their MTTF. This is calculated as the total running time on all failed and unfailed units divided by the total number of failures. MTTF is generally a less useful metric than, say, the reliability of a product at a specified time[7] (see also Section A3.3 of Appendix).

5. We tend to deal with complete systems near the end of the design cycle.

6. High availability, in addition to requiring high reliability, also requires high maintainability, where maintainability deals with the time required to repair the system. Improving reliability improves availability by increasing uptime; improving maintainability improves availability by decreasing downtime.

7. When the probability density function for life is highly skewed, (e.g., has a long right tail, as in many life data applications), a high MTTF does not necessarily result in acceptable reliability. For example, a product with an exponential distribution for lifetime and a MTTF of 25 years is still expected to experience 18% failures in its first 5 years.

Similar to MTTF for nonrepairable units, the reliability of *repairable* products and systems has frequently been characterized by their MTBF. Reliability improves as MTBF increases.

5.6 RELIABILITY EVALUATION OF A CONCEPTUAL DESIGN

5.6.1 The Challenge

Early design choices, such as selection of components for an electronic circuit, impact ultimate product reliability. It is, therefore, important to identify and confront, as early as possible, issues that may impact reliability, despite the fact that it may be difficult to do so while the design is still evolving and the product-operating conditions are not well established.

Design of high reliability products calls for the use of reliability management and engineering tools such as design reviews, risk analysis and mitigation, and so forth (see O'Connor 2002). We focus on some statistical aspects.

5.6.2 Reliability Modeling

The reliability of a product is assessed by viewing it as a system comprised of subsystems of assemblies and components, as in Table 3.3 of Chapter 3. A model is then developed that shows how the failure of individual components impacts system reliability.

Components in Series Many components in a system act in series (see Figure 5.1). Failure of *any one* component results in the failure of the entire system. A television set, for example, generally becomes inoperable with the failure of any one of its critical parts, such as its power supply, receiver, audio amplifier, or screen.

Components in Parallel Other components may be redundant and, thus, operate in parallel (see Figure 5.2). A system failure occurs only when *all* such components have failed. For many products, added reliability protection is gained by building redundancy into critical areas of the design. Typical examples are redundant sus-

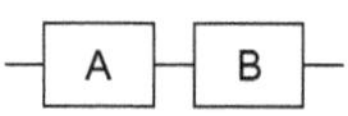

Figure 5.1 Schematic illustration of a system with two components in series.

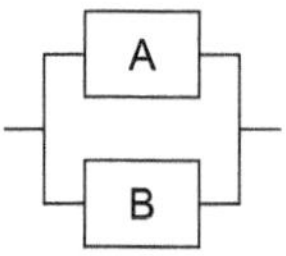

Figure 5.2 Schematic illustration of a system with two parallel components.

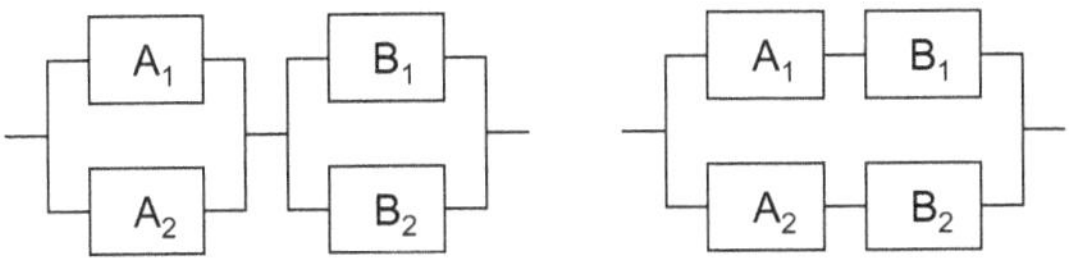

Figure 5.3 Schematic illustrations of systems with both series and parallel components.

pension cables in an elevator, dual tires on heavy trucks, and multiple fire alarm systems.

Components in Series and in Parallel Many systems consist of combinations of series and parallel components (see Figure 5.3). An aircraft engine, for example, requires the successful operation of each of various subsystems, such as its engines and landing gears. Some of these subsystems, such as the engines, may be regarded as redundant in a reliability analysis.

From Component Reliability to System Reliability Elementary probability theory lets us estimate the reliability of simple systems – assuming the components fail independently of one another – from estimates of the individual component reliabilities. This allows engineers to obtain an initial estimate of system reliability and, for example:

- Estimate the impact of the reliabilities of individual components and subsystems.
- Compare alternative designs and redundancy schemes.
- Assess the impact of using higher reliability components.

Assessing Component Reliability The preceding evaluations require information about component reliabilities. These might be obtained from component reliability handbooks (see Sidebar 5.2) or by reliability testing.

SIDEBAR 5.2: COMPONENT RELIABILITY HANDBOOKS

Handbooks and databases provide information on the reliability of electrical and electronic components, such as microcircuits, diodes, transistors, resistors, relays, and capacitors. Similar information, although less detailed, is also available on mechanical parts, such as springs, filters, pumps, electric motors, and compressors.

MIL-HDBK-217 (Reliability Prediction of Electronic Equipment) is the best known such source. It provides "base" failure rates for various types of components. An exponential distribution for life is generally assumed. This assumption may be reasonable for electronic components, which do not tend to wear out – but is less likely to be correct for mechanical components. Adjustment factors are provided for the effects of temperature, application stress, and various environments. The failure rates are based on data obtained from sources such as field history, laboratory testing, and physics-based models.

5.6.3 Example: Washing Machine Design

This example has been adapted from Doganaksoy, Hahn, and Meeker (2006a).

System Description A new washing machine has a reliability goal of 90% during its first 10 years of operation. The system consists of seven key parts: a motor, a main bearing, a brake, a cycle/timing control, a water temperature control, a hot water valve, and a cold water valve. We will consider each of these parts to be components; they could also be regarded as subsystems, which themselves can be broken down into their own constituent components for reliability estimation.

Reliability Model We will use R_1, R_2, R_3, R_4, R_5, R_6, and R_7, respectively, to denote the 10-year reliabilities (i.e., proportion surviving without failure) for the seven components. The machine fails as soon as one of these components fails. This "series system" is sketched in a "reliability block diagram" in Figure 5.4.

The goal that at least 90% of the washing machines operate for 10 years (i.e., $R_{system} \geq 0.90$) requires that all seven components operate failure-free for 10 years in at least 90% of all washing machines. Assuming that components operate and fail independently of each other, the proportion of washing machines that will survive 10 years can be obtained from the individual component reliabilities using the well-known formula from elementary probability:

$$R_{system} = R_1 \times R_2 \times R_3 \times R_4 \times R_5 \times R_6 \times R_7.$$

Reliability Allocation Based on experience with similar components in previous products, the design engineers assigned the individual minimum target values $R_1 = 0.970$, $R_2 = 0.990$, $R_3 = 0.990$, $R_4 = 0.985$, $R_5 = 0.985$, $R_6 = 0.980$, and $R_7 = 0.995$ to the 10-year reliabilities of the seven components. If these values are met exactly, the overall system reliability is:

$$R_{system} = 0.970 \times 0.990 \times 0.990 \times 0.985 \times 0.985 \times 0.980 \times 0.995 = 0.90.$$

Introducing Redundancy Assume now that redundancy is added by having two hot water valves (6a and 6b) that operate independently, each with reliability R_6 (see Figure 5.5).

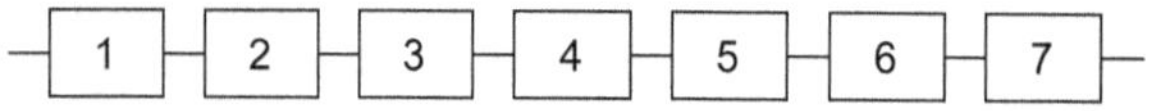

Figure 5.4 System reliability block diagram with seven components in series.

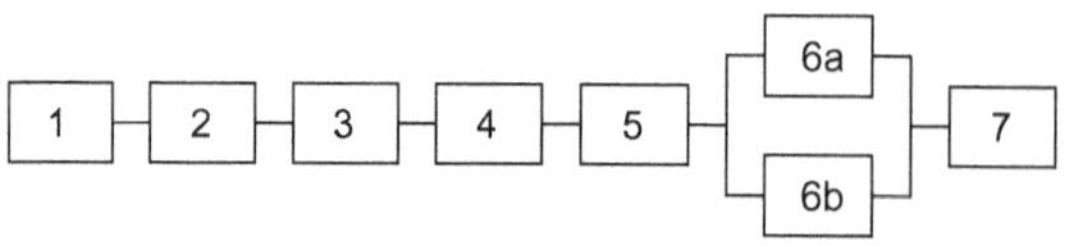

Figure 5.5 Reliability block diagram for washing machine with redundant hot water valve.

Then hot water valve reliability is determined, again from elementary probability, to be:

$$R_{\text{hot water valve}} = [1-(1-R_6)^2]$$

and

$$R_{\text{system}} = 0.970 \times 0.990 \times 0.990 \times 0.985 \times 0.985 \times [1-(1-0.980)^2] \times 0.995 = 0.92.$$

Adding a second hot water valve thus increases system reliability from 0.90 to 0.92.

5.6.4 Assumptions and Extensions

Types of Redundancy The type of redundancy in the preceding example is known as active redundancy, that is, redundant units are active and age simultaneously. Other common types of redundancy (that need to be recognized in the analysis) include:

- Stand-by redundancy. Redundant components are initially inactive, and do not age until activated, typically due to failure of the previously active component(s). Examples include the spare tire of an automobile and spare batteries for a laptop computer.
- k-out-of-n configuration redundancy: At least k out of n parallel components must be functioning.

The Assumption of Independence In the washing machine example, statistical independence of component lifetimes is assumed. This assumption would be incorrect if, for example, the deterioration of the main bearing, though not creating an immediate failure, leads to increased vibration of the washing machine, subjecting other components to higher stress, and accelerating their failures.

When the assumption of independence does not hold, system reliability cannot be obtained by simply multiplying the component reliabilities. Instead, a method that takes the dependence into consideration is required. For example, in studying the strength of a bundle of cables in a suspension bridge, a load-sharing model is used. This assumes that when one cable fails, the total load is distributed among the remaining cables. Knowledge of the nature of the interdependence is required to develop an appropriate model.

Uncertainty in the Estimates The component or subsystem reliability estimates used to estimate system reliability are obtained from various sources, such as past testing, that are generally themselves subject to uncertainty. This results in uncertainty in the system reliability estimates that, in practical applications, also needs to be evaluated.

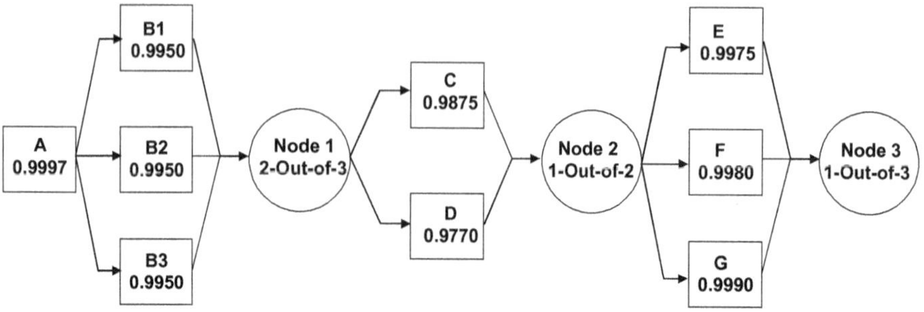

Figure 5.6 Graphical representation of a reliability block diagram by BlockSim.

Advanced Tools Hoyland and Rausand (2004, Chapter 6) describe advanced tools for system reliability assessment. Simulation is frequently used for complex assessments. Specialized software facilitates the evaluations (see Sidebar 5.3).

SIDEBAR 5.3: COMPUTER TOOLS FOR SYSTEM RELIABILITY ANALYSIS

The reliability of simple systems can be estimated using a simple spreadsheet computer program. More complex systems require specialized software. Such programs typically include graphical user interfaces to facilitate building complex system models.

Figure 5.6 shows the reliability block diagram of a system using BlockSim (from ReliaSoft Corporation). The nodes in this diagram describe the redundancy; interdependencies can also be modeled. The software then estimates system reliability either analytically or by simulation. For the example in Figure 5.6, system reliability is estimated to be 99.93%.

5.7 PRODUCT RELIABILITY DEVELOPMENT

5.7.1 Fundamentals

During product development, a major goal is to achieve the reliability targets set for the components, subsystems, and the system as a whole. Design engineers strive to understand how failures occur so that their causes can be addressed. Frequently, a failure takes place when the stress (or load) applied to a product exceeds its strength; for example:

- A car tire bursts when over-inflated.
- A valve breaks when its seal cannot withstand the water pressure.
- A transistor in an electronic circuit fails due to a voltage surge.

Also, the strength of a part may deteriorate over time due to the accumulation of previous exposure and stresses placed on the part.

A reliable design, therefore, requires a thorough understanding of:

- the key strength characteristics of the product (e.g., hardness, adhesion, and resistance);

- the effect of product design (e.g., chemistry, mechanical tolerances, and materials) on product characteristics;
- the nature of the stresses that the product will experience during normal use (e.g., vibration, pressure, and temperature) and how these vary over time; and
- the product use environment (e.g., temperature variations, voltage transients, altitude, lubricant contamination, and salt in atmosphere) and its fluctuations.

5.7.2 Understanding Failures

Reliability improvement requires an understanding of failures and what made them happen. Such understanding is acquired at different levels:

- Failure mode: This is often a description of the failure symptoms (e.g., car does not start, washing machine is leaking water, and bearing is overheating) and how these impact the operation of the product. Failure modes are usually easy to recognize, but such recognition alone is often insufficient to remove them.
- Failure cause: This requires responding to "what led to the failure?" (e.g., loss of lubrication in the bearing, deterioration of a connecting part, and corrosion of an internal switch). This assessment is generally made by a technician (but may also be a subject of speculation by the user). Such understanding can be a major step toward improving reliability.
- Failure mechanism: The ideal often is to identify and understand the fundamental physical, chemical and electrical processes underlying the failure. Frequently referred to as the physics of failure, such root cause analysis typically requires specialized knowledge of the impact of such factors as fracture, fatigue, creep, cracking, wear, corrosion/oxidation, or weathering. It may require further in-depth evaluation, such as a product autopsy.

An engine may, for example, fail to start (failure mode) because of water getting into the engine (the failure cause); this may lead to the corrosion of rotating motor parts or to an electrical short – two quite different failure mechanisms. Similarly, an underlying chemical degradation mechanism might assert itself as cracking or peeling of a coating (failure modes). Knowing the mechanism – although often hard to pinpoint – may provide the best chance of eliminating the failure.

5.7.3 Paths to Reliability Improvement

Designers aim to build products with *consistently long* (relative to customer expectations) life and, especially, to avoid "premature" failures. This is done by such measures as:

- Designing to eliminate the occurrence of failures or making them difficult to occur, using an understanding of the underlying failure mechanisms (e.g., by reducing the mismatch of thermal expansion coefficients between adjacent materials to reduce failures due to thermal cycling; by derating parts so that the applied stresses are below the rated value; or by using stronger materials).

- Slowing down the rate of failure by providing added protection (e.g., applying coating to combat the detrimental effects of corrosion).
- Building redundancy into the product.

Improved reliability might also be gained subsequent to design by:

- Requiring or strongly encouraging planned maintenance and periodic inspection. Some typical examples are scheduled automobile oil changes, periodic thermal barrier coating applications for turbine components, and replacement of filters in air conditioners.
- Making impending failures detectable (e.g., installing thermocouples to signal changes in temperature) and perhaps automatically shutting down operations to avoid or minimize damage.
- Planning "retirement" of the product before there is much chance of failure.

(See Section 9.9 for further discussion.)

5.7.4 Some Engineering Tools

Failure Mode and Effects Analysis (FMEA) FMEA is a systematic, structured method for identifying system failure modes and assessing their consequences. For example, an FMEA to study the reliability of a telecommunications relay repeater might assess all discrete devices in a circuit (e.g., ICs, capacitors, resistors, diodes, etc.) and their connections. The first step in an FMEA is to identify the possible failure modes of each of the components and their possible interactions. Then the severity, likelihood of occurrence and difficulty of detection of each failure mode is evaluated and is typically assigned a rating from 1 to 10. The resulting three numbers are then multiplied to obtain a risk priority number for that failure mode. (See Section 12.6.3 for a brief example and Wheeler 2005 for a discussion of FMEA.)

Modeling Tools These include the use of Computer-Aided Design (CAD) methods and analytical modeling techniques such as finite element analysis, computational fluid dynamics, and electronic circuit modeling. Such models often require knowledge of material properties that might have to be acquired by laboratory testing or from material handbooks.

Intensive Early Testing to Identify Failures To supplement more analytic approaches, testing is used to obtain, as early as possible, an improved understanding of known failure modes, discover unsuspected ones, and assess the associated causes and mechanisms.

Since testing at use conditions is often expected to result in few or no failures in a short time, tests are often conducted at high stresses. Highly Accelerated Life Testing (HALT), for example, is a type of overstress testing used by engineers for speedy discovery (and subsequent elimination) of key failure modes during product development (see Hobbs 2000). Test units are subjected to stresses (e.g., voltage, temperature, humidity, and vibration), often in combination, at much more severe levels than encountered in normal operation. HALTs are typically conducted on components, subsystems and, perhaps, a few initial prototype units of the complete product. They are performed principally for engineering evaluation to rapidly identify weaknesses in the design and may not require any statistical evaluations.

5.7.5 Statistical Experimentation

Statistically planned experiments are often used to quantify the effect of design factors (e.g., material composition, geometric characteristics, and processing conditions) on product life – much as for other product performance characteristics; these frequently result in censored data. For example:

- Zelen (1959) provides an early illustration of a factorial experiment dealing with the life of glass capacitors.
- Grove and Davis (1992, Section 8.2) report the results of an experiment that showed that weld strength (and thus the reliability of the part being manufactured) could be increased by changing the force with which the pieces of metal were held together during manufacture.
- Byrne and Quinlan (1993) describe an experiment which showed that the reliability of a mechanical speedometer cable can be improved by appropriate choice of wire braid and wire diameter.

Condra (1993) and Graves and Menten (1996) provide further discussion and illustrations of statistically planned experiments for reliability improvement.

Robust design is also important in reliability applications. For example, products need exhibit *consistently* high reliability under varying operating conditions, involving such noise variables as moisture, vibration, and ambient temperature (see Hamada 1995 and Phadke 1989, Chapter 11).

5.7.6 Reliability Estimation

Many statistically planned investigations during product design aim at assessing the reliability that can be expected for components, assemblies, subsystems, and eventually the final product or system. There is, typically, insufficient time to make such evaluations at normal operating conditions. For example, we might want a product to have a 10-year reliability of 95%, but we only have 6 months to assess whether a proposed design is likely to meet this requirement.

Three common approaches for addressing this challenge are use-rate acceleration, accelerated life testing, and degradation testing.

5.7.7 Use-Rate Acceleration

Concept Many products are not in constant use. Toasters, washing machines, photocopiers, printers, bicycles, and seat belt buckles are just a few examples. We capitalize on this through use-rate acceleration testing by increasing the use rate, thereby shortening the idle time. A toaster might, for example, be designed for 12-year failure-free operation, assuming twice-a-day usage. The equivalent of 12 years of normal use can be obtained in less than 90 days by running the toaster 100 times daily.[8]

Use-rate acceleration is based on the assumption that the increased cycling excites the failure modes seen in normal operation; that is, the failure results directly from product operation and not, say, chemical change over time. This assumption is reasonable for many, but not all, failure modes. Failures due to corrosion, for example, are likely to be dependent on elapsed time, rather than usage.

Washing Machine Motor Example In the earlier example in this chapter, the motor of a new washing machine was to be designed to have a 10-year reliability of at least 97%. Skilled engineers used top quality materials and state-of-the-art methods to correct reliability problems on previous designs. They also performed some HALTs, subjecting components and some prototype motors to intensive temperature cycling, vibration, and overvoltage conditions to discover, understand, and remove new or remaining failure modes. Failures were still expected to occur due to wearout of the bearing, but this was felt to be unlikely during the first 10 years of life.

The engineers, therefore, were confident they had developed a motor with the desired reliability. But had they? To assess this – at least under lab conditions – a life test was to be conducted for six months prior to the planned product release.

Use-Rate Accelerated Testing This situation lends itself to use-rate accelerated testing. Tests were conducted on special equipment on which the motors were subjected to mechanical loads resembling those experienced during a typical wash cycle. A sample of motors was to be run for 24 cycles per day, shutting down after each cycle for brief cool-downs. In this manner, 3.5 years of field operation were simulated in each month of testing – assuming a use rate of four washes per week. The assumption that failures are the result of the actual running of the motor, independent of elapsed time, seemed reasonable based on engineering considerations.

A test of 66 motors (the number of available test stands) for 6 months (equivalent to 21 years of field operation) was proposed. Computer simulation showed this plan to provide a reasonable degree of precision in estimating 10-year reliability.

The motors were lab prototypes, but needed to reflect as closely as possible the variability expected in production. Thus, the 66 motors were built at three different times, using multiple lots of material for each of the key assembled parts. Detailed records of the motors' manufacturing histories and performance were maintained.

Results after 6 Months The results after 6 months of testing (4,368 test cycles), converted into customer use years, are shown in Table 5.1. By that time, there were seven bearing failures; the remaining 59 units were unfailed. Most of the unfailed units were still running when the test was ended; a few had been withdrawn from testing earlier for various reasons unrelated to their failure proneness.

8. As before, age may be expressed in a variety of ways, such as number of hours of operation, number of start-ups, number of operating cycles; or whatever is most reasonable from physical considerations.

TABLE 5.1 Washing Machine Motor Test Data (Converted into Years of Operation Under Assumed Normal Operating Conditions)

Motor Lifetimes	Survival (Censoring) Times for Unfailed Units
9.6, 13.5, 16.8, 17.4, 18.7, 19.3, and 20.9	4.5, 6.7, 7.5, 10.6, 10.7, 17.8, 19.7, 17.5 (4 units), and 21 (48 units)

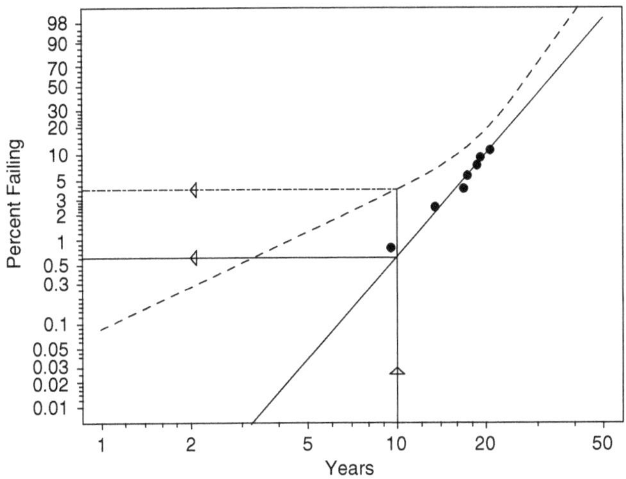

Figure 5.7 Fitted Weibull distribution for motor lifetimes and approximate upper 95% confidence bound (dotted curve).

Analysis[9] Both Weibull and lognormal distributions were fitted to the data. These models (see Appendix) are frequently used in applications and seemed reasonable on both theoretical and empirical grounds. The results for the two analyses were similar; therefore, only the Weibull analyses are shown.

Figure 5.7 is a Weibull probability plot (see Sidebar 5.4) of the data. The solid line in this plot is the Weibull distribution estimate of the percent devices failing as a function of years in service. The data were fitted by the method of maximum likelihood (see Appendix). The dotted curve gives approximate upper 95% confidence bounds on these percentages. The plotted points scatter around a straight line, supporting the assumption of a Weibull distribution within the range of the data.

SIDEBAR 5.4: A NOTE ON PROBABILITY PLOTS

A probability plot for an assumed distribution model is a plot of the cumulative distribution function (as estimated from sample data) whose axes are constructed so that the plotted points tend to scatter around a straight line *if* the assumed model is correct. For

9. Further details on this example will be provided on the book's ftp site.

censored data, only the noncensored observations are plotted, but the censoring is taken into consideration in determining plotting positions. In small or heavily censored samples, the plotted points may, due to sampling variability, show appreciable deviations from linearity even if the assumed life distribution is correct.

The generally best-known probability plot is for the normal distribution. Lognormal and Weibull probability plots are frequently used in life data analysis.

The estimated 10-year reliability (see Section A6 of Appendix) is 99.4% (percentage failing 0.6%), with a 95% lower confidence bound of 96% (upper confidence bound of 4% on percentage failing).[10] Further analysis showed that the desired 97% 10-year reliability could be demonstrated with approximately 92% confidence.

Resulting Action The results were sufficiently favorable to allow management to give the go-ahead for full-scale production of the motor under the assumption that this test realistically represented what would be encountered in actual operation.

All failed motors (and some unfailed ones) were taken apart and evaluated to obtain information to further improve reliability in the future. Also, 10 surviving motors were selected randomly for another 4,000 cycles of testing to obtain more precise reliability estimates. In addition, 25 units built during a 1-month period and 25 units randomly sampled from the first week of production were tested for varying times. If these had yielded unfavorable results (which they did not), start-up would have been halted.

5.7.8 Accelerated Life Tests (ALTs)

High-Stress Testing and Product-Aging Acceleration In many situations we cannot employ use-rate acceleration. For example, a refrigerator typically operates continuously and, therefore, it is not possible to accelerate its use rate. An alternative then might be to conduct an ALT.

An ALT involves high-stress testing and/or product-aging acceleration. High-stress testing requires increasing the stress (e.g., voltage or pressure) at which test units operate. A unit fails when its strength drops below the applied stress. High-stress testing causes units to fail sooner than normal stress.

Product-aging acceleration involves exposing test units to more severe than normal environments, such as higher than usual levels of temperature or humidity. This accelerates the physical/chemical degradation processes that cause certain failure modes (e.g., the weakening of an adhesive bond or the growth of a conducting filament across an insulator). Such testing tends to age the units faster and cause them to fail earlier.

10. This example illustrates the fact that demonstrating a certain level of reliability with high statistical confidence requires the *true* reliability to be appreciably higher.

Similar statistical approaches are used for conducting ALTs for high-stress testing and for product-aging acceleration. We will use the term "(high) stress" to refer to the accelerating variable for both situations.

Acceleration Models An important first step in planning an ALT is to identify one or more accelerating variables – and a physically appropriate acceleration model that relates stress to lifetime (see Sidebar 5.5) This model is then extrapolated to estimate the expected lifetime under the stress encountered at normal operating conditions.

SIDEBAR 5.5: SOME ACCELERATION MODELS

The selection of an acceleration model is guided by product domain knowledge and, if possible, should be based on the fundamental physics and chemistry of the application. When such knowledge is not available or is incomplete, models are developed from a combination of physical knowledge and experience. Well-known acceleration models include:

- The Arrhenius relationship describing the effect of temperature on the rate of a simple chemical reaction (which, in turn, can be expressed in terms of life). This model results in a linear relationship between the reciprocal of temperature in degrees Kelvin (Centigrade degrees + 273.2 degrees) and log lifetime.
- The inverse power relationship between life and voltage-stress acceleration. This model results in a linear relationship between log voltage and log lifetime.
- The Coffin–Manson relationship between temperature cycling and number of cycles to failure. This model results in a linear relationship between log lifetime and the range of the thermal cycle.

In addition to the acceleration model, a statistical model, such as a Weibull or lognormal distribution, is used to describe the distribution of lifetimes at each level of stress. (See references in Section 5.13 for further discussion.)

Conducting an ALT An accelerated life test is conducted at various stresses to obtain data to estimate the unknown parameters in the assumed model. Each test unit is run at a constant stress.[11]

Planning an ALT, in addition to defining the accelerating variable(s) and the acceleration model, requires specifying the test stress conditions, appropriately selecting test units, allocating these to the stress conditions, and determining test duration. This, in turn, requires understanding of the test facilities and their limitations and of the availability of test units, as well as the requirements of the study.

ALTs are contrasted with HALTs in Sidebar 5.6.

11. In an alternative approach, known as step-stress loading, test units are each subjected to successively higher levels of stress (see Nelson 1990, Chapter 10).

SIDEBAR 5.6: HALTs, ALTs, AND IRRELEVANT FAILURE MODES

Engineering-based HALTs and statistically based ALTs both involve testing at accelerated stresses. The major goal of a HALT is to rapidly identify and eliminate reliability problems. In contrast, an ALT typically seeks data to develop a model to estimate reliability at the normal operating condition and to compare the reliabilities of alternative designs and competitive products.[12]

ALTs, and especially HALTs, might generate failure modes, particularly at extreme stresses, that would not occur at normal operating conditions. Their impact on the fitted ALT model can be removed by statistical analysis – *if* the units experiencing such failures can be identified and are independent of the failure mode of interest. Especially when such identification cannot be made, test stresses should not be so severe as to result in irrelevant failure modes. In any case, careful engineering evaluation is needed to determine whether an observed failure mode at an accelerated stress presents a genuine risk at normal operating conditions.

Generator Insulation Example This example deals with a newly developed dielectric insulation for stator bars used to operate power generators. The insulation consists of a mica-based system bonded together with an organic binder. It is to be designed to have a reliability of 99% after 10 years of use at normal operating conditions.

Eleven months were available for testing the new insulation. If reliability is as expected, it is unlikely that any test unit would fail during this time at normal operating conditions. It is not possible to perform use-rate acceleration because the product is operated essentially continuously. Thus, an ALT is to be conducted.

Acceleration Variable and Model Insulation failures occur due to the failure of the organic material, causing a sudden reduction in strength. As a result, the insulation can no longer handle the voltage stress. Increased voltage speeds degradation, and thus failure, and was taken as the accelerating variable. From physical considerations and past experience, it was reasonable to assume an:

- inverse power relationship between voltage and lifetime; and
- a Weibull distribution for lifetime at each voltage.

Test Plan and Protocol Tests are conducted on special lab stands using short electrodes built to represent production bars. Sufficient units and testing facilities were available to test 90 electrodes.

Since the new insulation system was still under development, the 90 electrodes were prepared in the lab and, therefore, were not fully representative of future production. The results need, therefore, be validated by subsequent testing on regular size bars from normal production.

12. In the early stages of an ALT, the product often fails to meet the reliability goal; then, as in a HALT, the test units are scrutinized to gain understanding of the basic failure mechanism and to make improvements.

TABLE 5.2 Failure and Survival Times (in Hours) from Accelerated Life Test of Dielectric Insulation Material

75 vpm	95 vpm	105 vpm	115 vpm	125 vpm
7,800 + (15)	2,914	3,046	1,119	883
	6,650	4,384	1,406	996
	6,919	4,864	1,925	1,085
	7,800 + (27)	6,644	2,328	1,206
		7,510	2,465	1,664
		7,800 + (15)	3,399	1,819
			3,588	2,187
			3,854	2,400
			4,116	2,663
			4,262	2,938
			4,279	
			4,905	
			5,697	
			6,066	
			7,800 + (1)	

The numbers in parentheses designate the number of unfailed electrodes.

The stress on an electrode is measured in volts per mil (vpm); 60 vpm is the normal operating condition. Stress levels above 125 vpm were to be avoided since these were likely to lead to failure modes (such as deformation of insulation due to high heat generated by the high voltage stress) that do not occur at normal operating conditions.

The 90 electrodes were randomly allocated for testing for up to 11 months at five stress levels[13]:

- 15 electrodes at 75 vpm
- 30 electrodes at 95 vpm
- 20 electrodes at 105 vpm
- 15 electrodes at 115 vpm
- 10 electrodes at 125 vpm

The majority of test units at 125 vpm were expected to fail within 11 months. Few failures were expected to occur at 95 vpm and none at 75 vpm.

Results The data after 11 months from this ALT are shown in Table 5.2 and plotted in Figure 5.8. By that time, a total of 32 units had failed at four of the five test conditions; 58 units had not yet failed, leading to censored observations.

13. Five voltages were used to make the results comparable with those from earlier studies. The proportion of test units allocated to the different voltages was expected to result in a roughly equal number of expected failures at conditions other than 75 vpm. The units at 75 vpm provided potential added insurance of acceptable reliability (assuming none failed) and were to stay on test beyond 11 months. Simulation indicated that this plan would provide reasonable precision in estimating 10-year reliability at 60 vpm.

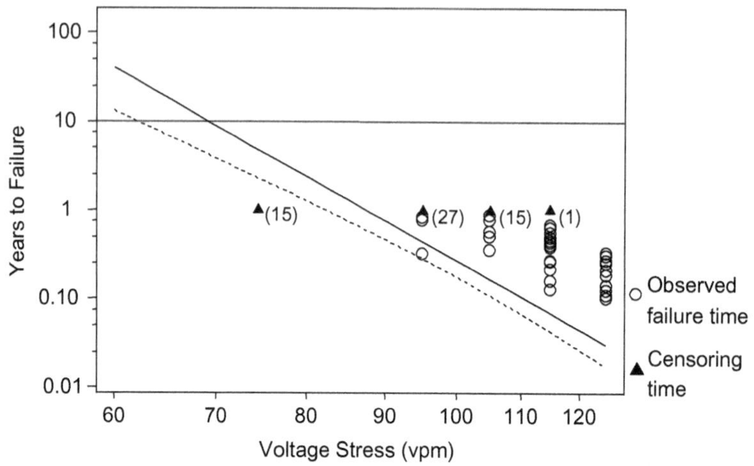

Figure 5.8 Results of high-stress test for dielectric insulation material, showing estimated first percentile of life distribution and associated 95% approximate lower confidence bound (dashed line).

Analysis[14] The data were fitted to the assumed model using maximum likelihood to estimate 10-year reliability at 60 vpm. The heavy line shown in Figure 5.8 is the estimated *first percentile* of the life distribution as a function of voltage. The associated 95% approximate lower confidence bound on this line is shown by the dashed line.

In particular, the approximate 95% lower confidence bound on the first percentile of the lifetime distribution at a voltage of 60 vpm is 13.2 years. Since this exceeds 10 years, the results suggest that the new insulation meets its reliability goal. This conclusion is subject to the validity of the assumed model throughout the region of interest[15] and to the assumption that the test correctly represents what will happen in operation on normal size units from regular production.

5.7.9 Degradation Testing

Concept Many failure mechanisms can be traced to an underlying degradation process. Degradation eventually leads to a reduction in strength or other change in physical state that can cause failure. Tire wear, as measured by tread, provides an example. Degradation measurements provide information on a continuous, rather than a go/no-go or attribute, scale. Thus, such measurements often yield more information than lifetime data. Degradation measurements are especially useful when the testing to date has resulted in very few or no failures.

Sometimes failure is defined, possibly arbitrarily, as the degradation exceeding or falling below some specified threshold value. A consumer might define a tire to

14. Further details on this example will be provided on the book's ftp site.
15. This is why it is important to base these assumptions on physical theory, as much as possible.

fail at "blowout." But the tire manufacturer or a regulator might say that a failure occurs when the tire's tread falls below a specified value at which the tire is no longer felt to be "safe." In such cases, it is especially appropriate to base evaluations directly on the degradation measurements themselves, rather than on the failed/not failed dichotomies derived from these.

The relationship between the amount of degradation and lifetime makes it possible to use degradation models and data to make inferences and predictions about reliability. This generally requires one to define a certain degree of degradation, when first reached, as constituting a failure. Degradation data also allow modeling physical failure mechanisms more directly than lifetime data, providing increased insights.

GaAs Laser Example Meeker and Escobar (1998) report a reliability study on a GaAs laser used in telecommunications systems. This type of laser has a built-in feedback circuit to generate a constant light output. Over time, a larger current is needed to maintain a specified level of light. Failure was defined as the time at which a 10% increase in current is first needed. Engineers desired to estimate the probability of failure at 5,000 hours (equivalent to about 20 years of normal operation) on a single-condition high-temperature test. The result was to be used to determine the degree of redundancy required in the systems in which the lasers are used. This assessment, moreover, was needed after only 2,000 hours of testing.

Degradation Measurements Tests were conducted on 15 lasers randomly selected from a lab prototype of the production process. Degradation, as measured periodically by the percent increase in operating current for up to 2,000 hours of testing on each of the test units, is displayed in Figure 5.9a for three of the test units and for all 15 units in Figure 5.9b.

Analysis None of the lasers exceeded a 10% increase in current by 2,000 hours, limiting the ability to estimate reliability at 5,000 hours using lifetime data only. Figure 5.9 suggests, however, that important information can be gained by analyzing the degradation data.

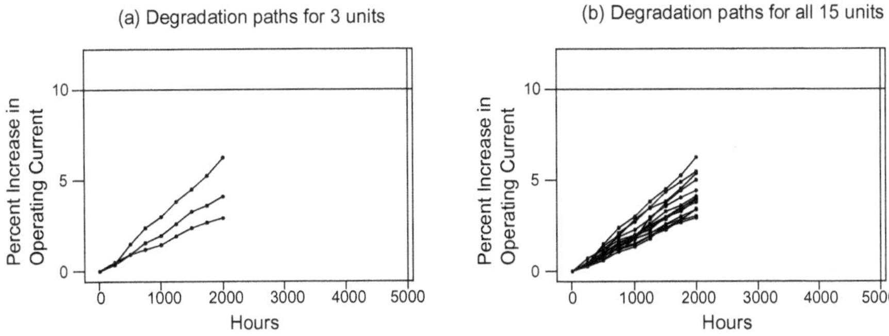

Figure 5.9 GaAs Laser degradation paths after 2,000 hours of testing.

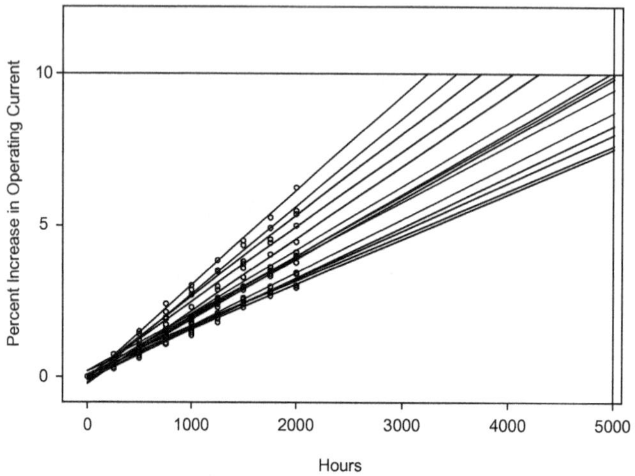

Figure 5.10 GaAs laser degradation paths at 2,000 hours extrapolated to failure (10% current increase) or 5,000 hours.

The basic approach is to generate "pseudo-lifetimes" by extrapolating the individual degradation paths of unfailed devices by, say, regression analysis, and using these extrapolations to predict the unit's lifetime. Figure 5.10 shows the 2,000-hour degradation paths extrapolated to time to failure or 5,000 hours, whichever came first, assuming that the linear trend noted in the first 2,000 hours can be validly extrapolated. Seven paths exceed a 10% current increase, resulting in "pseudo-lifetimes" of 3,229, 3,514, 3,742, 4,047, 4,282, 4,781, and 4,969 hours. The eight remaining units continue to be unfailed at 5,000 hours and were taken as censored observations.[16]

Figure 5.11 shows a lognormal distribution probability plot of these pseudo-lifetimes, together with a fitted lognormal distribution and the associated 95% confidence interval around the fitted line. The scatter of the pseudo-lifetimes around the fitted straight line suggests that the lognormal distribution provides a reasonable fit.

This approach led to the estimate of the probability of failing by 5,000 hours as 0.48 with an approximate 95% confidence interval of 0.24 to 0.72 under the previously stated assumptions.[17]

Limitation Degradation testing is not applicable for all situations. The so-called "catastrophic" failures are typically not the consequence of continued degradation, but due to a sudden outside event, such as a lightning strike. Also, degradation often

16. It was felt that a more conservative approach was to make the extrapolations only up to 5,000 hours, rather than until eventual failure. This was done so as to avoid too much extrapolation. In actuality, we did both analyses and compared the results (see next footnote).

17. This wide interval reflects the extrapolation and the fact that the analysis is based on a small number of failures. Linear extrapolation of *all* 15 units to failure resulted in an estimate of 0.54 for the 5,000-hour failure probability and a slightly narrower 95% confidence interval of 0.34 to 0.73.

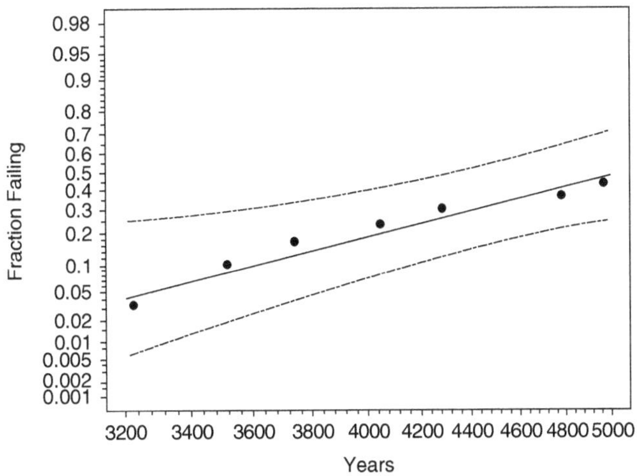

Figure 5.11 Lognormal probability plot for pseudo-lifetimes with fitted distribution and 95% confidence interval.

involves extrapolation, as in the example. This is always dangerous since it requires the assumption that the pattern that has been observed in the past, such as a linear trend, will continue into the future. (See Sidebar 5.7 for some further comments.)

SIDEBAR 5.7: SOME RAMIFICATIONS OF DEGRADATION TESTING

From Failure Mechanisms to Degradation Measurements

Establishing appropriate degradation measurements requires an understanding of the underlying failure mechanism. Corrosion, for example, is a major concern for applications involving metals. Oxidation is a type of corrosion that leads to formation of oxides (e.g., rusting of iron) on metal surfaces, consuming metal thickness. Thus, in studying corrosion, the amount of oxidation is often used as a measure of degradation. This might call for monitoring the weight change of the metal over time.

Types of Degradation Data

In some situations, such as the GaAs Laser example, a unit's degradation is measured continuously over time. In other cases, especially when readings are taken manually, measurements are taken periodically.

Still other applications (e.g., estimating burst strength of an angioplasty balloon) require a destructive process to measure degradation. In such cases, only a single degradation measurement is available for each test unit. This calls for staggering the times at which such readings are obtained on different test units (see Nelson 1990, Chapter 11, and Meeker and Escobar 1998, Chapter 21).

Degradation and Accelerated Testing

One can combine degradation testing with use-rate acceleration. In this case, the measurement obtained on the use-rate test is the degree of degradation of the test unit. Similarly, degradation testing might be built into an ALT.

5.8 RELIABILITY VALIDATION

Do not count your chickens before they are hatched.
—Popular saying

5.8.1 Goal

Much testing during early development is on the component or subsystem level. This generally does not permit evaluation of interactions among components and subsystems. Even when system testing is conducted, it is likely to be on prototypes rather than normal production. Also, it might be difficult to simulate the operating environment. For example:

- The generator insulation study was based on short electrodes (about 2 feet long) built in the laboratory to represent full-size stator bars (about 18 feet long). Testing was on special equipment under increased electrical stress. In actual operation, the full-size bars are subjected to additional thermal and mechanical stresses.
- For the washing machine motor, units built in the lab were tested outside the washing machine by applying an electric load simulating the load expected in normal operation.

These examples involved what we characterized in Chapter 1 as analytic studies. Any statistical inferences drawn from the results hold only to the degree that the test units and the test environment accurately represent those encountered in normal operation.[18]

Reliability validation aims to ensure, to the greatest extent possible, that the reliability goals will be met on scaled-up manufactured product under actual field operating conditions. The focus increasingly shifts to the evaluation of:

- product on the system level;
- sample units that represent future production as closely as possible; and
- use conditions that approximate what might be encountered in actual operation.

Thus, our discussion in this section parallels that on design validation and scale-up to manufacturing in Chapters 3 and 4.

For consumer products, such as CD players, dishwashers, toasters, etc., it is often possible to conduct validation testing on a relatively large number of units. For large products, such as generators, locomotives and aircraft engines, the testing that can be conducted is generally more limited, and due to cost considerations, often nondestructive. The resulting reliability inferences from small samples will likely have limited statistical precision. If reliability issues are identified during validation, these need be addressed immediately.

18. The testing described earlier in this chapter also involves analytic studies to varying degrees.

5.8.2 In-House Systems Testing

Generator Insulation Example: Description This test was to validate, in a short elapsed time, the design using full-size bars from manufacturing. Therefore, unlike the ALT conducted during development, the test was run at a single voltage of 125 vpm. A major goal was to determine whether the first percentile of the lifetime distribution at this condition was an improvement over the 100 hours attained on previous similar tests for the current insulation system.

The validation test needed to reflect, as closely as possible, the variability expected in normal production. Thus, it utilized 58 full-size bars built on the production line at three different times using different lots of materials for each major assembled part.

Discovery of a New Failure Mode The failures that occurred during this test were found, based upon an autopsy, to be due to one of two modes:

- Mode E (Early failure): Insulation defects due to a processing problem. This failure mode was apparently introduced in the switch to manufacturing and/or going to regular-size bars; it had not been experienced in the ALT.
- Mode D (Degradation failure): Degradation of the organic material; these failures typically occur late in life and were similar to those observed in the ALT. (The degree of degradation could not be measured prior to failure.)

There were 18 failures due to mode E and 27 failures due to mode D. Thirteen bars were still running when the data were analyzed. The full results are shown in Table 5.3.

TABLE 5.3 Voltage Endurance Test Results

Hours	Failure Mode	Hours	Failure Mode	Hours	Failure Mode	Hours	Failure Mode	Hours	Failure Mode	Hours	Failure Mode
2	E	53	+	144	E	236	E	303	E	350	D
3	E	64	E	157	+	241	+	314	D	360	D
5	E	67	+	160	E	257	+	317	D	369	D
8	E	69	E	168	D	261	D	318	D	377	D
13	+	76	E	179	+	264	D	320	D	387	D
21	E	78	+	191	D	278	D	327	D	392	D
28	E	104	E	203	D	282	E	328	D	412	D
31	E	113	+	211	D	284	D	328	D	446	D
31	+	119	E	221	E	286	D	348	D		
52	+	135	+	226	D	298	D	348	+		

E = Failure due to processing defect.
D = Failure due to degradation of the organic material.
+ = Unfailed bar.

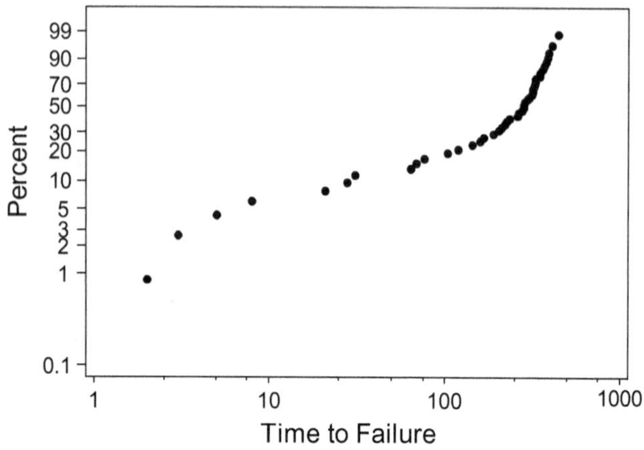

Figure 5.12 Weibull probability plot of voltage endurance data.

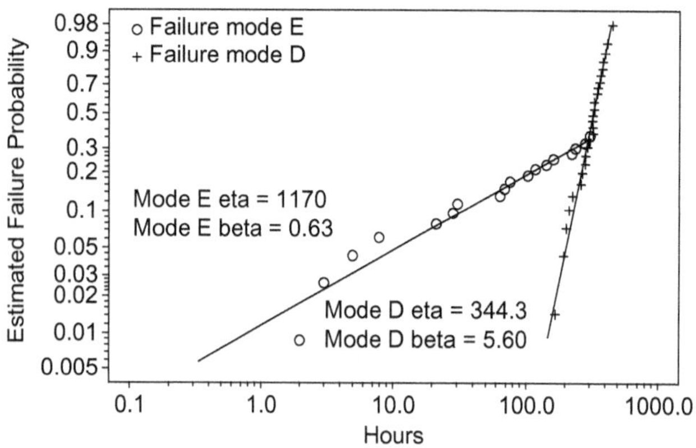

Figure 5.13 Separate Weibull probability plots of voltage endurance data for failure modes D and E (with maximum likelihood estimates of Weibull distribution parameters).

Initial Analysis[19] Figure 5.12 is a Weibull probability plot of the data in Table 5.3, ignoring the failure mode information.

The plot shows that a Weibull distribution does *not* fit the data since the points do not scatter around a straight line. A lognormal probability plot yielded a similar result. Thus, a single simple distribution does not seem to adequately represent the data. The reason is the presence of two failure modes. When these are known for each failed unit, as in this example, it is generally advisable to conduct *separate* analyses for each mode.

Separate Analyses Separate Weibull distributions were fitted for each of the two failure modes.[20] The results are shown in Figure 5.13. The scattering of the points around two straight lines indicate that each of these two failure modes, when considered separately, could be represented by a simple Weibull distribution.

19. Further details on this example will be provided on the book's ftp site.

20. The failures due to failure mode D were taken as censored observations in the analysis of failure mode E and vice versa. This analysis assumes that the two failure modes are independent.

The results of these separate analyses were then combined probabilistically to obtain system reliability estimates[21]; also, propagation of errors was used to obtain approximate 95% confidence intervals. The resulting estimate of the first percentile of the lifetime distribution was 0.84 hours with an approximate 95% confidence interval of 0.06 hours to 10.2 hours – a far cry from the desired 100 hours! Failure mode E clearly needed to be eliminated to meet the reliability requirement. This seemed doable from a physical evaluation of the underlying failure mechanism.

It was noted, moreover, from Figure 5.13 that if failure mode E were successfully removed, leaving only failure mode D, the estimated first percentile of the lifetime distribution would be 151 hours with an approximate 95% confidence interval of 118 to 195 hours – exceeding the 100-hour requirement.

Validating Fixes Unexpected failure modes discovered in validation testing are often evaluated and corrected, as in the example. Frequently, there may be little time left to verify that this has been done successfully and to ensure that new failure modes have not been introduced in the process. It is, however, highly desirable to conduct at least some further validation testing after making such fixes.

5.8.3 Beta Site Testing

Concept Any in-house testing, irrespective of how well constructed, cannot fully represent the diverse customer uses (and misuses) of the product and the variation in operating environments experienced in field operation.

One of our favorite examples is that of short circuits of ovens due to cockroaches accessing the circuit board – resulting in the malfunction of the oven (as well as the incapacitation of the invading cockroaches). This failure mechanism had not been envisaged by the product designers, and was not simulated by in-house testing. Once discovered, it was addressed by a minor design change. Bell Labs encountered a similar problem with their first underwater digital cable systems: sharks liked to bite the new cables (see Marra 1989).[22] The problem was reportedly addressed by burying the cable in shallow water.

Thus, as a final validation test, a sample of systems is put into the field prior to general product release and full-scale production in so-called "beta site testing." Typical examples are testing:

- washing machines in laundromats;
- refrigerators in high-temperature and high-humidity locations;
- car exterior body panels in hot, sunny climates to assess color fading from sunlight exposure; and
- locomotive operated with heavy loads in mountain terrains and tunnels.

21. The estimated probability of a system failure by any specified time was calculated as the estimated probability of a mode D and/or a mode E failure occurring by that time, based upon their separate fitted distributions.

22. The sharks had shown no interest in the old analog technology.

Such tests should be conducted under diverse field environments, especially ones that result in severe stress on the product in a short time. Ideally, such products are closely monitored to gain information about usage frequency and the operating environment – as well as performance and reliability (including the mode, cause, and mechanism of any failures). Degradation should also be measured, if possible.

Example: Laptop Computer Reliability Tracking System A new laptop computer was designed with careful consideration given to identifying and eliminating potential premature field failures. All new components were tested thoroughly and the machine itself was run through an extensive accelerated life test.

In addition, a 6-month beta site test was conducted before release for high-volume production. This involved building 300 computers over a 1-month period in a manufacturing environment and giving these to heavy-use employees in different company functions. A rigorous mechanism for reporting usage and malfunctions was developed. All failed machines were returned for in-depth evaluation.

Because few failures were expected in the relatively short time available, 30 machines were randomly selected for return after 1 week and also after 1 and 3 months and torn down to measure the degradation of key performance characteristics. Problems uncovered by this test were corrected before going to full-scale manufacture.

The 210 remaining computers continued to be carefully scrutinized to provide information about possible longer-term problems. This included returning 20 computers for further evaluation after every 6 months of operation.

Overall, the results provided added confidence of good product performance and reliability. A potential improvement is to imbed the computers with monitoring devices that continuously track degradation and provide information about usage intensity.

Limitations Beta site tests potentially provide more realistic information about performance than in-house tests. However, they often can give only limited information due to test time and samples size restrictions. Capabilities for monitoring usage might also be limited.

Manufacturers, moreover, are often reluctant to expose systems that might still have problems to customers for fear of the negative impact of poor reliability. This is why the computer manufacturer selected its own employees for such testing. Manufacturers of industrial products might similarly use one or a small number of customers with whom they have close and solid relationships for beta site testing.

5.9 PRODUCT SAFETY

5.9.1 The Concern

Product safety deals with the assurance that the use, or even presence, of a product does not lead to personal harm or injury. Potential fire hazards are a key safety concern in the design of many household products. So is the possibility of exposure to radiation. Unanticipated ways that imaginative customers might employ a product (e.g., using an electric iron to heat food and cleaning vegetables in a washing machine) can also create safety problems. Injury caused by inappropriate use is of particular concern to the makers of children's products. Toy manufacturers don't intend small pieces to be swallowed, but sometimes infants fail to read the cautionary labels.

An uncompromisingly high level of safety, in addition to high reliability, is essential for all products and the two areas are closely related.

5.9.2 Design Considerations

A common theme in ensuring product safety is assessment and prioritization of safety hazards so that they can be addressed proactively. As in other applications, careful engineering analysis with a keen understanding of the product use environment needs to precede any major statistical studies.

A proactive approach calls for designing and building products that are maximally robust to malfunction and misuse; for example:

- Minimize the temperature during operation of the outside parts of a toaster
- Design a dishwasher to turn off automatically if its door is opened
- Build a safety device into a firearm that must be deactivated before use
- Avoid parts in a toy that an infant might swallow

Disciplined product safety risk assessment and mitigation tools for product safety management have been developed. In the United States, military standards (e.g., MIL-STD 882) and industry guidelines provide a general framework. Underwriters Laboratories certification includes requirements specifically aimed at reducing safety risks. (See also discussion in Section 13.2.6 on food safety.)

Formal risk analyses to ensure product safety need to be conducted during design, especially in highly complicated situations, such as the construction of a nuclear power plant. A safety directed FMEA can be a major step toward proactively identifying safety concerns. In addition, a probabilistic model of product operation might be developed and evaluated, perhaps using simulation, to determine the probability of, say, injury or property loss. Such models are also used to assess the impact of features such as an automatic fire suppression system to mitigate the impact of potential safety hazards. (See Bedford and Cooke 2001 for further information on probabilistic risk assessment.)

For many products, failure in itself creates a safety hazard; tire blowout is a typical example. Thus, product safety is often addressed by reliability improvement. And even though eventual failure may – like the death of humans – be unavoidable, one can often work toward delaying its happening and lessening its adverse impact.

5.9.3 Statistical Considerations

Many of the key statistical issues dealing with product safety are similar to those for reliability assurance. What differentiates this area is the generally low occurrence rate of an incident (and the resulting sparse data), together with the high level of attention that even a single event merits (due to its serious consequences). A 1-in-100,000 probability of an electric iron causing a fire in normal use during its lifetime is still of extreme concern to manufacturers who sell millions of irons yearly. This is so irrespective of whether the fire was due to a rare product defect or the possibly inappropriate way in which the iron was used.

An in-house or beta site test of, say, 500 irons in this situation is, moreover, unlikely to result in any fires. Such testing might, therefore, include an engineering evaluation that takes irons apart after use to examine the wear of critical parts to detect possible defects. In addition, one might try to induce a dangerous type of failure resulting from possibly improper use through imaginative testing in a highly controlled environment.

Once a product is in the field, detection of safety issues may be easier due to the larger quantities involved (and the more realistic operating environment) – although this is precisely what we want to avoid. Nevertheless, we need to strive to build into the system ways of detecting and speedily reporting such problems in the unhappy circumstance that they do occur (see Section 9.6).

5.10 RELIABILITY GROWTH ANALYSIS

5.10.1 Concept

Reliability evolves continually during the design, development, validation testing, production, and field use of a product as design or manufacturing flaws are discovered and fixed to improve future product. This has led to the quantification and assessment of a product's reliability growth over time, especially for repairable systems, or so-called "reliability growth analysis[23]".

5.10.2 Duane Plots

Based on the analysis of failure data from five different products, Duane (1964) concluded that reliability improvement frequently follows a consistent pattern; namely a straight line relationship when the log of cumulative MTBF is plotted against the log of cumulative operating time, with later generation units tending to have better reliability than their predecessors. The slope of this line provides a measure of reliability improvement; typically encountered slopes range from 0.3 to 0.6. The steeper the slope, the more rapid is the improvement.

Example During the reliability test of a single prototype system, design flaws were identified after each failure and fixed.

System failures occurred after 52, 216, 496, 685, 1,007, 2,022, 3,019, 3,530, 4,067, and 5,653 hours. The times between successive failures seem to be getting longer, suggesting that reliability is improving over time.

23. Reliability does not invariably improve with time. It may remain unchanged when information on past failures is *not* used to make improvements and even deteriorate as a consequence of, say, inappropriate "cost reduction" measures.

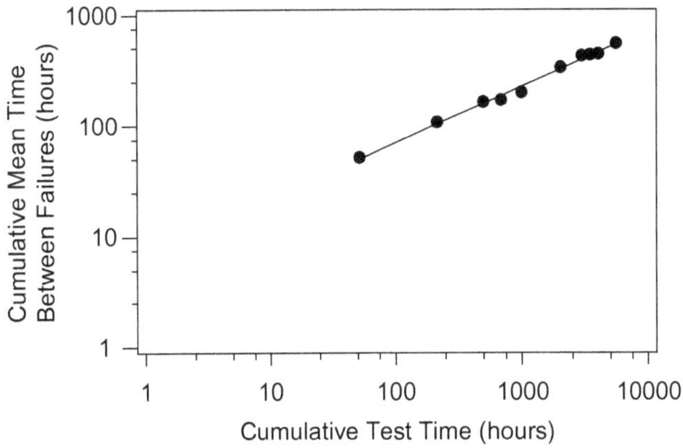

Figure 5.14 Duane plot of system failure data.

The Duane plot of the data, shown in Figure 5.14, suggests a straight-line relationship, as claimed by Duane, with an estimated slope of 0.51. In a further analysis,[24] the system MTBF after 6,000 hours of testing was estimated to be 1,158 hours and extrapolated to be 1,503 hours after 10,000 hours – as compared to an estimated MTBF of 465 hours after 1,000 hours of testing.

5.10.3 Applications

Reliability growth analysis is used to estimate the expected reliability of a system at some future time, and to decide when to release a product for full-scale manufacture or distribution. In the development of a new wind power generator, for example, pilot units were placed on a reliability growth test. Observed failures were identified, diagnosed and fixed; the design was modified to avoid the occurrence of similar failures in the future. Once a satisfactory MTBF was achieved the generator was approved for distribution into the field.

Software reliability (Section 5.12) is another important application area for reliability growth analyses. A key issue, again, is that of helping determine when the product in which the software is embedded can be released. Reliability growth analyses are also frequently used to report reliability improvement to management.

5.10.4 Extensions

Most reliability growth analyses are more complicated than the example and typically involve multiple systems.

The model proposed by Duane was empirically based. Later, Crow (1974) provided a theoretical framework for Duane's model[25] and developed statistical

24. Even though the Duane plot is based on cumulative MTBFs, to be most useful, the estimates are reported in terms of MTBFs (see Tobias and Trindade 1995, Chapter 12).

25. The Duane model is a particular type of a nonhomogeneous Poisson process, often referred to as a Weibull process (that should, however, not to be confused with a Weibull distribution for the life of nonrepairable products).

estimation methods. Various alternative models have also been suggested. (See O'Connor 2002, Chapter 12; Klion 1992; and Pecht 1995, Chapter 11 for further information on reliability growth models.)

5.11 PRODUCT BURN-IN

Another way of "improving" reliability experienced by customers has been through so-called "burn-in" of each manufactured unit. This calls for running *all* units for a designated time to weed out early failures (so-called "infant mortalities") – that is, to have frail units fail at the manufacturer's site, rather than in the field.[26] Statistical methods help establish optimum burn-in times to maximize the number of bad units removed, while minimizing the testing cost and the reduction in field life of the accepted units (see Kim and Kuo 1998).

Proactively removing failure modes during design and early in manufacturing should make burn-in unnecessary. However, burn-in still has some use today, especially for critical components and new products (Section 13.3.6).

5.12 SOFTWARE RELIABILITY

Modern systems of all kinds contain often complicated circuitry that depends heavily on embedded software for successful operation. Automobiles, appliances, televisions, communications networks, and spacecraft are just a few examples of such products. Reservation systems, banking, billing, and numerous other financial products and services also rely on software for their functionality. As a result, high software reliability has become a critical, although sometimes hidden, requirement of modern life.

5.12.1 Characteristics

Most software, unlike most hardware, generally does not degrade over time; failures are usually due to inherent faults present from the start. Such faults may, however, remain undiscovered until a specific set of inputs is used or a particular system state, such as heavy user load, is encountered. Also, for at least some software, it may not always be clear when a failure has occurred. Billing errors created by a software problem, for example, might only become known when a customer complains of being overcharged.[27]

Software problems vary in their criticality. Most of us have learned to live with the inconvenience of rebooting our computer to get around a software problem. Safety critical software failures – such as in medical, air traffic control, telecommunication or military systems – can, on the other hand, have serious, and even life-threatening, consequences and need to be addressed before product release.

26. For product burn-in to be useful, the life distribution of the product needs to have a decreasing hazard function (see Section A2 of Appendix).
27. Under-charges are less likely to be reported.

5.12.2 Software Testing

Identifying important failure modes and removing them prior to product release, typically calls for testing software at conditions, often identified by experts, that are expected to maximize the number of errors detected (and corrected) as speedily as possible. As indicated, reliability growth analysis is frequently used to determine when to release a product in which the software is embedded.

Statistical concepts and tools are used in the planning and analysis of software reliability studies (see Dalal and Mallows 1998 for an example).

5.13 FURTHER READING

As for other topics, we have only scratched the surface in this discussion of reliability assurance. Applications-oriented books in this area, emphasizing different topics, include those by Lawless (2002), Meeker and Escobar (1998), Nelson (1982, 1990), and Tobias and Trindade (1995). Also see Juran and Godfrey (1998, Section 48) and Ruggeri, Kenett, and Faltin (2008).

In addition, with Bill Meeker, we (Doganaksoy and Hahn) have since 1999, written at least one article yearly on reliability directed at practitioners for *Quality Progress* as part of the Statistics Roundtable series.

5.14 MAJOR TAKEAWAYS

- Product reliability, or quality over time, is a key concern for all products.
- The critical need for designing products with high reliability is receiving increasing recognition. This requires a team effort with reliability engineers and statisticians important members.
- The role of statistics in reliability assurance has evolved from fire fighting to validation testing to actively contributing to reliability improvement.
- The first step in a product reliability assurance program is to set clear and measurable reliability goals. In so doing, it is important to distinguish between repairable and nonrepairable products and to recognize that the subsequent analyses require concepts and tools different from those presented in most introductory courses on statistics.
- The next phase involves the evaluation of reliability of the conceptual design. This may require the construction of a probabilistic reliability model and its assessment.
- In product development, the next phase, we strive to ensure that the design meets or exceeds reliability goals. This ideally calls for an understanding of the underlying failure mechanisms, as well as failure modes and causes. It requires the use of various engineering tools, and may involve statistical experimentation.
- Empirical reliability estimates are obtained through statistically based tests on components, assemblies, subsystems, and eventually the final product or

system. Use-rate acceleration, accelerated life tests, and/or degradation testing are used to speed up the process.

- Validation, the final phase, involves further testing to ensure that product or system reliability goals are likely to be met under conditions that closely resemble normal manufacturing conditions and field operations. This typically involves in-house systems testing and, sometimes, in-field beta site testing.
- The need for ensuring product safety is a critical element of reliability assurance; statistical evaluations are often based on sparse data, at least early on, about events with potentially serious consequences.
- Reliability growth analyses are performed to track reliability improvement, especially for repairable products, over time and help determine when a system is ready for release.
- Product burn-in has been used to weed out early failures; a proactive approach to avoid such failures could make burn-in unnecessary.
- Software reliability assurance is characterized by the need to detect inherent faults in software.

APPENDIX: PRODUCT LIFE DISTRIBUTIONS AND CONCEPTS

In this appendix we review some concepts and statistical distributions commonly used in life data analysis of nonrepairable products (components).

A1: THE RELIABILITY FUNCTION

In dealing with life distributions for nonrepairable products two carryovers from elementary statistics are:

- $f(t)$: The probability density function (PDF) for product life.
- $F(t)$: The cumulative distribution function (CDF): The probability of failure (or the proportion of the population failing) by a specified time t. The $100p$'th percentile t_p satisfies $F(t_p) = p$. For example, the 10th percentile of a life distribution is the age by which 10% of units in the populations fail.

$R(t) = 1 - F(t)$ is the probability of survival to time t and is defined as the reliability function (RF).

A2: THE HAZARD FUNCTION (HF)

The HF[28] of a nonrepairable product is defined as

28. The counterpart of the HF for repairable products or systems is the rate of occurrence of failures, also referred to as the intensity function or recurrence intensity (see Nelson 2003).

$$h(t) = \frac{f(t)}{1 - F(t)}.$$

The HF characterizes the propensity of a unit that has survived to time t to fail in the next small interval of time. In particular,

- An *increasing* HF over time indicates that failures occur at an increasing rate as the product ages and suggests product wear-out. A severely increasing HF may call for critical components in a system to be routinely replaced at a specified age so as to decrease the likelihood of in-service failure.
- A *decreasing* HF over time indicates that failures occur at a decreasing rate as the product ages. This might suggest infant mortalities for some fraction of manufactured product, perhaps caused by a manufacturing defect. After the weak units have failed, the HF will stabilize or even begin to increase. It is important to understand and eliminate the failure mode(s) that cause early failures. Product burn-in (Section 5.11) can be a short-term corrective measure.
- A *constant* HF over time suggests that failures may be due to some random external mechanism, such as a lightning strike, that is generally independent of product age. Proactive replacement or burn-in does not benefit products with a constant HF.

A2.1. The Bathtub Curve HF

Many products experience multiple failure modes. Their overall HF might be characterized by the so-called "bathtub curve," shown in Figure 5.15. This suggests that the product population experiences a decreasing HF early in life (perhaps, resulting from one or more failure modes due to manufacturing defects), and then an increasing HF, later in life (due to different failure modes, perhaps caused by wearout) – with an approximately constant, and relatively low, HF in between. Human mortality is one "product" whose HF tends to follow a bathtub curve – although its shape is not symmetric.

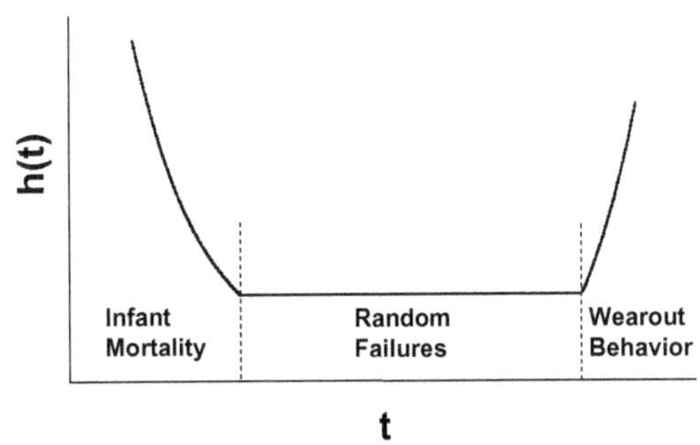

Figure 5.15 The bathtub curve hazard function.

A3: THE WEIBULL DISTRIBUTION

A3.1. Definition

The Weibull distribution is the proverbial workhorse of product life data analysis. Its CDF is

$$F(t) = 1 - \exp\left[-\left(\frac{t}{\eta}\right)^{\beta}\right], t > 0$$

where $\eta > 0$ is a scale parameter (also the 63rd percentile of the distribution) and $\beta > 0$ is a shape parameter. Figure 5.16 displays the CDF, PDF, and the HF of the Weibull distribution for various values of β.

A3.2. Justification and Application Areas

The Weibull distribution is one of the most frequently used distributions to model the lifetimes of products. Its use is justified on the following grounds:

- Theoretical: Extreme value theory shows that if failure takes place at the occurrence of the first of a large number of independent and identically distributed failure modes (i.e., the weakest link), the resulting life distribution, under certain circumstances, is Weibull (Meeker and Escobar, 1998, Section 4.8).
- Generality: The Weibull distribution can represent products with a decreasing HF ($\beta < 1$), an increasing HF ($\beta > 1$) and a constant HF ($\beta = 1$). It does *not* apply, however, for products that have *both* a decreasing and increasing HF, such as the bathtub curve HF (see examples in Sections 5.8.2 and 9.3.3).
- Practicality: The Weibull distribution has sometimes been found to provide a good fit to the lifetimes of various products, including insulation materials, steel bars, x-ray tubes, ball bearings, capacitors and ceramic parts. Furthermore, many statistical software packages include features for analysis of Weibull distributed life data.

A3.3. Special Case: The Exponential Distribution

The exponential distribution is a special case of a Weibull distribution with shape parameter $\beta = 1$ (see Figure 5.16). The scale parameter η of the exponential distri-

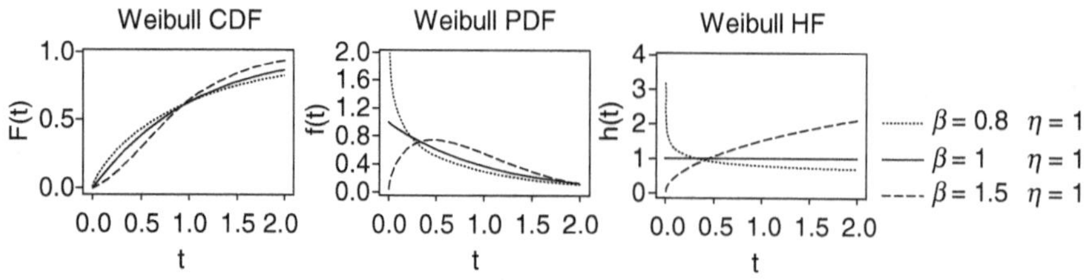

Figure 5.16 Weibull cumulative distribution function (CDF), probability density function (PDF), and hazard function (HF).

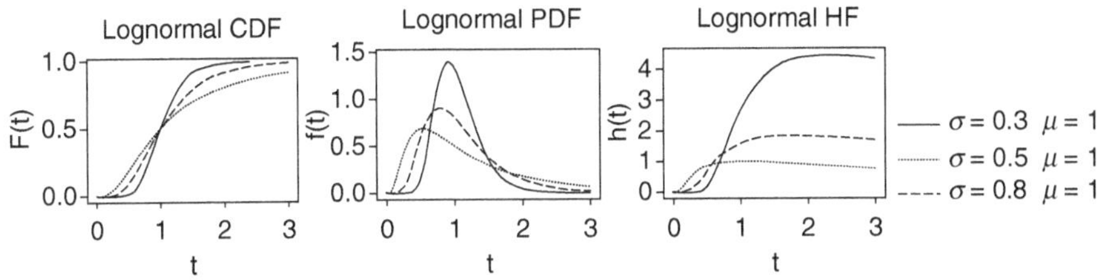

Figure 5.17 Lognormal cumulative distribution function (CDF), probability density function (PDF), and hazard function (HF).

bution is its mean (MTTF for life data). This distribution also has the important property that its HF is constant over time; in particular, its HF is $1/\eta$.

The exponential distribution can, therefore, be used to represent the life distribution for such phenomena as the number of years until outdoor equipment is destroyed by a lightning storm or the life of a plate that "fails" in a restaurant due to breakage. It has also been used to model the lifetimes of electronic components, which do not show wear and tear during their useful life.[29]

A4: THE LOGNORMAL DISTRIBUTION

A4.1. Definition

The lognormal distribution is also frequently used to represent product life. For a lognormal distribution, the logarithms of life are normally distributed. The parameters μ and σ are the distribution's scale and shape parameters, respectively, and are the mean and standard deviation of the log lifetimes.

Figure 5.17 displays the CDF, PDF, and HF for lognormal distributions with various values of σ. The lognormal distribution initially leads to an increasing HF over time, but eventually its HF decreases.

A4.2. Justification and Application Areas

The lognormal distribution has been justified theoretically as a model for life data by a central limit theorem. This states that a random variable that arises from the *product* of a number of identically distributed independent positive random quantities approximately follows a lognormal distribution. Thus, quoting Meeker and Escobar (1998, Chapter 4), "it has been suggested that the lognormal distribution is an appropriate model for lifetime caused by a degradation process with combinations of random rate constants that combine multiplicatively ... [and also] for fatigue crack growth in metals."

29. Because of its mathematical simplicity, the exponential distribution has sometimes been, at least implicitly, assumed to hold for all life data; this, in turn, has led to the popularity of MTTF. In practice, the lifetimes of most products do *not* have a constant HF and, therefore, do not follow an exponential distribution. Erroneously assuming an exponential distribution for lifetime can lead to seriously incorrect conclusions.

The lognormal distribution has been used successfully to represent the life distributions of various electronic components whose failures are driven by degradation.

A5: OTHER DISTRIBUTIONS

Meeker and Escobar (1998, Chapters 4 and 5) and Nelson (1982, Chapter 2) describe other distributions for product life and their applications. A three-parameter distribution, such as the three-parameter Weibull or lognormal, includes a threshold parameter that denotes the time prior to which failure cannot occur. Also, the extended generalized gamma distribution provides a highly flexible, although complex, model that includes Weibull and lognormal distributions as special cases.

A6: FITTING LIFE DISTRIBUTIONS

Which distribution to use is, in practice, often determined empirically by probability plotting, as illustrated in Section 5.7.7. (When more than one model seems to fit the data, each such model is often used and the results compared.) The unknown parameters of the assumed distribution(s) are then typically estimated from life data using the method of maximum likelihood.[30] These parameter estimates are then used to estimate the distribution's CDF, PDF, and HF.

For the example in Section 5.7.7, maximum likelihood analysis of the data in Table 5.1 resulted in the Weibull distribution parameter estimates $\hat{\eta} = 34.7$ years and $\hat{\beta} = 4.1$.[31] The fact that $\hat{\beta}$ appreciably exceeds 1 suggests a sharply increasing HF over time. These results, in turn, yielded the estimated 10-year reliability

$$\hat{R}(10) = exp\left[-\left(\frac{10}{34.7}\right)^{4.1}\right] = 0.994 \text{ or } 99.4\%.$$

DISCUSSION QUESTIONS

General

1. Suggest some other examples, possibly from your own experience, of poor reliability. How might the problem have been averted or mitigated by appropriate measures in the design, manufacture, maintenance, and/or use of the product or service?

2. What costs do manufacturers experience when products have poor reliability? How can such losses be quantified for management? What are some likely obstacles in convincing

30. Maximum likelihood is used since standard elementary methods, such as least squares regression, generally do not apply for censored data. The basic principle underlying maximum likelihood is to choose as estimates those values of the parameters from among all possible parameter values that make the observed data most likely.

31. It is more common to have a β of about 2.5 for bearing life data. The higher value was attributed to the well controlled production environment – hopefully to be replicated in high-volume production.

managers focused on short-term profits to invest in proactive reliability improvement and how might you address these?

3. Provide some further examples for which use rate acceleration, accelerated life testing and degradation testing, respectively, seem appropriate for reliability assessment. Explain.

4. Engineers frequently conduct autopsies on failed units. Deming urged that unfailed units also be evaluated. What are the advantages of conducting autopsies on both failed and unfailed units?

5. How would you define reliability for a CD player? Why might its assessment sometimes be ambiguous, and how can we help avoid such ambiguity? Suggest a program to help ensure high reliability for a newly designed CD player.

6. We state in a footnote in Section 5.8.1 that "the testing described earlier in this section also involves analytic studies to varying degrees." Elaborate on this comment.

7. You need to propose a reliability validation program for a newly designed motorcycle. To make a meaningful proposal, you must understand the product, its anticipated use, how it differs from other models, testing options, past evaluations, etc. What questions would you ask to develop an appropriate proposal?

8. Consider how much you rely on others for your own safety. For a particular situation, what do you think (or hope) manufacturers of products or providers of services do to attain high safety? What role does (or should) statistics play in these efforts?

Technical

1. We state that it "is incorrect to treat the survival times as if they were failure times, or to ignore the survived units altogether" (Section 5.5.1). Describe how you would handle data comprised of a mix of failure and survival times.

2. We state in a footnote in Section 5.5.3 that "a product with an exponential distribution for lifetime and a MTTF of 25 years is still expected to experience 18% failures in its first 5 years." Verify this statement.

3. Determine the 5-year reliability of the system whose structure is displayed below, assuming that all components act independently of one another.

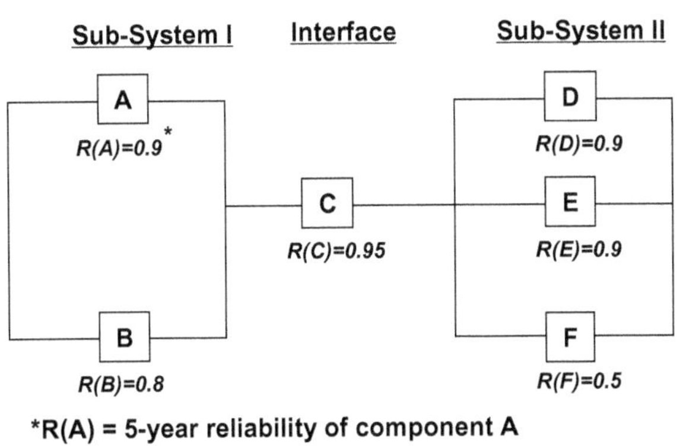

4. In the preceding question, assume that the reliability estimates for components A, C, and F and components B, D, and E were based upon random samples of size 15 and 25,

respectively. Thus, each of these estimates is subject to statistical uncertainty. Find an approximate 95% lower confidence bound on system reliability.

5. It is stated in Section 5.7.1 that "frequently, a failure takes place when the stress (or load) applied to a product exceeds its strength." Explain this concept probabilistically and indicate how it can be applied to accelerated testing and to degradation measurements.

6. In the washing machine motor reliability example, a use rate of four washes per week was assumed. In actuality, the use rate has a statistical distribution. Assuming knowledge of this distribution, how would you include this in your analysis?

7. In the motor reliability example, we indicate in Section 5.7.7 that a simulation analysis indicated that the proposed test program provided a "reasonable degree of precision in estimating 10-year reliability." A similar comment was made in a footnote in Section 5.7.8 about the generator insulation example. Describe how you would conduct such evaluations.

8. In the motor reliability example, we reported a 95% lower confidence bound of 96% on 10-year reliability and that "the desired 97% 10-year reliability could be demonstrated with approximately 92% confidence." How were these determinations made?

9. In Section 5.7.8, we state that the impact on the fitted ALT model of irrelevant failure modes at high stresses "can be removed by statistical analysis – if the units experiencing such failures can be identified." How would you do this?

10. In the generator insulation example in Section 5.7.8, more units were tested at the lower stresses (other than 75 vpm) than at the higher ones. Using an analytic or simulation approach, compare the stated test plan for assigning the 90 electrodes to the five test conditions with one that calls for equal allocation and also with other plans that might come to mind. Continue to assume that the major goal is to estimate 10-year life at 60 vpm.

11. We assert in Section 5.7.9 that "degradation testing might be built into an ALT." In this case, the measurement obtained in the ALT is each test unit's degradation at its accelerated condition. Elaborate on how you might analyze an ALT that uses degradation measurements.

12. In discussing validation testing for the generator insulation example in Section 5.8.2, we state that "propagation of errors was used to obtain approximate 95% confidence intervals" on system reliability. Elaborate on this approach.

13. How might statistical methods be used to help establish optimum burn-in times (Section 5.11)?

CHAPTER 6

MANUFACTURING QUALITY IMPROVEMENT: CONCEPTS

6.1 ABOUT THIS CHAPTER

Applications of statistics in business and industry have traditionally focused on product manufacturing. Although the role of statistics has significantly expanded to other arenas, statistical concepts and methods are still used extensively for quality improvement in manufacturing.

We begin this chapter by discussing the continuing need for manufacturing quality improvement, describing some special considerations, and advocating a disciplined approach. We then consider two key topics (1) quality assessment and improvement and (2) statistical monitoring using control charts.

The concepts are illustrated in the next chapter by an application dealing with the manufacture of stator bars for electric power generators. Some further manufacturing applications are considered in Chapter 8.

Our presentation will focus on product characteristics (e.g., weight, chemical composition, color, and strength). These might involve process measurements during production or at the end of production on the final product; they might reflect a particular feature of the product, a performance property, a measure of reliability, or an aesthetic attribute.

6.2 THE NEED FOR MANUFACTURING QUALITY IMPROVEMENT

Montgomery (2004) defines quality improvement as the reduction of variability in processes and products. Taguchi states the goal to be "on-target with minimum variance" (see Wheeler and Chambers 1992).

Forward-looking companies strive to build high quality and reliability into products, make them robust to variability, and anticipate and proactively eliminate manufacturing problems during design, as described in previous chapters. The increased expenses and delays associated with fixing problems during manufacture are thereby avoided.

The Role of Statistics in Business and Industry, by Gerald J. Hahn & Necip Doganaksoy

It is, unfortunately, still not possible to anticipate *all* the problems that might arise in manufacturing. Product quality may fail to meet expectations for such reasons as:

- An inherent problem that somehow was missed during design.
- Differences that still exist, despite earlier design scale-up evaluations, between the product development environment and the manufacturing plant.
- Inconsistencies in manufacturing procedures, such as in the methods used to set up and run equipment.
- Variability in the manufacturing environment, such as fluctuations in ambient temperature and humidity.

In addition, product quality might worsen over time due to such factors as changes in raw materials or parts, wearout of equipment or tools, deterioration of raw materials during storage, operator turnover and inadequate training of new operators, and various cost-cutting ventures by suppliers or by the manufacturer. Finally, we need to ensure that nothing that is done during manufacturing compromises the high level of reliability that, hopefully, was built into the product during design.

Achieving high quality in a manufactured product, as in design, calls for attaining both a desired level of performance and ensuring a high degree of consistency.

6.3 DIFFERENT INDUSTRIES

The role of statistics varies by type of business and manufacturing operation. One important differentiation, discussed in Sidebar 6.1, is between the discrete parts and process industries.

SIDEBAR 6.1: DISCRETE PARTS AND PROCESS INDUSTRIES

Many products are built by assembling discrete parts. Some examples are motors, light bulbs, bicycles, automobiles, and aircraft engines. Typical manufacturing operations include cutting, casting, machining, and assembly of parts.

The process industries, in contrast, generally involve moving, mixing and separating bulk materials and often involve chemical operations. Examples are the plastics, textile, food, pharmaceutical, and petroleum industries.

Some products require *both* discrete parts assembly and product processing. For example, semiconductor manufacturing involves processing during wafer fabrication and, subsequently, discrete parts assembly and packaging of chips cut from wafers. Also, computers and appliances include plastic parts that are created by a processing operation.

Traceability of individual components is often possible for discrete parts, but difficult for processing operations; these typically require drawing raw materials from storage silos in powder, pellet or liquid form.

A further differentiation (especially for process industries) is between manufacturing operations that yield a continuous flow of product and those that produce product in batches.

Finally, it is useful to distinguish between job-shop operations and repetitive production. A job shop typically builds a specialized product to fill a specific customer order. Repetitive product operations generally yield one or more standard products and frequently involve high-volume production.

6.4 CONSIDERATIONS IN MANUFACTURING QUALITY IMPROVEMENT

6.4.1 Improving Target-Is-Best Characteristics

As previously noted, "target-is-best" characteristics generally have two-sided specification limits; to be "acceptable," these characteristics need fall within the bounds defined by such limits. Achieving a high level of performance requires centering the distribution of the characteristic on its target value *and* minimizing variability.

Figure 6.1 illustrates four different situations for target-is-best characteristics. The curves can be thought of as smoothed histograms constructed from data on product measurements that demonstrate:

a. On-target within-spec performance that can be further improved by reducing variability.

b. Off-target within-spec performance that can be improved principally by shifting the mean, and secondarily by reducing variability.

c. On-target out-of-spec performance, *requiring* reduction in variability.

d. Off-target out-of-spec performance, *requiring* reduction in variability; further improvement can be gained by shifting the mean.

6.4.2 Improving Higher- (or Smaller-) Is-Better Characteristics

As previously observed, characteristics for which the goal is the higher-the-better or the smaller-the-better typically have one-sided lower or upper specification limits, respectively. Achieving a high level of performance requires consistently

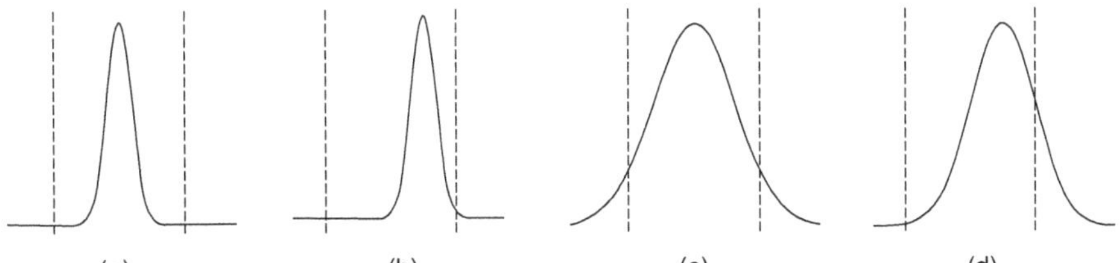

(a) (b) (c) (d)

Figure 6.1 Four situations for target-is-best characteristic (showing two-sided specification limits).

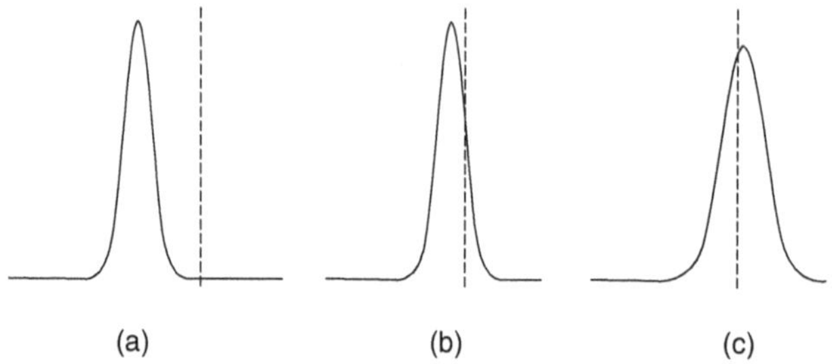

Figure 6.2 Three situations for smaller-is-better characteristic (showing upper specification limit).

maintaining high (low) values exceeding (falling below) the lower (upper) specification limit.

Figure 6.2 shows three situations for the smaller-the-better characteristics that demonstrate:

a. Within-spec performance that can be further improved by shifting the mean and/or reducing variability.

b. Out-of-spec performance, *requiring* shifting the mean and/or reducing variability.

c. Out-of-spec performance *requiring* shifting the mean; further improvement can be gained by reducing the variability.

6.4.3 An Example

Figures 6.1 and 6.2 again illustrate that achieving mean performance and reducing variability are *both* important. As a further example, consider the filling of 1-gallon bottles with liquid. We require the average liquid content to be sufficiently high so that the vast majority of bottles contain at least 1 gallon. But to minimize material costs, we want that average to exceed 1 gallon only slightly; this requires minimizing variability.

6.4.4 Gaining Insights from Data

Figures 6.1 and 6.2 are idealizations. Plots of the data, coupled with process understanding, often suggest idiosyncrasies that warrant further study. To illustrate, Figure 6.3 shows histograms of measurements for three product characteristics.

Closer Look at Case A Case A shows an apparent mixing of two distributions, suggesting two underlying processes. Questioning revealed that the product came from two production lots. Plots for each lot (Figure 6.4) appeared to separate this mixing and suggested the need to search for the cause for the difference.

Closer Look at Case B Case B shows an abrupt cutoff of the distribution at 100. This plot was constructed upon receipt of the parts. It turned out that parts with measurements below the lower specification limit of 100 had been removed by the

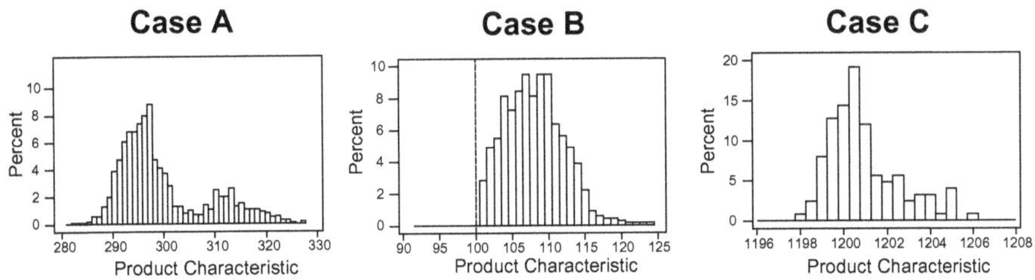

Figure 6.3 Histograms for measurements of three product characteristics (dashed line for Case B is lower specification limit).

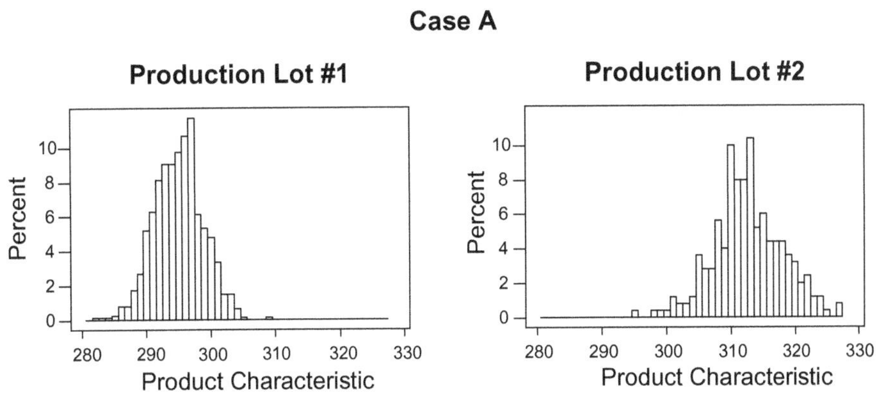

Figure 6.4 Histograms by production lot.

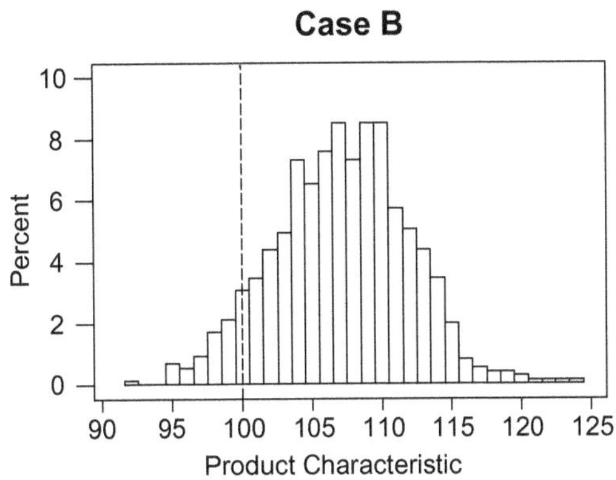

Figure 6.5 Histogram including out-of-spec measurements.

supplier prior to shipment. Thus, the histogram was constructed from only previous within-spec product. Figure 6.5 shows a histogram with measurements on rejected parts included.[1]

1. A histogram similar to Case B, but with a spike at 100, would be obtained if the instrumentation is unable to read values below 100 and the resulting truncated measurements are then incorrectly assigned the value 100.

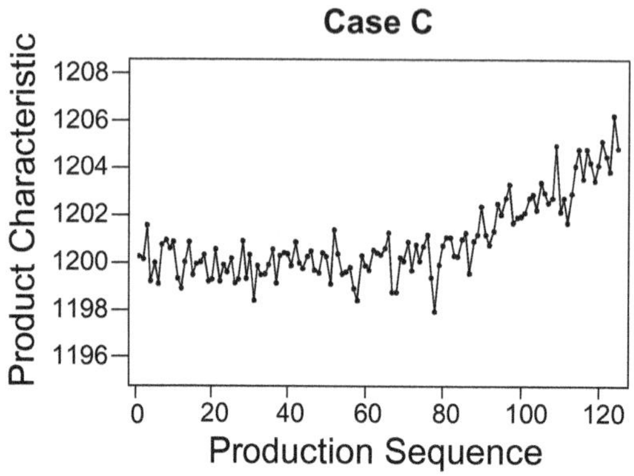

Figure 6.6 Time series plot for measured product characteristic.

Closer Look at Case C Case C suggests a distribution with a slight skew to the right, but no readily apparent anomalies. A plot of the same data against time is shown in Figure 6.6. This shows a clear pattern, starting after approximately the 80th unit.

Much data are obtained over time, and warrant plotting against time. Constructing histograms of time-ordered data without such further plots may obscure important information.

6.4.5 Handling Multiple Characteristics

Figures 6.1 and 6.2 deal with a single product characteristic. For most products, one is concerned with multiple characteristics, each of which is important. These may be correlated favorably or, as in the child car seat example of Chapter 4, unfavorably. The need to consider multiple product characteristics, and especially those that are correlated in an unfavorable manner, often adds significantly to the challenge of quality improvement.

6.5 DISCIPLINED APPROACHES TO MANUFACTURING QUALITY IMPROVEMENT

Manufacturing quality improvement, like design, requires a disciplined approach. Solutions based on "gut feel" may yield quick fixes to the immediate problem, but often fail to address the root cause(s). It is then often only a matter of time for the problem to resurface. Disciplined (and databased) approaches to manufacturing problems aim to identify and address the causes of problems.

Some well-known approaches for manufacturing quality improvement are the PDSA (plan, do, study, act) paradigm (Section 2.2.5) and Six Sigma (Sidebar 6.2); see also Juran's Improvement Program (Juran and Godfrey 1999, Section 5). These approaches have much in common and involve the use of statistics at varying levels. They are most powerful when viewed as part of a continuous learning process directed at improvement – rather than one-shot attempts at fixing broken products.

SIDEBAR 6.2: SIX SIGMA FOR MANUFACTURING IMPROVEMENT

Six Sigma was initially targeted at helping improve manufacturing processes through the following five stages (DMAIC):

- Define (D): Define the problem to be addressed, including determining customer impact and potential benefits.
- Measure (M): Identify the critical-to-quality characteristics (CTQs) of the product. Process improvement usually targets these characteristics. This step includes verifying the capability to measure the CTQs, estimating current process capability, and setting goals for improvement.
- Analyze (A): Understand the root causes of failure to achieve the desired level of performance and/or consistency in the CTQs, thereby identifying the key process variables that cause defects.
- Improve (I): Quantify the influence of the key process variables on the CTQs, determine specification limits for these, and modify the process to stay within such limits, thereby reducing the defect levels in the CTQs.
- Control (C): Ensure that the modified process maintains its achieved level of performance and that the key process and product variables are, and continue to be, within acceptable limits.

See Breyfogle (2003) for further discussion. Six Sigma has recently been linked closely with lean manufacturing (Section 8.6.2).

The stages of DMAIC relate to the discussion in this chapter as follows:

- The Define stage applies to initial definition of the need to improve, as well as assignment of the team and development of the project plan.
- The Measure stage applies principally to manufacturing capability and stability assessment and addressing measurement error.
- The Analyze stage applies to identifying and analyzing causes for deviations from target and variability.
- The Improve stage applies to acting on our findings to reduce deviations from target and variability and to validate that we have done so.
- The Control stage applies to statistical monitoring.

Also Steiner and MacKay (2005, 2006) propose and illustrate a more specific strategy for reducing variation in manufacturing processes based upon their experience in the automotive industry.

6.6 QUALITY ASSESSMENT AND IMPROVEMENT

6.6.1 Manufacturing Capability and Stability Assessment

The difference between performance and capability is opportunity, and that drives improvement.

—Lynne Hare

The first step in the assessment and potential improvement of a manufacturing process is to gain an understanding of existing manufacturing capability and stability. Close scrutiny of available data presents a good starting point; often this needs to be supplemented with carefully collected new data.

Percent Yield The (estimated) percentage of units for a manufactured product that meets specification limits for all required characteristics is frequently referred to as product "percent yield."

Percent yield plotted against time (e.g., day, week, and shift) of manufacture often provides a useful summary to present to management, especially when there are multiple performance characteristics. This may be supplemented by a Pareto chart – a simple chart that lists problems according to their frequency of occurrence.[2]

Percent yield may, however, be uninformative in helping understand and remove the causes of quality problems, and in helping improve quality. If, moreover, measurements are dichotomized as acceptable and unacceptable (or good and bad), valuable information may be lost. Such dichotomization, for example, does not permit acceptable units at or close to target to be differentiated from those that just miss being unacceptable.

Variables Assessment Measurements on individual characteristics, or variables data, are generally preferable for seeking quality improvement. With advances in measurement technology, most instrumented measurements provide readings on a continuous scale. This even includes such features as color, appearance, and surface defects that were traditionally evaluated subjectively.

Process Stability Assessment A process is said to be stable if the measurements on the process tend to fluctuate around a constant mean with constant variability *over time*.[3] If a process is unstable, the identification and removal of the causes for this instability needs to be the focus of an improvement project.

Breaking Down Variability Statistical methods can be applied to appropriately gathered data to break down total variability, measured by the standard deviation, into its constituent components (e.g., part-to-part, machine-to-machine, and line-to-line), as per the example in the next chapter.

Short-Term and Long-Term Variability The variability observed for a performance characteristic in the short-term is generally less than that over longer time periods. This is due to such factors as variation in raw materials, environmental conditions, equipment differences, and personnel changes that occur over time. A common practice is to obtain separate estimates of "short-term" and "long-term" variability.

2. This chart stems from the principle suggested by an Italian economist that for many phenomena, 80% of the consequences stem from 20% of the possible causes.
3. See forthcoming discussion on statistical monitoring.

Process Capability Evaluation Process capability deals with the margin between the tails of the distribution of a characteristic and its specification limits. A process with a large margin is said to be capable. Figures 6.1a and 6.1c illustrate centered processes with high and low process capabilities, respectively.

Process capability indices have become popular metrics that relate product specifications to performance for individual product characteristics. A variety of process capability indices have been proposed; C_p and C_{pk} appear to be the two best known and most frequently used.

The Process Capability Index C_p The C_p metric compares the process specification range to the variation in the process (as measured by its standard deviation σ). It is calculated for characteristics with two-sided specification limits as:

$$C_p = \frac{\text{Upper Spec Limit} - \text{Lower Spec Limit}}{6\sigma}.$$

The denominator measures the total spread; 99.7% of the observations are contained within a $\pm 3\sigma$ range around target for a normally distributed centered characteristic. For such characteristics, C_ps of 0.8, 1.0, and 1.5 correspond to reject rates of 1.64%, 0.27%, and 0.0007%, respectively.

For processes with one-sided lower and upper specification limits (but with a specified target value), C_p is calculated, respectively, as:

$$C_p = \frac{\text{Target Value} - \text{Lower Spec Limit}}{3\sigma}$$

and

$$C_p = \frac{\text{Upper Spec Limit} - \text{Target Value}}{3\sigma}.$$

The Process Capability Index C_{pk} The C_{pk} metric, unlike C_p, takes the deviation of the process mean μ from target into consideration. It is calculated for characteristics with two-sided specification limits as:

$$C_{pk} = \text{smaller of} \left\{ \frac{\text{Upper Spec Limit} - \mu}{3\sigma}, \frac{\mu - \text{Lower Spec Limit}}{3\sigma} \right\}.$$

For characteristics with a one-sided upper (lower) specification limit, C_{pk} is calculated by using the first (second) bracketed item alone.

Estimation of μ and σ The values of μ and σ used to calculate C_p and C_{pk} are generally estimated from the results of a process capability study. The short-term standard deviation is generally used to estimate σ. If the long-term standard deviation is used instead, the resulting metric may be better referred to as a performance, rather than a capability, index.

Other Capability Indices Capability (and performance) indices, other than C_p and C_{pk} have been proposed in the literature (see Kotz and Johnson 2002, and Pearn and Kotz 2006).

Limitations Despite their popularity, process capability indices have distinct limitations. Their interpretation is difficult for nonnormally distributed performance characteristics or when the process exhibits instability over time. Also, they depend upon the method used to estimate σ and the specification limits themselves. In many situations, they are less direct, and perhaps less informative, than an estimate of the percentage within specifications[4] or of the process variability. Gunter (1991), Hare (2007), and Montgomery (2004, Section 7.3) provide further discussions and critiques of process capability indices.

6.6.2 Addressing Measurement Error

We now extend our discussion of measurement error in Section 3.4 to manufacturing processes.[5]

Estimating True Product Variability We are typically interested in quantifying and reducing the *actual* variability of key product characteristics since this is what customers experience. However, we generally observe only the *measured value* of a product characteristic; this deviates from the *true value* due to measurement error. Measurements, thus, consist of two components, that is:

$$\text{Observed (or Measured) Value} = \text{True Product Value} + \text{Measurement Error.}$$

We are unable to separate the true product value and the measurement error for *specific observations.* However, estimates of the standard deviations for measurement error and for the *total observed variability* can be used to estimate the *true product variability*. In particular, let:

- $\sigma_{OBSERVED}$ denote the standard deviation of the observed product measurements;
- $\sigma_{MEASUREMENT}$ denote the measurement error standard deviation; and
- $\sigma_{PRODUCT}$ denote the standard deviation of the true product values.

Then, if measurement error and product variability are uncorrelated, it follows (from the simple rule for the variance of the sum of two uncorrelated random variables) that:

$$\sigma^2_{OBSERVED} = \sigma^2_{PRODUCT} + \sigma^2_{MEASUREMENT}.$$

In practice, we generally use estimates $\hat{\sigma}_{MEASUREMENT}$ of $\sigma_{MEASUREMENT}$ and $\hat{\sigma}_{OBSERVED}$ of $\sigma_{OBSERVED}$ to obtain an estimate $\hat{\sigma}_{PRODUCT}$ of $\sigma_{PRODUCT}$, that is:

$$\hat{\sigma}_{PRODUCT} = \sqrt{(\hat{\sigma}^2_{OBSERVED} - \hat{\sigma}^2_{MEASUREMENT})}.$$

Discussion If measurement error is an appreciable part of the observed product variability, this usually needs to be addressed before proceeding, perhaps through a

4. One possible application might be in situations in which the observed sample has no out-of-specification observations.

5. We assume that the measurement systems have previously been established – an important task in itself.

special team effort. As indicated in Section 3.4, there are generally multiple sources of variability that contribute to $\sigma_{MEASUREMENT}$ (e.g., sample preparation, test equipment, and operator performing measurement). A GRR study might be conducted to quantify these individually.

There are typically also multiple sources that contribute to $\sigma_{PRODUCT}$ (e.g., raw materials, manufacturing process, and machine operators). Doganaksoy (2000) illustrates the assessment of measurement error in such situations.

6.6.3 Reducing Deviations from Target and Variability

Happy is he who had been able to learn the cause of things.
—Virgil

To develop an effective plan for correcting deviations from target and reducing variability, we need to understand how the variability came about. This requires us to:

- Identify the likely contributing factors.
- Quantify the impact of each of these factors.
- Find the root causes.
- Improve the process.
- Validate that the change is working.

These steps are closely intertwined.

Identify As a first step, a team of experts from manufacturing, design, and other parts of the organization identify the variables that might create deviation of the mean from target and/or variability. The results are often conveniently conveyed by a fishbone diagram or a Pareto chart (see Table 2.1 of Chapter 2).

Quantify The next step is to break down and quantify *separately* the impact of each of the (key) components (e.g., raw material batches, shift operators, equipment, and measurement error) of deviation from target and process variability. Such assessments often require careful gathering and recording of data from the process during normal operations. Individual components of variation can then be estimated using analysis of variance methods. Graphical methods also provide useful insights (see Hamada, MacKay, and Whitney 1993, Mentch 1980, and Snee 1983).

Find Root Causes Quantifying (and displaying) the major sources of variability often provides a path to, but generally does not identify per se, the underlying root causes of variability. Knowing, for example, that variability between batches of raw material is the largest source of variation provides focus – but does not explain *what* it is about the raw materials that causes the variation. This might be due to preprocessing of the material, the relative amounts of ingredients, or some other factor or combination of factors. Similarly, if the machining operation is the major source of variation, the underlying culprit could be the stamping force, curing temperature, machining defects, or wear of machine parts over time. Thus, isolating the root causes of deviations from target and of variability is generally required to achieve improvements.

This often calls for an assessment of the relationships between process variables and product characteristics. Sometimes, regression analyses[6] are conducted on historical data from operations. Often it will be necessary to conduct a statistically planned investigation to get the needed data.

Improve Once there is an understanding of the root causes of deviations from target and of variability for the key product characteristics, action can be taken to make improvements.

Correcting for Deviations from Target If mean performance is off target, we would aim to shift the mean accordingly. In some applications – often involving target-is-best performance characteristics – this can be accomplished by a simple dial adjustment. For example, the mean amount of product per container for a packaged product may be adjusted by resetting the feeding machine. Similarly, in the car seat example in Chapter 4, we were able to center viscosity on its target value by changing the relative amounts of polymers during manufacture.

In other applications, it is more difficult to shift mean performance. These often involve higher-is-better or smaller-is-better performance characteristics. Reducing emission from an automobile and increasing the strength of a material or the life of a tire are two typical examples.

Reducing Variability Variability is reduced by acting on the identified root causes. Reducing variability in setup might, for example, require standardized and perhaps more stringent procedures, improved documentation, or better operator training. Variability in raw materials may be addressed by reconsidering specifications, negotiating with suppliers, or making the product more robust to such variation.

Validate The validation of a change generally involves both checking on whether the desired improvements have been attained and making sure that the change does not have a deleterious impact on other characteristics.

Validation, moreover, is required not only for changes made to gain quality improvement, but also for design and process changes made by manufacturers and their suppliers for a variety of reasons, such as to:

- Enhance product features or performance by making, what are presumed to be minor, design or processing improvements.
- Reduce costs by upgrading manufacturing equipment or using a less expensive material.
- Address events beyond their control; for example, replacing a supplier who has gone out of business or meeting new environmental regulations.

6. In situations involving processes with complex dynamics and multiple variables, time series models and multivariate techniques may be used.

Assertions that changes will not impact product performance, quality, or reliability detrimentally (and might even improve things) need be demonstrated empirically, unless there is overpowering physical evidence.

The manufacturer of an automatic transmission system for an automobile is, for example, faced with frequent proposals for minor changes and improvements. Such changes might be examined together periodically in an off-line designed experiment, and then assessed online in a scale-up evaluation, as discussed in Section 3.7, before full-scale online implementation.

Sometimes a change must be made immediately. Its impact might then be evaluated initially online by alternating between the new and old process over some period of time and observing the results before possible permanent implementation (see Hahn 1982).

6.7 STATISTICAL MONITORING

There's nothing stable in the world; uproar's your only music.
—John Keats

6.7.1 Basic Concept

It is important to distinguish between the Voice of the Customer and the Voice of the Process ... A control chart defines the Voice of the Process.
—Donald Wheeler

Our discussion to this point has focused on building products that meet their target values and on minimizing variability. A further challenge is to ensure that the desired performance persists over time and to seek added opportunities for improvement. Statistical process monitoring is a key tool for accomplishing this.

Statistical monitoring was popularized by Shewhart with his introduction of the concept of statistical control (now often referred to as statistical process control or SPC) and the traditional control charts that bear his name. Control charts raise a flag when there is statistical evidence that a process is behaving differently from the past. Shewhart and Deming advocated control charts as a means for developing knowledge about processes and improving them in line with their proposed PDSA cycle.

Control charts are based on the concept of two types of variation in a process: variation produced by chance (or common) causes and variation due to special (or assignable) causes. Chance causes are inherent in the system and act consistently over time. A process that exhibits variation produced by chance causes alone is said to be in statistical control. In contrast, special causes produce variation in addition to that from chance causes and suggest some inherent change in the system. Special causes of variation result in an "out of control" situation. Control charts allow one to determine whether and when special causes are present. If they are, they need to be identified and removed if they have a detrimental impact.[7]

Control charts are discussed in detail in texts such as Montgomery (2004), Ott, Schilling and Neubauer (2005), and Ryan (2000). See also the original expositions by Shewhart (1931 and 1939); Wheeler (1999) and Wheeler and Chambers (1992) for practitioner-oriented discussions; and Hare (2001) for a broad perspective.[8] We provide only a brief overview.

6.7.2 Areas of Application

> **Millions of people have proven, by their own experience that Shewhart's control charts work.**
> —Donald Wheeler

Control charts are employed to monitor end-of-line product performance. They are also very useful in assessing the consistency over time of the many process factors that contribute to final product performance, such as raw material characteristics, equipment settings, ambient conditions, etc. Control charts during upstream processing help us identify the root cause of inadequate product performance and gain understanding when end-of-line control charts show changes – or, preferably, to trigger advance action to avoid such changes.

Control charts are also used at different stages of multistage processes, as in the manufacture of semiconductor products, and for assessing the continuing consistency of a measurement system. In addition, control charts are now employed extensively in applications beyond manufacturing, as will be illustrated in subsequent chapters.

6.7.3 Traditional Control Charts

The $\bar{X}$ ("X bar") chart for the process mean, the R chart for the range (i.e., the difference between the largest and smallest observation) and the S chart for the standard deviation are frequently used control charts. Control charts on medians, on proportions, or % defectives (p charts), and on counts (c charts) – such as the number of defects on the surface of a product – are also popular. In addition, control charts on single observations (X charts) are used in many applications, especially in the chemical and process industries.

6.7.4 Control Chart for the Process Mean

Description Figure 6.7 displays control charts on the *mean* of a process; these are referred to as $\bar{X}$ charts. Each plotted point is the mean of a characteristic measured on

7. Much of our discussion implies that a special cause has a detrimental impact, as is often the case. Special causes can, however, also have beneficial impact. In such cases, they should again be identified and, in this case, incorporated into the system, if possible.

8. Statistical process monitoring has also become a highly researched topic – see the *Journal of Quality Technology, Quality Engineering* and *Technometrics* for recent developments.

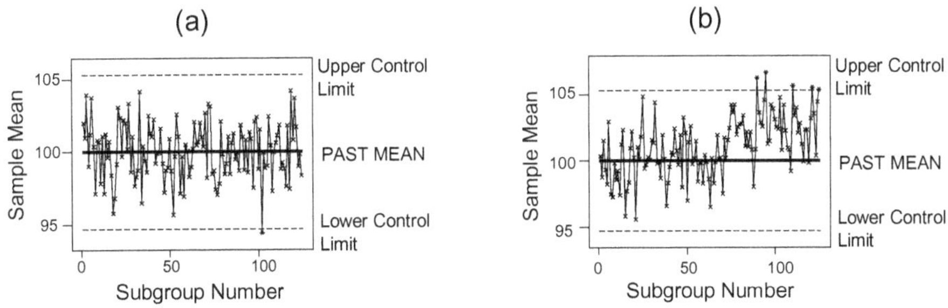

Figure 6.7 Control charts on mean for processes (a) for which there appears to be no change in the mean over time and (b) for which there is a change in the mean over time.

n items (typical values for n are 3 and 5), forming a so-called "rational subgroup," (see Sidebar 6.3). The points are plotted in time order of production.

SIDEBAR 6.3: RATIONAL SUBGROUPS

Nelson (1988) defines a rational subgroup as "a sample in which all of the items are produced under conditions in which only random effects are responsible for the observed variation." Thus, the variation of the readings in the subgroup provides a measure of common cause variability. This is in contrast to the special cause variation that we desire to isolate, identify, and frequently remove. The determination of rational subgroups is an important part of developing a control chart (see references cited in Section 6.7.1).

The centerline of the $\overline{X}$ chart is the historical mean of the characteristic being plotted. The two outer lines in Figure 6.7 represent the lower and upper statistical control limits. These are often three standard error limits (i.e., $3\sigma/\sqrt{n}$), removed from the mean, where σ is the common within-group variability, which itself is often evaluated by an R or S chart.[9]

Most sample means are expected to fall within the limits if the underlying process remains unchanged. Occasional points outside the limits need to be studied carefully to find the possible cause(s). Multiple points falling outside the limits suggest that there is a process change, and a special cause that needs to be studied and acted on. Such action is facilitated by preparing a list, in advance, of possible special causes of variation.

Statistical control limits deal with maintaining the degree of variability established over time, that is, the inherent variation produced by the process. They should not be confused with specification limits on individual units.

Process with no Mean Change Over Time Figure 6.7a is an $\overline{X}$ chart for a process for which there appears to be no change in mean performance over time; all

9. Setting the control limits is critical to the performance of the charts. Too narrow limits increases the false alarm rate and too broad limits may fail to detect important changes. Even though three standard error limits are typically used, other factors may be appropriate in specific applications.

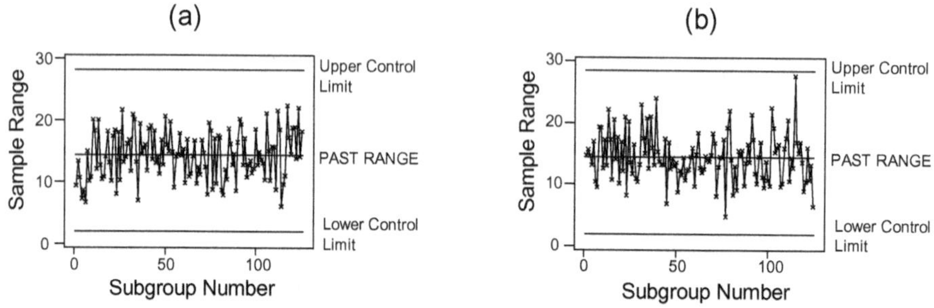

Figure 6.8 Control charts for range for processes shown in Figure 6.7.

points except one are within control limits. The conditions under which this point was obtained need be studied carefully to search for a possible cause and its removal.

An Out-of-Control Process Figure 6.7b shows a process for which a change in the mean occurred somewhere between subgroup 75 and 90 due to some, presumably, special cause. This is evidenced by various plotted points after subgroup 90 falling outside the control limits and also by various "runs," or sequences, of points above the established process mean.[10] The control chart suggests that the process mean has shifted (or drifted) to a different level from what it had been historically.

The process owners are alerted immediately; they need to identify the (root) cause of the change and act thereon. The reason for the change might be the unintended consequence of a specific action (e.g., a new raw material supplier and equipment upgrade); or it may be inadvertent and unexpected (e.g., improperly set machines and defective raw materials). Upstream control charts can help make this determination.

6.7.5 Control Chart for the Process Range

Figure 6.8 provides *R* charts on the *range* of each of the rational subgroups for the processes shown in Figure 6.7. These are constructed in a manner similar to the $\bar{X}$ chart.

In both cases, the charts did not show any change in variability over time and no out-of-control points. One point in Figure 6.8b, however, comes very close to the upper limit and warrants study.

6.7.6 More Advanced Control Charts

Many different types of control charts have been developed and applied in the years since Shewhart's pioneering work (see Montgomery 2004, Part III). Perhaps, best known among these are the exponentially weighted moving average (EWMA) chart and the cumulative sum (CUSUM) chart. The plotted points on these charts take into consideration past history, rather than just the most recent results. They are

10. Such runs can be formally evaluated by a statistical "runs test."

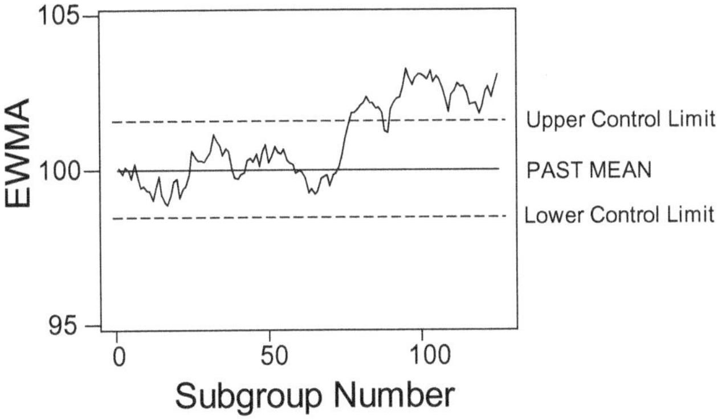

Figure 6.9 EWMA chart on data plotted in Figure 6.7b.

especially appropriate in dealing with single, rather than grouped, observations and when one might expect a drift in the process (due, say, to steady deterioration of a tool or machine), rather than a sudden change.

Figure 6.9 shows an EWMA chart for the data shown in Figure 6.7b. This chart shows more directly, and perhaps earlier, the change in the mean.

Although more powerful statistically than the traditional Shewhart charts for many situations, more advanced charts are also more difficult to explain and management may be less familiar with them.

6.7.7 Control Charts for Multiple Characteristics

In practice, we are frequently concerned with multiple characteristics. In some situations, and especially for process variables, separate control charts are maintained for each variable. It is not unusual for a complex manufacturing process to have tens, or even hundreds, of control charts. This can also increase the number of false alarms, thereby undermining the credibility of the system. Appropriate modifications in the control limits are, therefore, required (see Mason and Young 2001).

Frequently, and especially for end-of-line product performance, multiple characteristics are correlated, as previously indicated. Multivariate control charts help address such cross-correlations and often provide faster detection of process changes and a reduction in false alarms. A particular challenge in such applications is identifying the individual variable responsible for an alarm (see Doganaksoy, Faltin, and Tucker 1991). Special methods, such as principal component analysis have also been proposed for control charting multiple characteristics (see Bisgaard and Kulahci 2006).

When multiple grades of a product are manufactured at the same time, we need to differentiate global from local process changes. A change in the characteristics of a common raw material, for example, affects all product. Equipment wear on individual manufacturing lines has a more localized effect, impacting only the product(s) manufactured on that line. (See Doganaksoy, Schmee, and VanDeven 1996 and Doganaksoy and VanDeven 1997 for further discussion of these topics.)

6.7.8 Addressing Autocorrelation

In Shewhart's days it was often difficult and expensive to sample product and, as a result, only occasional measurements, such as on three or five samples per shift, were taken. In contrast, much modern instrumentation provides high volumes of data and, sometimes, an almost continuous flow of information through electronic or sensor-based equipment. Consecutive observations usually share, to a great degree, the same sources of random variability, resulting in autocorrelated data, that is, observations that are strongly correlated with one another over short periods of time.

Autocorrelation is especially common in the chemical and process industries. This is because disturbances, such as changes in raw materials, tend to have lasting effects on the process, even though these may be slowed by, for example, mixing tanks. Similarly, ambient conditions that impact the process typically persist over some period of time and vary smoothly.

Autocorrelation, if not handled properly, can create false alarms and need be considered in constructing control charts. See Bisgaard and Kulahci (2007) for an introduction to control charting autocorrelated observations using time series models.

6.7.9 Remembering the Fundamentals

There is a need to advance control charts (and the associated theory) and to adapt them to today's complex manufacturing environment. We must, however, not lose sight of Shewhart's original purpose of using such charts for continuous learning about processes and improving them.

The fundamental goal of control charting is to identify and understand special causes of variation and then to remove these, leading to permanent improvements. If, for example, the cause of an out-of-control signal is found to be a defective batch of material, it does not suffice just to replace the bad batch with a good one. A process needs to be established to determine why the batch was bad and then to act to avoid similar bad batches in the future. Control charts, thus, provide signals for human intervention and call for those responsible for the process to determine the reasons for observed changes and act accordingly.

It may, admittedly, be difficult to remove some special causes of variation, and sometimes these need to be accepted as part of the process. For example, *some* variability in raw material or setup might be both inevitable and sometimes acceptable – and may itself warrant monitoring over time (see Hahn and Cockrum 1987 and Hare's 2001 concept of "structural variation").

Finally, the importance of having control charts at strategic points in the process warrants emphasis. This makes them part of a proactive system to determine the underlying causes of problems that might lie upstream in the process and expedites their identification.

6.8 MAJOR TAKEAWAYS

- Statistics continues to play an important role in helping improve the quality of manufacturing processes.

- The specific approach varies for discrete parts versus process industries, continuous versus batch operations, and job-shop versus repetitive production.
- For target-is-best characteristics, the goal is to maintain consistently on-target performance within two-sided specification limits.
- For higher-is-better (smaller-is-better) characteristics, the goal is to maintain consistently high (low) performance, exceeding (falling below) the one-sided lower (upper) specification limits.
- A disciplined approach for manufacturing improvement, such as provided by PDSA or Six Sigma, is required for effective quality improvement.
- The key phases of quality assessment and improvement are
  - Manufacturing capability and stability assessment.
  - Addressing measurement error.
  - Reducing deviations from target and variability.
- Statistical monitoring is employed strategically and systematically, through the use of control charts, to signal significant process changes, identify and remove special causes of variation, and provide the path to permanent improvement.

DISCUSSION QUESTIONS

General

1. We state in Section 6.2 that "we need ensure that nothing that is done during manufacturing compromises the high level of reliability that, hopefully, was built into the product during design." Elaborate and suggest some examples of how elements of the manufacturing process might compromise reliability.
2. Undesirable variability is not limited to manufacturing processes. Consider one or more of your personal experiences (e.g., baking brownies, playing golf, running the marathon, and brewing coffee). How would you quantify variability? What are the likely root causes and how can these be quantified? Summarize by a fishbone diagram.
3. Provide examples of short-term and long-term variability and the underlying causes.
4. What are some advantages and limitations of process capability indices in general, and the two indices presented in Section 6.6.1? Compare such indices to the Taguchi Loss Function discussed in Section 3.3.3.
5. Develop a control chart system to monitor your weight. Discuss the type of control chart to be used, the recommended sampling scheme and protocol, the tracking of root causes of potential changes, and the routine evaluation of results.
6. Before an appliance is shipped, a trained operator carefully inspects it for visual defects (scratches, etc). For this application:
   - How would you develop a consistent definition of a visual defect?
   - How would you design a process capability study?
   - How would you assess and monitor measurement error?
   - How would you set up a control chart system to monitor change?
   - How would you proceed in identifying, quantifying and removing root causes of defects?

Technical

1. Elaborate on how the differences between discrete parts and process industries impact the specific manner in which a statistical approach is likely to be applied.
2. How would you expect the application of statistics to differ between manufacturing operations that yield a continuous flow of product and those that produce product in batches?
3. How would you expect the application of statistics to differ between a job-shop and repetitive product operations?
4. Consider again the Case A, B, and C histograms presented in Section 6.4.4. How are these impacted by measurement error?
5. Measurement error is often proportional to the magnitude of the process value being measured. How might this impact the results of a study and what might you do to address this?
6. Explain the rationale behind control limits of $\pm 3\sigma/\sqrt{n}$ for an $\bar{X}$ chart.
7. The performance of a control chart is frequently evaluated with regard to its ability to detect real process changes and to avoid false alarms. How do these relate to Type 1 and Type 2 error risks in statistical hypothesis testing?
8. In what kind of situations would you expect an EWMA or a CUSUM chart to outperform a traditional control chart? When do you think a traditional control chart would be better?
9. For what kind of situations might you prefer an EWMA over a CUSUM chart?
10. Describe a specific situation that would result in autocorrelated data and describe a control chart that you might use in its assessment. How does this scheme handle the autocorrelation?

CHAPTER 7

MANUFACTURING QUALITY IMPROVEMENT: EXAMPLE

7.1 ABOUT THIS CHAPTER

This chapter presents a case study to illustrate some of the concepts and methods for product quality improvement described in the previous chapter.[1]

Czitrom and Spagon (1997) give further case studies involving the use of statistics for quality improvement in the semiconductor industry; these also are relevant for other industries.

7.2 GENERATOR STATOR BAR EXAMPLE

7.2.1 Description of Process

This application deals with the manufacture of stator bars for generators. These are placed in a magnetic field to generate electrical current that flows through the conductors of the bar. Due to their critical role in the operation of the generator, the bars must meet several key requirements.

The first step in stator bar manufacturing is the assembly of a series of conductor strands to form a bar about 15 feet in length. Figure 7.1 shows a bar assembled with 24 conductor strands. The conductors are then consolidated into a solid bar structure by applying a resin mold to the surface of the bar in a hot press. Four bars are processed simultaneously in each press run. Our discussion focuses on improving this step of the process.

The press machine is used for more than 20 different bar designs. Changing the setup from one design to the next requires replacement of the metal plates and is time-consuming. Thus, it is desirable to conduct as many runs of the same design in a setup as possible.

7.2.2 End-of-Line Evaluations

At the end of each hot-press molding run ("end-of-line" for our purposes), measurements are made on each of the four stator bars of their:

1. Technical and calculational details of various statistical analyses are to be posted on the book's ftp site.

The Role of Statistics in Business and Industry, by Gerald J. Hahn & Necip Doganaksoy

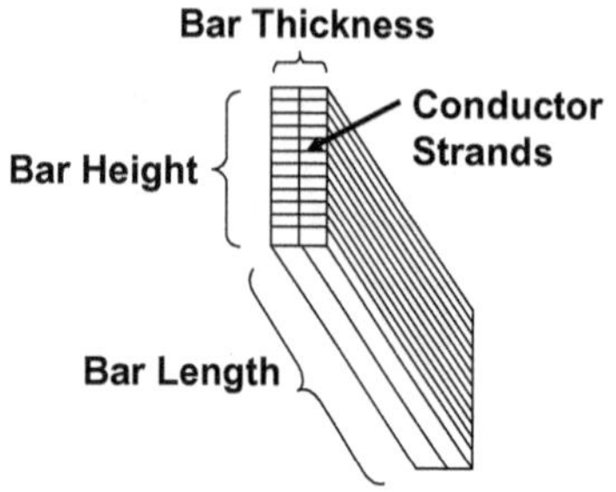

Figure 7.1 Schematic of an initial bar assembly.

- Electrical properties: An instrumented test for electrical shorts and other insulation defects.
- Dimensional conformance: A pass/fail gauge assessment (due to its ease of use) of bar height and thickness.
- Molding: A check for visible defects, such as streaks, twisting and delamination.
- Bar shape: A coordinate measuring machine assessment of bar shape.

Bars that fail to meet specifications on any of these tests are reworked and then inspected again.

7.2.3 The Challenge

The manufacturer has been building bars for a long time. We focus on a new type of bar that has been in production for nearly 3 months. This bar had been carefully designed and successfully transitioned to Manufacturing. To date, 780 bars have been manufactured, representing 15% of total bar production and involving 56 setups.

Rework of bars that did not meet specifications, though an expense and nuisance, had, in the past, not received much attention. A combination of circumstances changed this:

- The new design produced more initially defective product than the previous ones, resulting in appreciable productivity losses due to the required rework.
- Some nonconforming bars were accepted in error, creating costly problems downstream.
- Repaired stator bars were more likely to fail prematurely in the field, adding further costs and customer dissatisfaction.

An enterprising new manager recognized that reducing initial stator bar rejects would yield important savings and initiated a study to learn how to improve the quality of the new type bar. The resulting information was also expected to help improve other designs.

7.3 MANUFACTURING CAPABILITY AND STABILITY ASSESSMENT

7.3.1 Percent Yield

A total of 92 or 11.8% of the 780 bars, involving 118 defects, did not meet one or more of the requirements and had to be repaired over the 3-month period. It was desired to reduce this rejection rate drastically.

Operators had recorded the data on rejects from the hot-press molding operation by defect type on a check sheet, shown in Table 7.1 for a typical week.

A Pareto chart of the defects by type, over the 3-month period is shown in Figure 7.2.

This chart shows that failure to meet dimension (i.e., thickness) specifications was the most important source of rejects—accounting for 49/780 or 6.3% of all bars.

TABLE 7.1 Cheek Sheet by Defect Type for Typical Week

Fiscal Week 12							
	Day						
	Monday	Tuesday	Wednesday	Thursday	Friday	Saturday	Total
Number of bars	16	18		12		14	60
Number of defective bars	//	///		/		///	9
Reason	Number of Defects						
Dimensions	//	//		/		//	7
Molding	/	/				/	3
Bar shape				/			1
Electrical						/	1
Other							0
Total	3	3		2		4	12

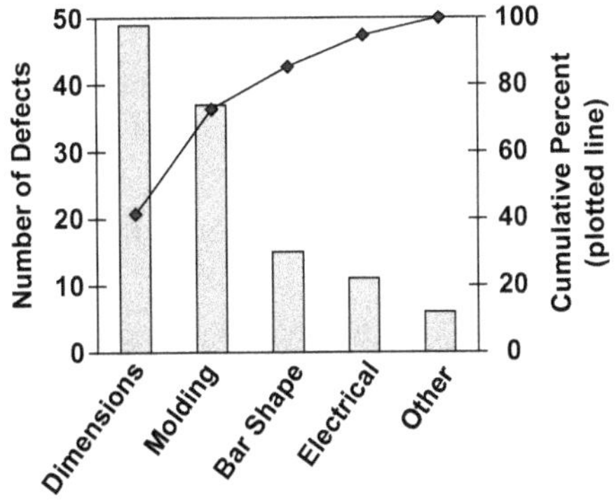

Figure 7.2 Pareto chart for defects for the hot-press molding operation.

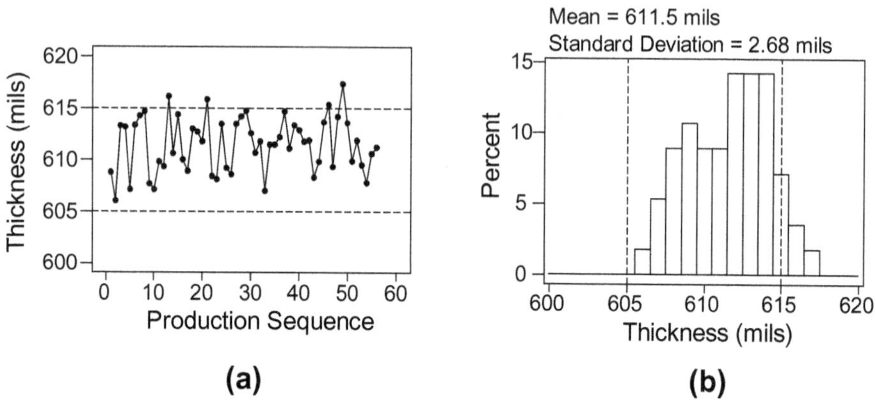

Figure 7.3 (a) Time plot and (b) histogram of thickness measurements of 56 production bars.

Some of the other causes for rejection were also expected to be alleviated by addressing the thickness problem; we therefore focus on this defect type.

A time plot (not shown here) of weekly defect rates due to failure to meet dimensional specifications did not suggest any trends over the 3-month production period.

7.3.2 Variables Assessment

The preceding evaluations are useful in quantifying the magnitude of the problem; however, they provide little insight. Variables data would be more informative.

The manufacturing team had the foresight, at the introduction of the new design, to establish a procedure by which a bar was randomly selected from each setup and measured for thickness and other characteristics.

The thickness measurements on the resulting 56 bars are plotted against production sequence in Figure 7.3a. Figure 7.3b is a histogram of the data. The target thickness for this bar was 610 mils with ±5 mils specification limits; these are shown as dashed lines.

Examination of Figure 7.3 led to following observations:

- Fraction defective bars: Four bars were above the upper specification limit, corresponding to 7.1% of the 56 bars. This is in line with the 6.3% defects, based on end-of-line thickness inspection, for all 780 bars.
- Mean: The mean bar thickness of 611.5 mils was 1.5 mils above the target value.
- Variability: The observed variability, $\hat{\sigma}_{OBSERVED} = 2.68$, indicated appreciable variation (relative to the specification limits).
- Time trend: The process appeared to be stable over the 3-month period.[2]

2. This conclusion was further supported by a moving R chart (not shown here).

7.3.3 Process Capability Evaluation

The data resulted in the estimate

$$\hat{C}_{pk} = \text{smaller of } \left\{ \frac{615-611.5}{3(2.68)}, \frac{611.5-605}{3(2.68)} \right\} = 0.44.$$

This, as well as the reject rate of 6.3%, was unacceptable.[3]

7.4 ADDRESSING MEASUREMENT ERROR

The results shown in Figure 7.3 are attributable not only to the pressing operation (the process), but also reflect measurement error. It is, therefore, important to obtain an assessment of how much of the observed variation and deviation from target is due to the process alone. This cannot be inferred from Figure 7.3 since each bar was measured for thickness only once.

7.4.1 Measurement Monitoring

The manufacturer had established a process to monitor the measurement system, using a standard bar with a nominal thickness of 610 mils. This bar was submitted twice a week for two thickness measurements on the same day to be taken by whichever one of three operators designated for this purpose happened to be available. (The three operators were randomly selected from the larger population of operators that took such measurements.)

7.4.2 Graphical Assessment

Figure 7.4 shows a time plot of the 48 thickness measurements on the standard bar taken over 24 days during the 3-month period under observation, with same-day readings joined by vertical lines.

If there were no offset error and no variability in the measurements, we would expect all thickness readings in Figure 7.4 to be 610 mils. The variation shown in the figure is therefore due to the measurement system. Visual examination suggests that:

- The measurements appear to be centered on or near the nominal value of 610 mils (sample mean of 610.1 mils for the 48 observations), suggesting no discernible bias in the measurement system.[4] Thus, the previously observed deviation of 1.5 mils of the mean thickness of production bars from target thickness cannot be attributed to measurement bias.

3. In this example, $\hat{\sigma}_{OBSERVED}$ reflected the long-term, and not the short-term, variability. Thus, $\hat{C}_{pk}$ might perhaps be better designated as a performance index.

4. Although hardly necessary, one can also show that this value does not differ significantly statistically from the 610-mil target.

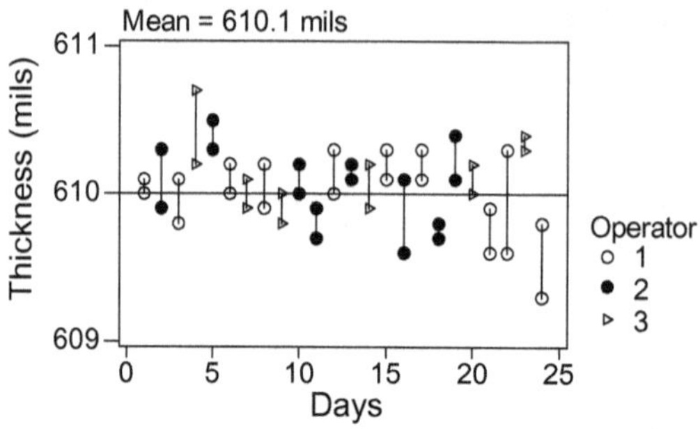

Figure 7.4 Measurements by three operators on standard bar with nominal thickness of 610 mils.

- There is moderate variability in the measurements taken on the same day.
- There may, in addition, be some important day-to-day variability.
- There seems to be little, if any, difference between operators.
- There do not appear to be any trends in the mean and the variability during this time period; that is, the measurements appear to be stable over time.

7.4.3 Formal Analysis

Analysis of variance can be used to break down the total estimated variability in measurement error into three constituent components: $\hat{\sigma}_{OPERATORS}$, $\hat{\sigma}_{DAYS}$, and $\hat{\sigma}_{READING}$, designating, respectively, the estimated variation attributable (solely) to operators, days, and individual readings (by the same operator on the same day). For example, $\hat{\sigma}_{OPERATORS}$ denotes the variability due to operators alone, that is, the variability that would remain if that due to the other two sources of variation were eliminated completely. Specifically,

- $\hat{\sigma}_{OPERATORS} = 0$
- $\hat{\sigma}_{DAYS} = 0.15$ (95% approximate confidence interval: 0.08 to 0.64)
- $\hat{\sigma}_{READING} = 0.22$ (95% confidence interval: 0.17 to 0.31)

(See Burdick and Graybill 1992, Section 5.4).

Although the point estimate for the variability in readings by the same operator was higher than the day-to-day variability, the 95% confidence intervals for these two estimates overlapped appreciably; a larger investigation would be required to get a more definitive comparison.

The total measurement error standard deviation (for a single measurement) is then estimated to be

$$\hat{\sigma}_{MEASUREMENT} = \sqrt{\hat{\sigma}^2_{OPERATOR} + \hat{\sigma}^2_{DAYS} + \hat{\sigma}^2_{READING}} = 0.27.$$

7.4.4 Impact

The histograms displayed in Figure 7.5 illustrate the impact of measurement variability on the total observed product variability. Figure 7.5a is a repeat of Figure

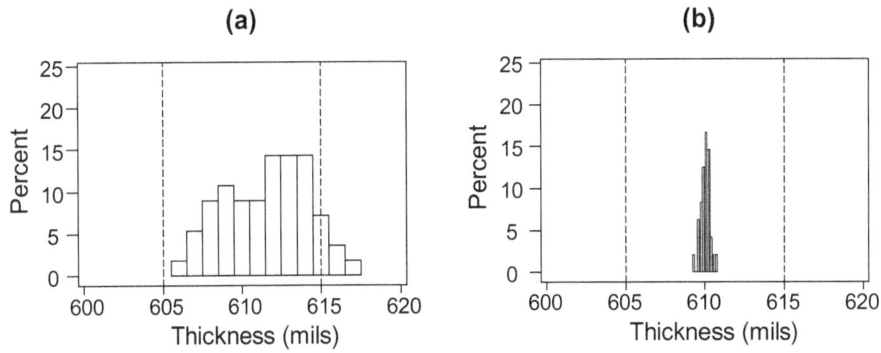

Figure 7.5 Histograms of thickness measurements on (a) 56 production bars and (b) 48 readings on standard bar (the dashed lines are specification limits).

7.3b, showing the observed thickness measurements on the 56 randomly selected production bars. Figure 7.3b shows the 48 readings on the standard bar over the approximate 3-month period. This figure shows the variability in the observed thickness measurements for the production bars to be much larger than that of the repeat measurements on the standard bar.

More formally, statistically removing the estimate $\hat{\sigma}_{MEASUREMENT} = 0.27$. from the previously obtained estimate $\hat{\sigma}_{OBSERVED} = 2.68$, we obtain $\hat{\sigma}_{PRODUCT} = \sqrt{2.68^2 - 0.27^2} = 2.67$ – a very small reduction.

Both the graphical and the statistical evaluations, therefore, suggest that the measurement variability is a small contributor to the total observed variability, and thus, that a substantial part of the observed product variability is real. This was also suggested by the estimated GRR index (Section 3.4.2) of

$$100\frac{(5.15 \times 0.27)}{(615-605)} = 13.9\%.$$

7.5 REDUCING DEVIATIONS FROM TARGET AND VARIABILITY

The process capability study indicated the need for reducing the percentage of defective product by both removing the offset error (mean of 1.5 mils above target) and reducing the variability in product thickness ($\hat{\sigma}_{PRODUCT} = 2.67$). This was to be accomplished by improving the manufacturing process.

7.5.1 Identify

A team of process experts developed the fishbone diagram in Figure 7.6. This breaks down the major factors believed to impact the offset error and variation in pressed bar thickness during hot-press molding into four categories: methods, measurements, equipment, and materials.

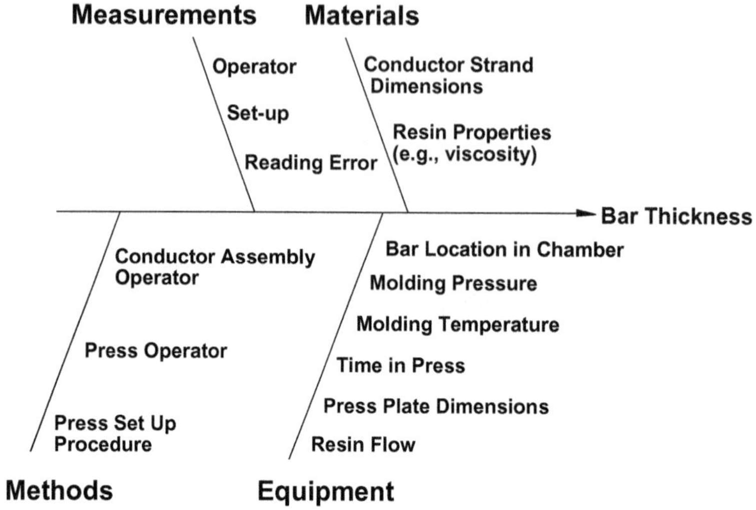

Figure 7.6 Fishbone diagram for the hot-press molding operation.

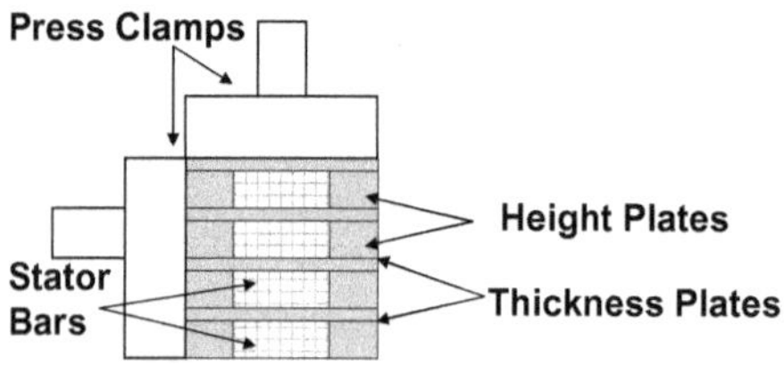

Figure 7.7 Hot-press molding operation.

Measurement error had been shown not to be a key contributor. The focus was, therefore, on the process itself, that is, the other three "bones" of the diagram (see Sidebar 7.1).

SIDEBAR 7.1: MORE ON THE HOT-PRESS MOLDING OPERATION

Figure 7.7 is a schematic of the press operation. The four bars are placed into press chambers and the operator applies metal plates to the sides of each bar. The melted resin is pushed into the bar chambers for molding under elevated temperature and the clamps are pressed. The material is then cooled and the clamps pulled back.

It seemed clear that the positioning and dimensions of the metal plates in the press chambers were major determinants of bar height and thickness; a too thick (thin) plate could lead to consistently under (over) sized bars. Thus, offset error can be corrected readily by adjusting the dimensions of the plates and this is what was done. This simple solution was not expected to have a deleterious impact on other characteristics—an assertion that, however, needed to be verified.

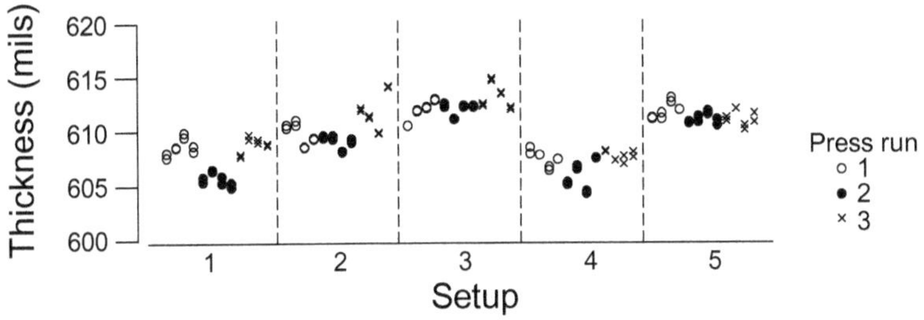

Figure 7.8 Multi-vari chart for bar thickness data.

There seemed to be no simple explanation for the variability in the thickness measurements. Some of the causes suggested in Figure 7.6 were dismissed as potentially unimportant from physical considerations. After some lively discussion, it was decided that further data were needed to break down the total variability in part thickness into individual components attributable to press setups, press runs within each setup and bars within a press run.

7.5.2 Quantify

After making the correction to remove offset error, a special study was conducted to obtain the needed data to break down the total process variability into individual identifiable components.

Description of Study Thickness measurements were obtained on bars selected during normal production from:

- five press setups, chosen at approximately equal intervals over a 1-month period
- three press runs from each setup
- four bars from each press run[5]

The thickness of each bar was measured at two different times.

Results of Study Figure 7.8 displays the results of this study in a "multi-vari chart" (see Snee 1983). This chart uses different symbols for the three press runs within each setup, and different positions on the abscissa for the four bars within a press run in each setup.

Analysis of Results The mean thickness of 609.8 mils was close to the target of 610 mils, indicating that the plate thickness adjustment had succeeded in removing the offset error.

Examination of Figure 7.8 suggests that setup is the most important source of variation, followed, perhaps by press run. The mean and variability, moreover, seemed to remain constant over time.

5. Such a study is known as a "fully nested design" in that press runs are "nested" within setups, and bars are nested within press runs (see Box, Hunter, and Hunter 2005, Chapter 9).

TABLE 7.2 Estimates of Variance Components for the Stator Bar Thickness Example

Source	Estimate of Variance	Estimate of Standard Deviation	Percent Contribution to Total Variance
Set-up	5.00	2.23	71.2
Press run	1.14	1.07	16.3
Bar	0.83	0.91	11.8
Measurement Error	0.05	0.22	0.7
Total	7.02	2.65	

A quantitative evaluation can be obtained by estimating the standard deviations attributable to each of the four components of variation using analysis of variance. The results are shown in Table 7.2.

Note that:

- The total observed standard deviation of stator bar thickness, estimated to be 2.65, is very close to the value of 2.68 obtained in the process capability study.
- The measurement error standard deviation, now estimated to be 0.22, is the smallest source of variability (and leads to an estimate of the standard deviation for actual stator bar thickness of 2.64 (i.e., $\sqrt{7.02-0.05}$)).
- The variability due to press setup is the biggest contributor to the total observed variability in stator bar thickness. The complete elimination of variability between press runs would reduce the estimated total observed standard deviation from 2.65 to 1.42 (i.e., $\sqrt{7.02-5.00}$). (The estimated standard deviation for actual total bar thickness would be reduced to 1.40.)
- The variability among press runs and bars within a press run, contributed only a relatively small amount to the total variability in stator bar thickness. Elimination of both these sources of variability reduces the estimated total process standard deviation from 2.65 to 2.25 (i.e., $\sqrt{7.02-1.14-0.83}$).

These results clearly suggest that in order to improve this process, we need to focus on reducing the variability between setups.

7.5.3 Find Root Causes

The findings of the special study helped the team focus its efforts on the major source of variability in the thickness measurements—that between press setups—and to search for exactly what was creating such variability. Physical considerations and past experience pointed to four factors: pressure (during pressing operations), molding temperature, time in press, and resin viscosity.

Each of these factors could impact thickness appreciably if permitted to vary over a sufficiently broad range. For this reason specification limits had been set initially on each characteristic and these appeared to have been met by Manufacturing. The impact of each factor within its normal manufacturing range, therefore, needed to be evaluated.

Nature of Study It was decided to conduct a statistically planned experiment since only limited data were available to make the desired assessments. The design used was a 2^4 full factorial experiment. This called for 16 setups, involving all combinations of two conditions for each of the four variables. Interactions among these variables were expected to be important and this plan allowed these to be evaluated. The high and low settings of each of the four variables were set close to the extremes encountered in manufacturing. These were also near their specification limits.

Major Results Graphical evaluations and an analysis of variance (not shown here) of the results of the study indicated an interaction between resin viscosity and press temperature. This is shown in Figure 7.9 by a plot of the mean thickness at each of the four combinations of press temperature and resin viscosity in the study. The plot shows that the thickness of a pressed bar is highly sensitive to press temperature when the viscosity of the resin is low, but not when the resin viscosity is high. This finding was supported by fundamental knowledge about the material and process.

7.5.4 Improve

Correct for Deviations from Target The offset error was corrected by adjusting the dimensions of the plates used in the pressing operation. The effectiveness of this action was validated by taking measurements on a sample of bars built immediately after making the change and by the special study.

Reducing Variability The special study indicated that reducing the variability in stator bar thickness required the resin supplier to maintain resin viscosity consistently at the high end of the original specification range.

The best way of achieving this would have been for the supplier to switch to a different formulation. This might have been possible during product design, but was no longer practical. Reduction of variability was, instead, achieved by adjusting the specs on resin viscosity (requiring the supplier to make some process adjustments) and by blending resin lots.

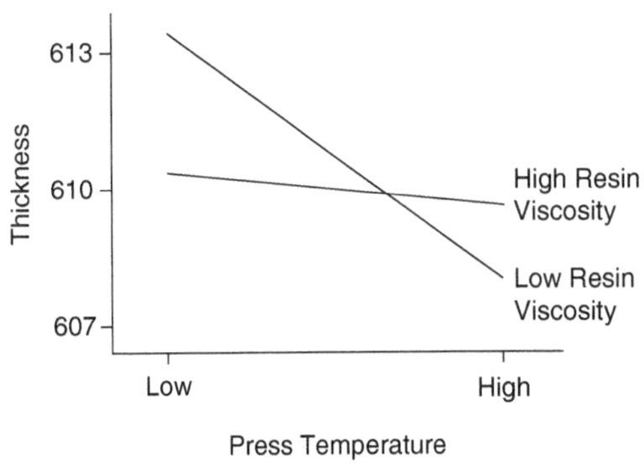

Figure 7.9 Temperature and resin viscosity interaction plot for stator bar study.

7.5.5 Validate

Thickness measurements on a subsequent sample of bars showed that $\hat{\sigma}_{OBSERVED}$ had been reduced from 2.65 to 1.90. This result—together with the centering of the process—reduced the percentage of stator bars failing to meet dimensional specifications from 6.3% to 0.9%. The changes also helped improve other performance characteristics and the total bar reject rate was reduced from 11.8% to 2.7%. The lessons learned from these assessments were also used in designing future stator bars.

7.6 STATISTICAL MONITORING

Several control charts were used to monitor the continuing performance of the process:

- Raw material properties: Process experts helped the resin supplier install a system of control charts to monitor resin consistency in the short run, and to identify the root causes of variability in the long run.
- Measurement system: The measurement-monitoring program, using the standard bar, was continued with sampling reduced from twice a week to once a week. The results were plotted on $\overline{X}$ and R charts.
- Process variables: Measurements were taken at the beginning of each new setup on press tool dimension and on various press parameters (e.g., temperature, pressure). These were plotted on EWMA charts for timely detection of gradual deterioration.
- Product performance: Measurements on key characteristics, such as thickness and height, were taken on a single bar randomly sampled from each setup. The results were plotted on an X chart. In addition, p charts on end-of-line yield were used to provide summary information to management.

An automated system was developed to implement these control charts. The engineer responsible for addressing the problem was alerted immediately of all out-of-control situations. All those involved with the process, including management, could go on line at any time to obtain up-to-date results.

DISCUSSION QUESTIONS

General

1. Which of the Basic Quality Tools enumerated in Table 2.1 of Chapter 2 were utilized in the case study of this chapter? What kinds of situations lend themselves especially well to the use of each of these tools?
2. We state that "a bar was randomly selected from each setup and measured for thickness and other characteristics" (Section 7.3.2). How would you select such a bar randomly?

3. How would you go about quantifying the financial benefits of the improvement achieved in the case study presented in this chapter?

Technical

1. In a footnote in Section 7.4.2 we state the sample mean of 610.1 mils for the 48 observations on the standard bar "does not differ significantly statistically from the 610 mil target." Perform this analysis and obtain a two-sided 95% confidence interval on the mean for the standard bar.

2. Verify the 95% confidence interval on the sources of variation for the measurement error study (Section 7.4.3).

3. Critique the design of the measurement error study and suggest possible improvements.

4. What are the degrees of freedom associated with each of the variables in the analysis of variance shown in Table 7.2?

5. In the special study that led to the data displayed in Figure 7.8, two repeat measurements were obtained on each bar. Suppose that these measurements were taken at two predetermined locations on the bar. How would this change the analysis of the resulting data?

6. In the special study, the total standard deviation from actual total bar thickness was estimated to be reduced from 2.64 to 1.40 upon removal of setup error. How was this estimate calculated?

CHAPTER 8

PRODUCT MANUFACTURING: FURTHER APPLICATIONS

8.1 ABOUT THIS CHAPTER

We begin this chapter with a brief description of engineering process control and how, when coupled with statistical process control, it can provide a "one-two punch" for quality improvement. We then briefly discuss materials and parts quality assurance, final product assessment and productivity improvement.

8.2 INTEGRATING ENGINEERING PROCESS CONTROL AND STATISTICAL PROCESS CONTROL

> **When steering a boat we should certainly not allow for every wave, but, if we want to stay off the rocks, it will be necessary to compensate for the current.**
> —George Box

Statistical process control (SPC)[1] focuses on the monitoring and improvement of a process. Typically, an out of control signal initiates a search for special causes and their elimination. In this section, we consider integrating SPC with engineering process control (EPC)—also referred to as automated process control—a quite different strategy for quality (and productivity) improvement that has been applied successfully in many operations, especially in the chemical and process industries.

8.2.1 EPC

EPC deals with ongoing online adjustments of one or more process variables so as to control the level and reduce the variability of key performance or output variables.

1. In this discussion, for convenience, we equate SPC with control charts; SPC is actually a broader topic, encompassing process improvement and involving the use of basic quality tools, such as those introduced in Section 2.2.5.

The Role of Statistics in Business and Industry, by Gerald J. Hahn & Necip Doganaksoy

We routinely exercise the principles of EPC in such daily tasks as taking a shower and driving a car. In a shower, we adjust the mix of cold and hot water to attain a comfortable temperature. We make such adjustments, both at start-up and in response to external disturbances, such as the use of water elsewhere in the building. In so doing, our goal is not to eliminate the disturbances—which sometimes may be difficult, impractical or even impossible—but to compensate for their effects.

Chemical and process operations employ EPC to regulate flowmeters, feed rates, temperatures, pressures, and so forth throughout the plant so as to control product performance in response to one or a sequence of readings. To do so, the relationship between one or more compensating variables and the process outputs needs to be established. This knowledge is then used to make appropriate process adjustments to return the process to target, if needed, and to reduce variability in performance.

Applications of EPC generally require:

- Definition of the output or performance variable(s) to be controlled
- Determination of the control variable(s)
- Establishing the relationship between the control variable(s) and the variable(s) to be controlled
- Development of a control rule
- Exercising the control rule

Some applications involve multiple controls and output variables, complex relationships, and issues involving systems dynamics, such as delays and transients.

8.2.2 Feedback Control

Feedback control calls for periodically (or continuously) measuring deviations from target of the performance variable(s) and using these deviations to adjust the process so as to reduce future deviations. In taking a shower, we feel the water initially; if it is too cold or too hot we adjust the dial until the desired temperature is attained. If the water temperature subsequently changes due to external disturbances, we make further adjustments. It takes time for the temperature to reach the desired level; we therefore need resist the temptation to overadjust.

Similarly, in the programmed logic for a hot water thermostat

- The temperature of the water in the tank is measured.
- The measured temperature is compared with the target (or set point) temperature.
- A control algorithm is used to decide whether the heater should be on or off.

Feedback control can, thus, be thought of as a form of constantly playing "catch-up."

8.2.3 Feedforward Control

Feedforward control involves measuring potential noise variables or external disturbances that are likely to impact one or more outputs *in the future* and adjust the control variable(s) accordingly. Feedforward control is used when we take initial measurements on a part about to be processed, and then adjust the processing conditions accordingly. In chemical processing, feedforward control may involve adjusting the process based upon measurements on incoming materials, such as unrefined petroleum.

Feedforward control is more proactive than feedback control. It may also be harder to implement due to the difficulty of identifying external disturbances and predicting their impacts.

8.2.4 Joint Use of Feedback and Feedforward Control

Sometimes feedback control and feedforward control are used together. One might make adjustments based on *both* past information about the process, such as recent performance, and information on factors that might impact future performance, such as measured properties of raw materials.

The U.S. Federal Reserve System's Board of Governors' periodic adjustment of the prime interest rate to impact future economic outcomes provides an, arguably less scientific, example of the joint use of feedback and feedforward controls. These adjustments are based upon a study of both current economic conditions and projections of future economic behavior using leading indicators and sophisticated economic models.

For further discussion and some cautionary notes on the use of EPC, see Kirby, Leitnaker, Moore, Ranney, and Sanders (1990).

8.2.5 EPC versus SPC

Short-Term Considerations Continually adjusting a process appears to contradict the basic teachings of SPC. In his famous funnel experiment, Deming (1986) demonstrated that adjusting a process that is in statistical control leads to increasing, rather than reducing, variability. Deming assumed serially independent observations on a process that exhibited stable random variation around a constant mean.

Deming's conclusions are, of course, correct given his assumptions; these, however, are not always met in practice. Many processes tend to wander from target due to such factors as changing raw material properties, equipment wear and tear, decaying catalyst, or changes in environmental conditions. In such cases, SPC's methodical approach may, in contrast to EPC, seem slow to react. In a follow-up to the funnel experiment, MacGregor (1990) showed how process adjustments and EPC can, for *some* processes, *reduce* variability.

The degree to which EPC is useful depends on the specifics of the situation. These include the dynamics of the process, measurement error, the ability to adjust the process and the cost of making adjustments.

EPC often works and is used extensively in the chemical industry because of the presence of autocorrelated observations. These make future process observations more predictable and helps explain why SPC and EPC found their roots in the parts and processing industries, respectively.

Long-Term Considerations A drawback of EPC is that it will make the necessary adjustments, without usually identifying, the root causes that made them necessary. Consequently, the underlying disturbances, and their impact, often are not fully addressed. Moreover, EPC may diminish the urgency to tackle issues directly and to make long-term improvements. The problems, however, are likely to worsen—to the point that corrections using EPC may no longer be effective or even be possible.

Assume, for example, that the special cause creating a disturbance is a leaky water tank. EPC might provide process adjustments to counteract the impact of the leak. These may be effective up to a point, but the leak may worsen over time. The tank may, in an extreme case, burst open altogether, creating much greater difficulties.

Thus, we need to get a handle on the underlying assignable causes of variability so as to ensure, as a minimum, that they do not get to the point at which they no longer can be corrected for by short-term adjustments. That is where SPC comes in. Its goal is to determine underlying problems, find their root causes, remove them where possible, and work to avoid similar problems in the future.

Admittedly, not all problems get worse over time. Environmental conditions, beyond our control, might, for example, swing back and forth. We strive to address these by robust design, per our earlier discussion—but may only be partially successful. So a second line of defense is to adjust the process in response to the environmental fluctuations—while being on the watch for new sources of variation that need be addressed.

8.2.6 Integrating EPC and SPC

There is clearly a place for *both* EPC and SPC. The power and limitations of these approaches when used individually has led to a drive to use them together to reap the benefits of each. The basic idea is to employ EPC to reduce predictable short-term variability using feedback and/or feedforward techniques. The adjustments are then themselves closely monitored using SPC to detect and then remove special causes of variation. This dual approach has been developed extensively by Box and Luceño (1997), and Del Castillo (2002); Montgomery (2004, Chapter 11) provides an introduction.

8.3 POLYMER RESIN EXAMPLE

8.3.1 Description of Process

This example of integrating EPC and SPC is based on Vander Wiel, Tucker, Faltin, and Doganaksoy (1992), who provide additional detail. It deals with the manufacture of a polymer resin used for consumer products.[2]

2. This example involves a batch process; many EPC applications are on continuous processes.

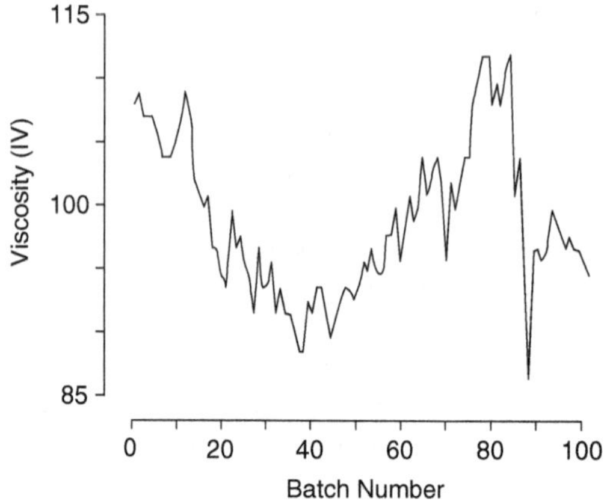

Figure 8.1 Batch viscosity for 106 batches.

Intrinsic viscosity (IV), a key performance characteristic of the polymer, is measured on each batch. The amount of catalyst added to the reaction is considered to be a promising control variable to keep IV at or near its target value of 100 and to reduce variability.

8.3.2 Past Results

Figure 8.1 displays IV measurements for 106 batches produced over a 30-day period for a catalyst addition of 50 gallons per batch.[3]

Figure 8.1 shows that IV tends to wander over time from the target IV of 100, indicating lack of statistical stability in the process if left unadjusted. Such variability creates problems in downstream manufacturing, and ultimately for the customer.

The data also suggest strong autocorrelation. This is not surprising, since it is not possible to fully empty a reactor between batches, resulting in left over "heel material." Furthermore, economic and environmental considerations necessitate the recycling of previously unused material into the storage tanks; this can induce "memory" into the process.

8.3.3 Feedback Control (EPC)

Analysis indicated that the amount of catalyst added to the batch reaction could, indeed, serve as a compensatory variable to control IV.

The relationship between catalyst added and the resulting IV was fitted statistically, taking the autocorrelations of the observed IV into consideration. This led to the simple control rule:

3. In actuality, the process was already being controlled by varying the amount of catalyst added based upon the IV readings of previous batches. The results shown in Figure 8.1 were obtained by statistically adjusting the IV values actually obtained to those expected *if* catalyst addition had been held at 50 gallons.

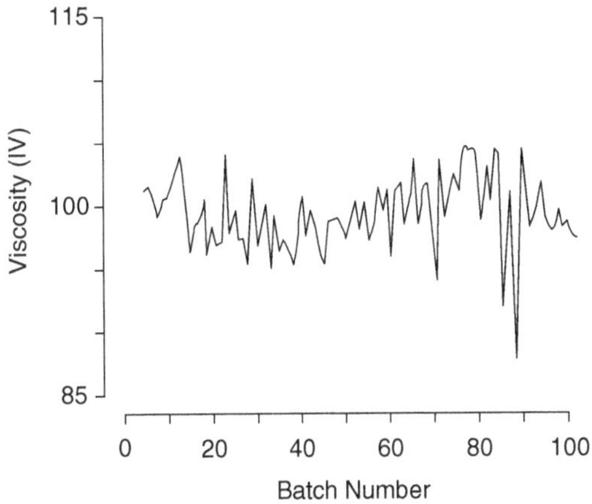

Figure 8.2 Predicted batch viscosity using catalyst addition from control rule.

Catalyst deviation from nominal 50 lbs. to be used for current batch	=	0.80 (Catalyst deviation from nominal 50 lbs. in previous batch) − 0.40 (IV deviation from target of 100 in previous batch)

The preceding rule, along with the fitted relationship between IV and catalyst added, was then applied to the data shown in Figure 8.1. The predicted IV as a consequence of applying this rule is shown in Figure 8.2.

This plot shows an appreciable decrease in IV variability, compared to Figure 8.1. It also showed a spike for the 83rd batch that needed to be investigated.

8.3.4 Superimposed Control Charts (SPC)

A series of CUSUM control charts to detect changes in *both* the IV measurements and the catalyst adjustment were superimposed on the preceding control rule. These showed the process drifting out of control. The root cause was tracked to a leaky catalyst feed valve that allowed residual catalyst to remain in the charge line. This could not be caught by EPC alone and would eventually have made the adjustments ineffectual.

8.3.5 Combined Scheme

Figure 8.3 is a schematic diagram of the combined EPC/SPC procedure.

The combined implementation of EPC and SPC resulted in elimination of all out-of-spec material within 3 months. This in turn led to a revised control rule that called for smaller changes from batch to batch in the amount of catalyst added. Continuation of this approach is expected to ensure further reduction in short-term variability and rapid identification of any future changes.

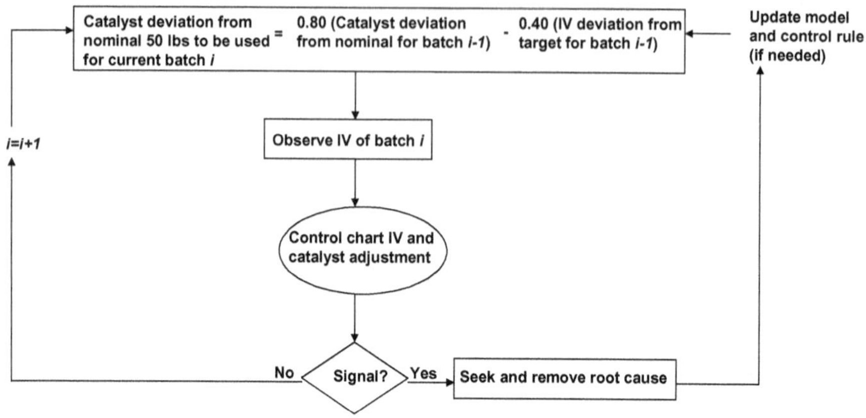

Figure 8.3 Flow diagram of joint application of EPC and SPC to the polymer resin example.

8.4 MATERIALS AND PARTS QUALITY ASSURANCE

8.4.1 Ensuring Supplier Quality

The quality and reliability of many products depend heavily on the quality and reliability of raw materials, components, and, sometimes entire assemblies, obtained from suppliers.[4] One recently highly publicized example is that of excess lead in paint used on toys.

Detailed quality and reliability specifications on supplied parts and materials are established and validated during product design. Standards ISO 9000 and ISO 14000, developed by the International Organization for Standardization, are directed toward standardizing and documenting quality management systems. They are especially relevant for global supplier qualification (see Evans and Lindsay 2007).

Ensuring that specifications *continue* to be met requires an ongoing effort. This calls for close interaction with suppliers, and sometimes on-site presence at their facilities. Statistical concepts contribute by helping:

- Define requirements quantitatively.
- Ensure that suppliers have appropriate quality monitoring and improvement processes in place.
- Have suppliers provide pertinent and timely quality data, including end-of-line audit measurements on shipped product.
- Insist on suppliers making known any changes in the design or the manufacturing process—and require thorough assessments to ensure that such changes do not have a detrimental impact on product performance, quality and reliability.

4. A "supplier" may be another organization within our own company, possibly offshore.

8.4.2 Materials and Parts Acceptance Sampling

Acceptance sampling of incoming material or parts has traditionally been used to decide whether or not a submitted lot or shipment is acceptable, based upon inspection of a random sample. This procedure has been deemphasized[5] in recent years in favor of a more proactive approach that emphasizes working with suppliers to build quality into their products. It, however, is still used to deal with some critical parts or material, or as part of a larger system to evaluate supplier quality and to foster continuous improvement.

Acceptance Sampling Plans An acceptance sampling plan specifies the size of the sample to select from a submitted lot and lot acceptance/rejection requirements.[6] These are usually established to meet specified initial criteria, such as average outgoing quality (AOQ), acceptable quality level (AQL), and lot tolerance percent defective (LTPD).

The U.S. government issued a series of acceptance sampling plans (MIL-STD-105 for attributes and MIL-STD-414 for variables) for use in procurement by its armed forces. The plans were subsequently replaced by civilian counterparts (ANSI/ASQ Z1.4 and ANSI/ASQ Z1.9), administered (and changed to varying degrees) by the American National Standards Institute and the American Society for Quality.

Extensions Sometimes, limits tighter than the product specifications are used in the so-called "narrow limit gauging." This makes the assessment more sensitive and provides early warning of impending and perhaps slowly emerging problems (see Ott, Schilling, Neubauer 2005).

In some applications, instead of sampling each lot, one might sample occasional lots, and perhaps, step up sampling if unfavorable results are obtained. See Doganaksoy and Hahn (1994) for an illustration of such "audit sampling," involving multiple product characteristics.

Further Reading Schilling (1982) provides an overview of acceptance sampling (see also Schilling 1998 and Ott, Schilling, and Neubauer 2005).

8.5 FINAL PRODUCT ASSESSMENT

8.5.1 100% Inspection

Final product inspection to identify and remove unacceptable product was, at one time, a key element of what was then known as statistical quality control (Section 2.2.3). "Inspecting quality into the product" has, in recent years, been replaced by identifying and removing the causes for defects upfront, preferably during product design.

However, there still remain situations for which 100% inspection of final product is used and may be appropriate, such as:

5. Deming (1986, page 133) states categorically that "... the day of such plans is finished."
6. In some cases, instead of accepting or rejecting a lot, it might be subjected to 100% inspection.

- For critical components, for example, the parts of an aircraft engine.
- When it is easy and inexpensive to inspect each manufactured unit (generally by an automated process) to remove unacceptable units without harming good ones. In the semiconductor industry, a new design is sometimes introduced even though only a fraction of the initially manufactured product may be acceptable. The good units are then sold, while efforts continue to improve product yield.[7]
- When the product is to be sorted into various grades, based on measured final characteristics. The semiconductor industry again provides an example.
- Defective product can be readily repaired. This was the case for dimensional conformance in the generator stator bar example of Chapter 7.

In addition to removing unacceptable units, 100% inspection should also serve as a source of information for improving the product. This requires consistent and systematic recording of the reasons for rejection, at as deep a level as possible (perhaps on a sampling basis), to help find the root causes.

8.5.2 Product Acceptance Sampling

In some situations, 100% inspection, though warranted by the criticality of the application, is very expensive. For other characteristics, it might weaken or destroy good units. For example, a child safety gate is required to withstand a push/pull test with 31.5 lbs. of force by a robot arm for 10,000 cycles. Units that pass this test may be severely weakened.

Instead of 100% inspection, one might in such cases employ acceptance sampling. Product is then sampled from each production lot (or a sample of such lots) and inspected. The procedure and some of the considerations are similar to those discussed in Section 8.4.2.

8.5.3 Final Product Testing in Context

Acceptance sampling, and even 100% inspection, does not guarantee the desired level of conformance due to errors associated with inspection and measurement systems. And the comfort that such inspection provides as a last line of defense should certainly not result in reducing the pressure to improve the product proactively. Or quoting Ott, Schilling, and Neubauer (2005), these tasks should be much "more than a police function" and instead provide information for future product improvement.

8.5.4 Past Product Retention

In some applications, it is useful to retain carefully selected samples of past product to compare with current product. For example, color chips from a new batch of

7. One hundred percent inspection might also be used at various process steps to remove potentially bad product and avoid additional production costs (see Section 13.3).

material may be compared with those from early production. Such comparisons can help address customer claims that the product has changed over time. They are especially useful when new diagnostic or measurement methods have evolved over time. Past material might, for example, then provide a standard to assess the drift in measurement methods.

Retention of past material is not always appropriate. The properties of the material might, for example, change over time (a phenomenon which may in itself warrant assessment). Also, manufacturers are concerned with the costs of storage and the diversion of product.

8.6 PRODUCTIVITY IMPROVEMENT

Productivity improvement calls for operating a production process more efficiently and cost-effectively. Some typical productivity metrics are throughput (production output over time), equipment utilization rate, machine downtime, and machine (and worker) idle time.

It is important to ensure that illusionary improved productivity is not gained at the expense of quality and reliability. Improving quality and reliability often also improves productivity. For example:

- Higher upfront quality can lead to earlier product shipment.
- The "right" process settings at new batch start-up can improve equipment utilization of a batch manufacturing operation.
- Increased confidence in product quality can reduce the required testing.

8.6.1 Productivity Characterization and Evaluation

You can't address it if you can't quantify it! That statement goes for productivity improvement, just as for quality improvement.

Typically, we need to quantify productivity under existing (and possibly alternative) operating conditions, analyze the data, and present the results in an easy-to-comprehend manner. A first step in reducing unutilized machine time in a manufacturing operation might, for example, be to track such time and its potential root causes and present this information graphically. Pareto charts are often used to highlight the major causes of production downtime (e.g., equipment breakdown, machine setup, waiting time for test results, etc.). The results are then used to prioritize efforts. A planned investigation might be called for, especially if the impact of a new approach is to be evaluated.

8.6.2 Lean Manufacturing

Lean manufacturing aims to identify process activities that do not directly add value to the final product (e.g., storage, transportation, rework, duplicate data entry, and testing). These are sometimes referred to as constituting a "hidden factory." The goal is to keep such work at a minimum—or, when possible, to eliminate it altogether.

Value stream maps are used to differentiate value added from non-value-added process steps (see Manos 2006). Process improvement ideas may be generated through Kaizen, a team-brainstorming activity to help remove waste (see Manos 2007).

Lean manufacturing (sometimes referred to simply as "Lean") gained momentum through its integration with Six Sigma, resulting in "Lean Six Sigma." This calls for applying Six Sigma concepts and tools, including statistical methods, to implement lean manufacturing. Reducing rework, for example, often requires a statistical approach to help identify and quantify its cause(s). See Breyfogle (2003), George, Rowlands, and Kastle (2003), George, Rowlands, Price, and Maxey (2004), and Snee and Hoerl (2007) for further information on the interplay of lean, Six Sigma, and statistics.

8.6.3 Discrete Event Simulation

Basic Concept Discrete event simulation is often used to model systems that involve a chronological series of events (e.g., parts processing in a job-shop operation or serving customers in a bank) to assess performance and to compare alternative modes of operation or proposed improvements over time. It is especially useful in studying processes that involve queues or waiting times.

The approach typically calls for building a computer model that provides a mathematical description of the process, including its random elements. The model is then exercised many times using streams of random values that closely resemble the statistical distributions of the underlying random elements[8] and the performance analyzed. The process is repeated for different scenarios and strategies and the results compared to identify the "best" strategy.

Job-Shop Example A job-shop builds a product to order—for example, a metal cutting operation makes parts from individual specifications. Upon receipt, an order enters a waiting line and is processed using a "first come, first served" priority rule. Different types of machines are required for different jobs at different workstations—some machines can be used for some orders, but not for all. The waiting time between receipt and completion of an order depends on such factors as the rate of arrival and backlog of orders, the availability of machines and operators and the time to process an order—each of which has random elements.

Management seeks an optimum policy for running the shop—one that meets or exceeds customer shipment and other requirements at a minimum cost. Among the many questions this poses is determining the number of machines of each type to purchase. Increasing this number results in processing orders faster, but also increases investment costs.

The Rationale for Simulation Making the evaluation directly under actual operating conditions by trial and error (e.g., steadily adding machines) is slow and

8. As in our earlier discussion of simulation, we will be principally concerned with randomly occurring events (Monte Carlo simulation). Discrete event simulation is also used to model deterministic systems.

impractical; it may result in disruption of operations, initial late deliveries, and an overinvestment in machines.

A computer simulation of the process is conducted instead. This mimics the actual operation of the job shop for a specified period of time, such as a month. It allows the performance of the process, including its ability to provide product on schedule and the total cost of production, to be compared under different machine purchase strategies.

Initial Steps We first develop and program a mathematical description of the process. In this case, the model describes the incoming flow of jobs and their processing by different machines over time. The time required to process the job is a key output. Such modeling requires an in-depth understanding of the operation.

The required inputs to the model are identified and estimates of their statistical distributions obtained. These include the distributions of incoming orders and the processing times using different machines. Machine breakdowns, lead times, and other random variables are also computer-generated from assumed models.

The adequacy of the simulation model is assessed before proceeding further. For an existing process, this might involve comparing the simulated results with the actual results at current operating conditions.

Exercising the Simulation The simulation process is summarized in Figure 8.4. It involves "running" the job-shop on the computer for the specified time period many times for each of the machine purchase strategies to be evaluated and comparing the results.

More specifically, for a particular strategy, job "arrival times" are randomly generated. The jobs are "assigned" to and "processed" by available machines for randomly generated times. The results are used to determine performance characteristics for a particular simulation over the specified time period. The

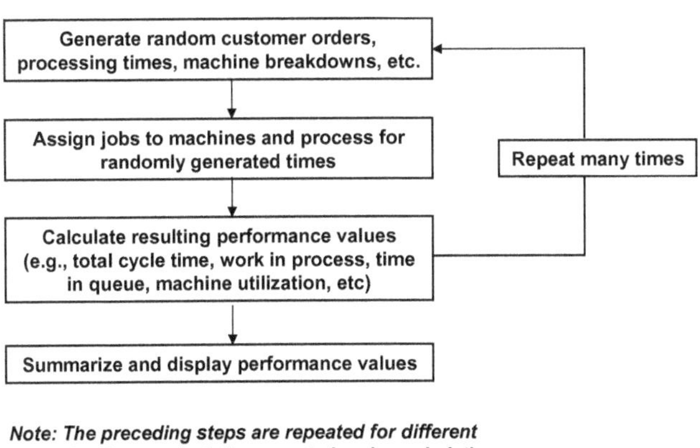

Note: The preceding steps are repeated for different scenarios (e.g., optimistic, expected and pessimistic distribution of job arrival distributions) and strategies (e.g., number of machines).

Figure 8.4 Flow chart of job-shop simulation.

preceding is repeated many times, using newly generated random inputs, to get a statistical distribution for each of the performance characteristics, for example, % jobs completed on time, % utilization of each machine type, and added costs (due, for example, to operators working overtime), for a particular strategy. The results are summarized and displayed in, say, a histogram. The process is then repeated for each of the scenarios and strategies to be evaluated and compared for different strategies.

Further Comment Once a simulation model has been developed, it can be re-exercised as conditions change or even routinely. In the job-shop example, an anticipated increase in business volume might call for rethinking the required number of machines. If the machines are rented, rather than purchased, the model might be exercised weekly, using updated input information.

8.7 MAJOR TAKEAWAYS

- Engineering process control involves ongoing online adjustments of one or more process variables so as to control the level and reduce the variability of the key performance variables. It has been applied successfully to many operations, especially in the chemical and process industries, that involve autocorelated outputs.
- EPC, unlike SPC, typically does not result in the identification and removal of root causes of problems. It is, therefore, advantageous to apply EPC and SPC *together*. EPC is used to correct off-target performance and reduce predictable short-term variability using feedback and/or feedforward techniques. The results and adjustments are then closely monitored using SPC to detect and then remove special causes of variation.
- Ensuring that the parts and materials received from suppliers continue to meet specifications is an important part of quality assurance. This traditionally involved acceptance sampling—an approach that is now deemphasized in favor of more proactive methods.
- Final product inspection is similarly being deemphasized in favor of upfront improvement. There are, however, still some situations that call for 100% inspection (e.g., for critical parts) or acceptance sampling of final product. The results should be used for future product improvement.
- For some applications, it is useful to retain carefully selected samples of past product to compare with future product.
- Productivity improvement calls for operating a production process more efficiently and cost-effectively. Approaches for productivity improvement include productivity characterization and evaluation, lean manufacturing (identifying and removing process activities that do not directly add value), and discrete event simulation.

DISCUSSION QUESTIONS

General

1. We stated in Section 8.2.5 that "In his famous funnel experiment, Deming (1986) demonstrated that adjusting a process that is in statistical control leads to increasing, rather than reducing, variability." Describe Deming's funnel experiment and discuss its relevance.

2. We also state that "In a follow-up to the funnel experiment, MacGregor (1990) showed how process adjustments and EPC can, for *some* processes, *reduce* variability." Describe MacGregor's experiment and why it provides results different from Deming's funnel experiment.

3. Consider Deming's (1986) claim (footnote in Section 8.4.2) that the day of acceptance sampling plans is "finished." What was Deming's justification for this claim and what is he assuming? Do you think he was correct?

4. We state in Section 8.5.4 that "in some applications, it is useful to retain carefully selected samples of past product to compare with current product." Suggest some situations for which this might be the case.

5. Suggest one or more processes (e.g., doctor's office) which might be made "leaner" and indicate how statistical concepts and methods might be applied in so doing.

6. Suggest a process that warrants evaluation by discrete event simulation. Provide details of the random elements and how you would conduct such a simulation.

Technical

1. In the polymer resin example (Section 8.3.3), we state that "the relationship between catalyst added and the resulting IV was fitted statistically, taking the autocorrelations of the observed IV into consideration." Elaborate on how this might be done.

2. Elaborate in the polymer resin example on the details of constructing superimposed CUSUM control charts to track changes in the IV measurements and the catalyst adjustments. Why would one want to track both?

3. We state in Section 8.4.2 that tabulations of acceptance sampling plans "are usually established to meet specified initial criteria, such as average outgoing quality (AOQ), acceptable quality level (AQL), and lot tolerance percent defective (LTPD)." Describe these criteria and indicate the type of situations for which each is most applicable.

4. Provide details on narrow limit gauging (Section 8.4.2) and why this approach "provides early warning of impending and perhaps slowly emerging problems."

5. How might you go about developing the required input distributions for the job-shop discrete event simulation example in Section 8.6.3?

CHAPTER 9

PRODUCT FIELD SUPPORT

9.1 ABOUT THIS CHAPTER

The proof of the pudding is in the eating.
—Popular saying

"Out of sight-out of mind" is not an appropriate strategy in dealing with product shipped to customers. Even though a product has been successfully built and sold, our interest and concern continues. We need to work towards the complete satisfaction and delight of customers – and, thereby, help ensure that past customers will become future customers and tell others about our "wonderful" product and service. In addition, we need to use our experience on existing products to build even better products in the future. For some products, moreover, servicing provides added business opportunities.

We begin this chapter by discussing the continuing need for and goals of tracking the reliability of product in the field and describe its use for nonrepairable and repairable products, respectively. We then consider the use of field data to assess product safety, track customer satisfaction, conduct competitive evaluations and make advertising claims. We conclude by describing some exciting opportunities dealing with proactive product servicing.

9.2 RELIABILITY DATA TRACKING

9.2.1 The Need

In a perfect world, marketing has fully identified customer needs, engineering has developed a flawless high-reliability product, and manufacturing has succeeded in building such a product consistently. Thus, the traditional "design, build, ship, fix" approach has been replaced by a proactive "do it right the first time" mindset.

Unfortunately, we do not yet live in a perfect world, and despite our best efforts, some problems still get overlooked. Others may arise as a consequence of unforeseen changes in the use environment. There may, in addition, be strong business pressures – that cannot be completely rebuffed – to release the product prematurely to meet customer needs before reliability has been fully validated. This is especially the case for components or assemblies that are part of a larger system for

The Role of Statistics in Business and Industry, by Gerald J. Hahn & Necip Doganaksoy

which delay in delivery can hold up the building, testing, and release of the entire system.

Companies, therefore, need to continue to scrutinize their products, collect appropriate data about performance, reliability and customer satisfaction over time, and carefully evaluate such data. The field provides the ultimate testing ground and yields the most realistic, though not the timeliest, information about product performance. The worst situation is when a customer is the first to bring a problem to our attention.

A reliability problem in the field is one that results in the product failing to perform its intended function over time, as experienced (or perceived) by the customer. Failures may appear to be sudden to the user, but they may be the consequence of degradation over time that may or may not be evident. The so-called dead on arrival (DOA) units that do not function properly on their first use are especially undesirable.

9.2.2 Goals

Some major goals of tracking the reliability of products in the field are to:

- Identify reliability problems as early as possible and speedily feed back this information to design engineering and manufacturing to make changes to avoid similar problems in the future. The resulting improvements may be immediate,[1] such as adding a protective coating to ameliorate a wear problem, or they may require fundamental changes to be incorporated into the next design.
- Proactively mitigate the harmful impact of potential failures on units already built. This may call for measures ranging from providing consumer warnings about improper product use to selective recall.
- Provide comprehensive product reliability assessments to management and others to use, for example, to prioritize reliability improvement efforts, plan future products, and develop advertising claims.
- Evaluate and update earlier reliability predictions and estimated warranty costs.

9.3 RELIABILITY DATA TRACKING FOR NONREPAIRABLE PRODUCTS

9.3.1 Laptop Computer Hard Drive Example

The Product We illustrate data tracking for nonrepairable products by considering further the new laptop computer subjected to beta site testing described in Section 5.8.3, and subsequently released into the field, resulting in yearly sales of

1. Such design changes need to be thoroughly validated to ensure their effectiveness and that they do not create other problems.

over a million units over approximately a 4-year period. The basic product design remained the same over this time, even though small changes and improvements were made (and validated) periodically, based upon field results and customer feedback, and to reduce costs.

Our evaluations will focus on the computer hard drive and is adapted from Doganaksoy, Hahn, and Meeker (2006b). Similar analyses were conducted on other subsystems.

Description of Data The product had a 1-year warranty. Consumers also had the option of buying an additional 3-year extended warranty and about 50% did so. Those availing themselves of this option are not a random sample of all purchasers; they might, for example, be heavier users. It was felt, nevertheless, that reliability estimates from units on warranty were relevant – even though, most likely, somewhat pessimistic.

A field failure tracking and reliability assessment system was developed prior to product release to obtain lifetime data on all units under warranty.

Management Overview The product has been in the field for 4 years. Figure 9.1 shows the number of hard drive field failures per 1,000 units for each calendar quarter since product introduction. The plotted points are the total number of failures that occurred during each quarter (times 1,000), divided by the total number of units under warranty that quarter.

This frequently used type of report provides a comprehensive summary for management, and gives useful information for estimating replacement part requirements. Casual examination of Figure 9.1 suggests field reliability is improving over time.

Segmented Data Analysis Figure 9.1 has some serious drawbacks as a tool for reliability analysis – it ignores the mix of ages of units in service. After 1 year, all units have been in operation for 1 year or less. After 4 years, there is a mix of new units with short operational times and older units that have been in service up to 4 years. Figure 9.1, thus, does not separate two effects of high interest: the change in quarterly failures per 1,000 units for product built at different times versus the change over the lifetime of a unit.

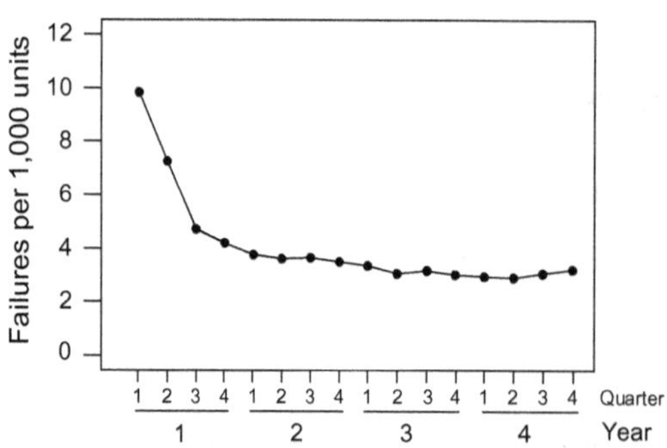

Figure 9.1 Quarterly failures per 1,000 units for the computer hard drive.

Thus, a segmented data analysis that breaks down the results into these two constituent parts is needed. This requires advance planning to get the data required to perform these (and likely subsequent) analyses (see Section 9.5). Fortunately, a system to provide the needed information had been developed and was in place. This included unique identification of each computer and its key parts (including the hard drive); consistent reporting of purchase date and of all failures (including date and type) and replaced parts; and information on such factors as customer type and geographical region.[2]

Improved Data Displays Table 9.1 provides a comprehensive summary of the hard drive failures per 1,000 units during each quarter of life for units built in each production quarter. For example, units built in the third quarter of production year 1 experienced 3.04 failures per 1,000 units during their sixth quarter of life.

The tabulation is triangular in form because the age of a unit has to be equal to or less than the number of quarters since manufacture.

These results are plotted in:

- Figure 9.2 to show the failures per 1,000 units versus product age for each of the 16 production quarters.
- Figure 9.3 to show the failures per 1,000 units versus production quarter for each of the 16 product age groups.

Discussion of Results Examination of Figure 9.2 indicates that:

- For units built during the first quarter of production, the number of failures per 1,000 units was especially high during the first quarter of product use. These numbers dropped sharply during the second quarter and still further in the third quarter. Units built during the second quarter showed a similar but less severe pattern. These results reflect a serious manufacturing defect – and its elimination by redesign on units built after the fourth month of production.
- For units built in subsequent production quarters, the failures per 1,000 units during the first quarter of use still tended to exceed those in the second quarter, and continued to decrease into the third quarter. This reflected early life failures due to a combination of manufacturing defects.
- There was a slight but consistent upward spike in the failures per 1,000 units during the fourth quarter of life for units built in almost all production periods. The slight downward trend resumed, however, in the fifth quarter. The spike was (after some further data analysis) attributed to closer scrutiny near the end of the 1-year standard warranty period by customers who did not have an extended warranty.
- After about the fifth quarter of life, the failures per 1,000 units remained relatively constant until an age of about 3 years (see, especially, first six plots).
- There seemed to be an increase in the failures per 1,000 units after about 3 years of life, suggesting product wearout. At this point, moreover, some

2. Data on use frequency and intensity, though highly desirable, could not be obtained.

TABLE 9.1 Quarterly Hard Drive Failures per 1,000 Units by Age and Production Quarter

Age (in Quarters)	Year and Quarter of Manufacture															
	Year 1 Quarters				Year 2 Quarters				Year 3 Quarters				Year 4 Quarters			
	1	2	3	4	1	2	3	4	1	2	3	4	1	2	3	4
1	9.85	7.50	5.46	4.75	4.29	4.29	5.07	4.42	4.37	4.18	4.67	4.60	3.93	3.81	4.27	4.36
2	7.09	5.06	4.21	3.31	3.53	3.45	3.34	3.58	2.86	3.48	2.97	2.82	2.89	2.93	2.76	
3	3.63	3.37	3.53	3.04	3.30	2.64	3.46	3.27	2.70	2.44	2.37	2.41	2.40	2.85		
4	4.45	4.54	4.02	3.68	4.02	3.55	2.98	2.88	2.75	2.92	3.40	2.92	2.62			
5	3.09	3.63	3.33	3.34	2.56	2.57	2.10	3.06	2.23	2.82	2.13	2.22				
6	3.16	3.44	3.04	2.55	2.72	2.29	2.27	2.52	2.21	2.51	1.83					
7	3.35	3.13	3.16	2.86	3.28	2.72	2.18	2.17	2.27	2.56						
8	4.00	3.53	2.81	2.43	2.43	3.21	2.12	2.99	2.65							
9	3.48	2.72	3.46	2.96	2.23	2.32	2.43	3.09								
10	3.57	3.53	3.19	2.97	2.22	3.20	2.88									
11	4.02	3.36	2.86	2.84	3.22	3.19										
12	3.49	3.95	3.72	2.92	3.00											
13	4.16	3.23	3.68	3.57												
14	4.48	3.40	4.23													
15	4.44	4.62														
16	4.86															

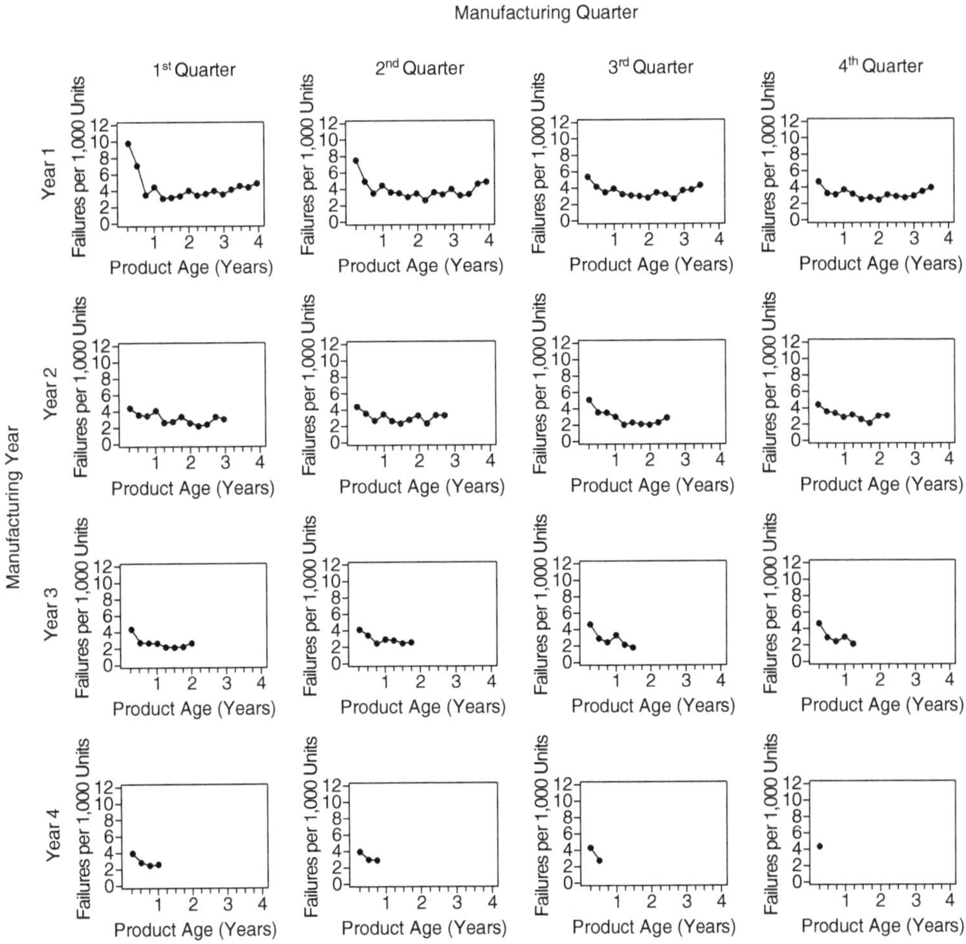

Figure 9.2 Hard drive failures per 1,000 units versus product age for each of the 16 production quarters.

customers might be replacing their computers, especially upon failure – thus resulting in under-estimates (increasing with time) of the true number of failures per 1,000 units.

Figure 9.3 shows similar results as Figure 9.2. In addition, these plots indicate that the failures per 1,000 units for each quarter of life tended to decline modestly over the production periods. For example, during the fourth quarter of life, the failures per 1,000 units for units built during the 13th production quarter was about half of that for those built during the first production quarter. This reflects the impact of various small design and manufacturing improvements.

The preceding information allowed the responsible engineers to gain a better understanding of failures and act accordingly to make both immediate and long-term improvements. This required further statistical analyses by product subpopulation and failure mode.

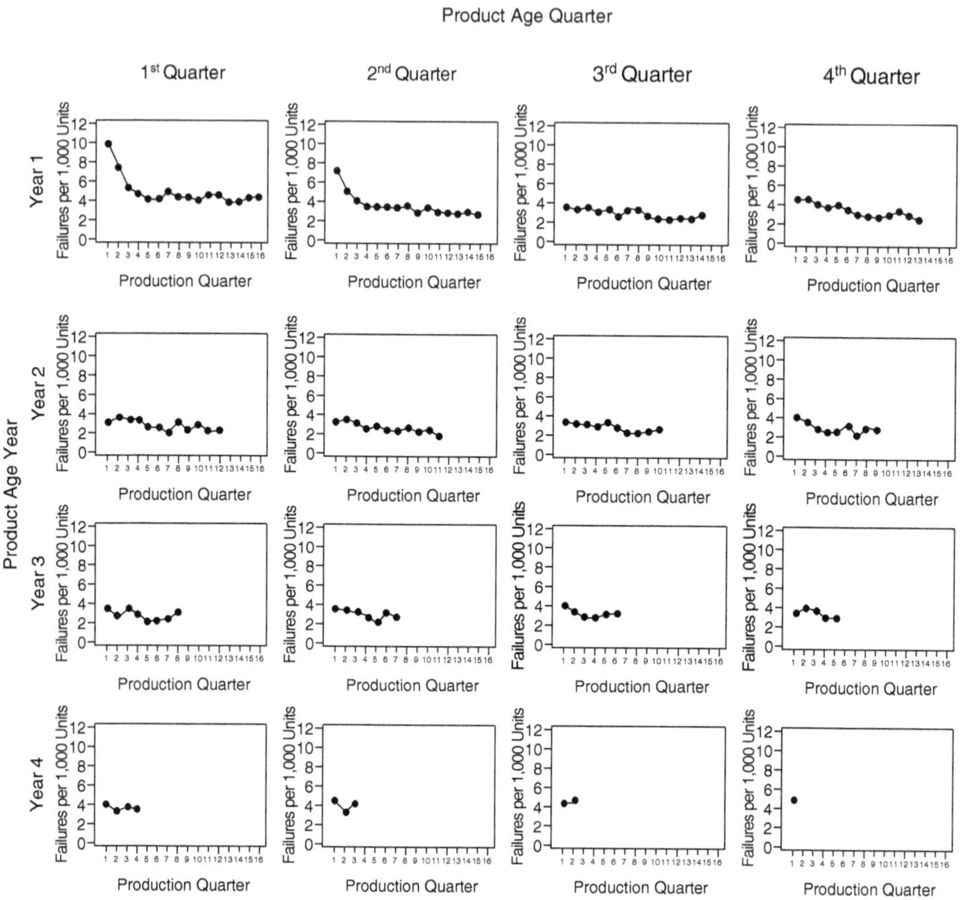

Figure 9.3 Hard drive failures per 1,000 units versus production quarter for each of the 16 product age groups.

A Dynamic Reporting and Corrective Action System The major benefit of the reliability tracking system is not the retrospective evaluation after 4 years, presented here, but the dynamic information it yields much earlier. The system provides ready access to current reliability estimates for:

- The entire system or a specified subsystem or component
- All units or a specified subset of units
- All failure modes or specified failure modes

For example, a design engineer might request 3-month, 1-year, 3-year, and 5-year estimates[3] of failures per 1,000 units for the hard drive for all units sold in a particular region of the United States built during each of the first 3 years of production, and for these years combined.

3. The 5-year estimates require extrapolating the data, using an assumed model for time to failure, such as the Weibull or the lognormal distribution.

Also, the system automatically provides early warning if the estimated failures per 1,000 units for any subsystem, component or failure mode exceeds a stated threshold and/or is changing significantly from that attained earlier. The system, thus, gives information similar to that of a statistical control chart applied to field failures.

In addition, the system provides periodic summary reports (monthly, unless requested otherwise) to management and responsible engineers. The management reports tend to deal with the entire product, serving as a reliability report card. Reports to engineers are more detailed, focusing on sub-systems, components and individual failure modes.

9.3.2 Subpopulation Segmentation

The preceding example involved segmenting data by production quarter. Segmentation was introduced in Section 2.2.5 as one of the basic quality tools. We now discuss this concept more generally.

The Challenge All products are not created equal – some are more likely to fail than others. Differences in performance and reliability may, for example, be due to variability in raw materials and components, and in manufacturing processing conditions.

A failure may, moreover, be due to a defect that occurs in only *some* production lots. For example, cracking of the plastic casing of a laptop computer was traced back to one of several assembly plants. It was then found that the wrong type screw was used in assembly over an approximately 1 month period at this plant. This resulted in the grease on some screws leading to chemical degradation of the plastic casing. Isolating the problem facilitated root cause identification and steps to ensure its avoidance in future product. More immediately, it led to identifying previously built vulnerable computers for corrective action.

Similarly, different units experience different use environments. A problem may be accentuated by and limited to occur only under extreme ambient conditions, such as severe heat or cold. Similarly, the performance of a dishwasher may depend upon the characteristics of the local water supply. In such cases, one would want to focus immediate corrective action on product in the most vulnerable geographical regions. The long-term answer, however, is to eliminate the problem in future units, perhaps by designing a sufficiently robust product whose performance is insensitive to the use environment.

Nature of Analysis Segmentation is a "divide and conquer" strategy that breaks down the product population into meaningful subpopulations, doing separate analyses on each, and acting on the resulting information.

The subpopulations are determined by what might, from physical and empirical considerations, be likely differentiating factors. For the laptop computer hard drive, segmentation was based somewhat arbitrarily upon production quarter. This choice should be trumped by manufacturing knowledge. The times at which changes are introduced on line might, for example, have been a better criterion. In other applications, segmentation might be based upon such factors as parts supplier, the geographical region in which the product is being used, or customer type – or a combination of these.

In the short run, segmentation may result in the speedy and accurate isolation of field problems to well identified subsegment(s) of the total product population, in identifying the most susceptible units, and in taking corrective action. Segmentation may, for example, help determine whether a recall is needed, and if so, what segment of the product population needs to be recalled. By isolating a problem to, hopefully, a relatively small part of the population, we may be able to manage an otherwise extremely costly problem without inconveniencing customers who are not impacted by the problem.

9.3.3 Further Segmentation Example: Aircraft Engine Bleed System

Initial Analysis The following example is from Meeker and Escobar (1998), using data from Abernethy, Breneman, Medlin, and Reinman (1983). It deals with a system that bleeds off air pressure from an aircraft engine to operate, for example, a compressor.

Data were available on bleed systems from 2,256 engines from aircraft operating from various bases, involving 19 failures. Figure 9.4a shows a Weibull distribution probability plot of the lifetime data. This plot indicates that a simple Weibull distribution does not provide an adequate representation for lifetimes since the plotted points do not scatter around a single line. Instead, the line seems to change slope at around 600 hours. This suggests more than one failure mode and/or product population.

Segmented Data Analysis Examination of the data revealed that 10 of the 19 failures occurred at Base D, one of the bases at which the aircraft were stationed.

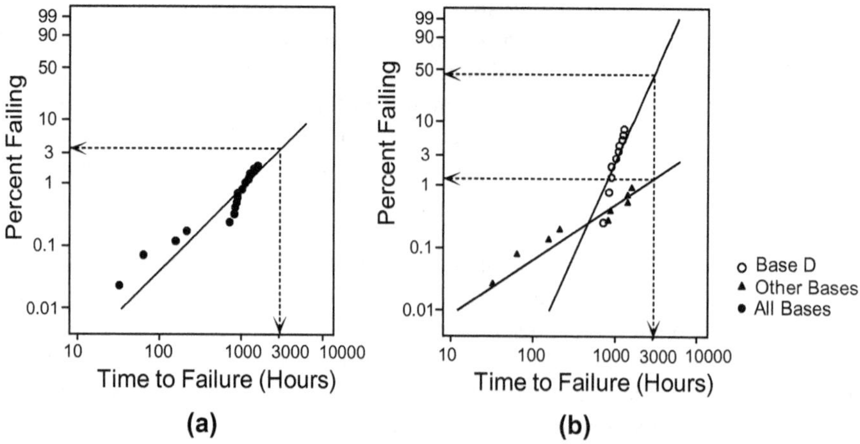

Figure 9.4 Weibull probability plot for lifetimes for (a) all bleed systems and (b) separate Weibull probability plots for lifetimes for bleed systems from base D and for those from all other bases.

Separate Weibull probability plots for lifetimes for the systems at Base D and for those at all other bases were therefore prepared and are shown in Figure 9.4b.

The data in each of these two plots now do scatter around straight lines, suggesting that simple Weibull distributions provide adequate representations if one considers Base D and the other bases separately. Moreover, the probability of failure by 3,000 hours is estimated to be 0.467 for the systems at Base D from the plot, as compared to 0.013 for the systems at the other bases.[4]

Further investigation revealed that the early failure problem at Base D was caused by salty air (Base D was near the ocean), and a change in maintenance procedures was implemented there. This resulted in removing the previous major failure mode at that base.

The segmentation methods used in addressing this problems bear some similarity to the analysis of multiple failure modes discussed in Section 5.8.2 (see Sidebar 9.1).

SIDEBAR 9.1: SEGMENTATION AND MULTIPLE FAILURE MODE ANALYSES

The bleed system example and the in-house generator insulation test discussed in Section 5.8.2 both illustrated situations for which the life data in totality could not be described adequately by a single, simple distribution. In the case of the bleed system, the data were segmented into sub-populations for separate analyses.

In contrast, in studying individual failure modes in the motor generation insulation example, all the data were considered, but observations from failure modes other than the one under consideration were taken as censored.

In both situations, the results of the individual analyses can subsequently be combined to obtain an omnibus analysis for the entire population.[5]

9.3.4 Some Further Comments

Segmentation analyses typically do not per se provide "cause and effect" conclusions. The difference in failure probabilities between Base D and the other bases could have been due to any or a combination of many different factors. The determination that the underlying cause was salty air involved an engineering assessment of failed parts.

We have discussed segmentation in the context of reliability data tracking for nonrepairable products. Segmentation of data, however, is useful in many other situations. For example, a chemical cure process showed inconsistent results. To gain improved understanding, the data were segmented and plotted in various ways; one of these was by shift. The resulting plot showed that two of the shifts were

4. The two plots in Figure 9.4b have quite different slopes, indicating different Weibull distribution shape parameters. Base D exhibits an increasing hazard function ($\hat{\beta} \approx 3$) over time, suggesting wearout (see Sections A2 and A3.2 of Appendix to Chapter 5). The hazard function for the other bases is close to constant ($\hat{\beta} \approx 0.9$).

5. For segmentation, this requires knowledge of the proportion of units in the population belonging to each subpopulation.

providing satisfactory product, but that the night shift was not. In a search for the underlying cause of this difference, a late night visit revealed that the third shift operators frequently turned off the plant air conditioning. This resulted in increased humidity, which in turn impacted subsequent product performance unfavorably. After correction of the problem, it was decided to control chart performance segmented by shift.

9.4 RELIABILITY DATA TRACKING FOR REPAIRABLE PRODUCTS

9.4.1 Introduction

Our discussion in this and much of the Chapter 5 focused on nonrepairable products or components; that is, the life of the unit or part under consideration ends when the unit fails. In contrast, many systems are repaired when a failure occurs, often by replacing the failed component or assembly.

9.4.2 Repair Recurrence Data

Repairable products, unlike nonrepairable ones, lead to multiple event data, such as repair times, on the same unit, resulting in what is known as "recurrence data[6]."

9.4.3 Example: Locomotive Braking Grids

Description Doganaksoy and Nelson (1998) provide the following example. A manufacturer of locomotives received orders from a railroad for 15 and 18 locomotives respectively. At the time of analysis, the locomotives from the first order had experienced about 700 days of service in the field, versus about 450 days for those from the second order.

Each locomotive has six braking grids. These are used in place of the power input to the electric motor to slow the locomotive. The motor then begins to act as a generator, powered by the inertia of the locomotive. The electrical resistance of the braking grid provides a load to the motor. This electrical load translates into mechanical resistance for the motor and, in turn, the wheels connected to it, causing the locomotive to decelerate.

In preparation for filling a third order, it was desired to assess and compare the reliability of the braking grids from the two orders, using the recorded data of the times (in days of operation) when the braking grids were repaired on each of the 33 locomotives.

Event Plot The repair (or failure) times for each locomotive are displayed in an "event plot" in Figure 9.5. Each line in this plot shows the history of a locomotive; the *X*s display the days in service when a braking grid was repaired. The length of a

6. Recurrence data are also encountered in other applications, such as the study of the occurrence of diseases or of customer repeat purchases.

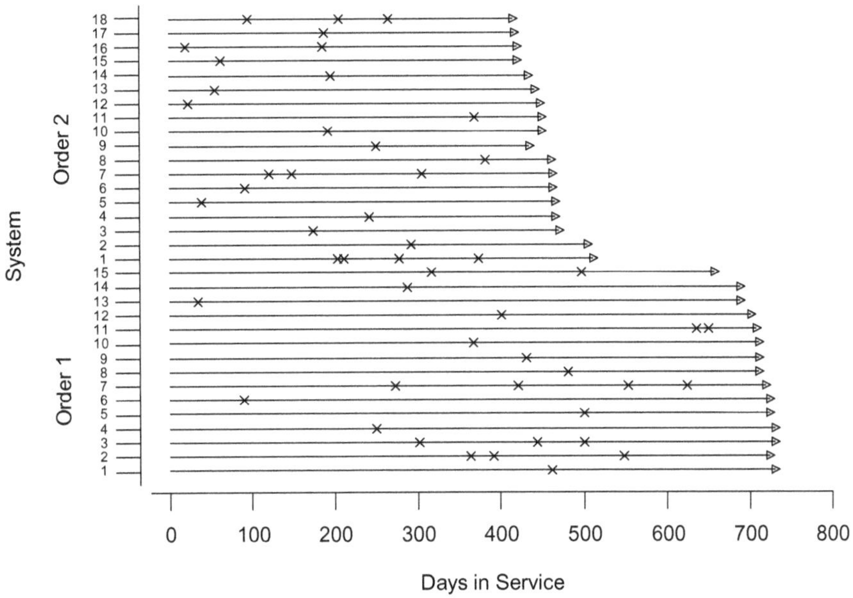

Figure 9.5 Event plot of braking grid repair times.

line shows the locomotive's time in service. For example, the 15th locomotive from Order 1 has been in service for 657 days. Its braking grids were repaired after 317 days and again after 498 days in service.

Examination of this plot indicates that in the first 400 days of service the 18 locomotives from Order 2 experienced a higher rate of braking grid repairs than those from Order 1 (26 repairs on 18 units versus 11 repairs on 15 units).

The Mean Cumulative Function (MCF) Plot The MCF of a product at age t is defined as the mean number of failures per unit up to that time. Many questions about the reliability of a repairable product, using field repair recurrence data, can be answered by estimating its MCF.

Since in the example, the running times for the locomotives within each order are almost the same, a simple estimator of the MCF at age t is the sum of all repairs by age t divided by the total number of units in service.[7] For example, a total of six repairs were needed during the first 300 days for the 15 locomotives in order 1. Accordingly, an estimate of the MCF at 300 days is 6/15 = 0.40 repairs per locomotive.

Figure 9.6 shows plots of the estimated MCFs of the braking grids against days in service for each of the two locomotive orders. These plots show a clear differentiation in the estimated MCFs between the two orders, with Order 2 requiring

7. When running times differ, due to, say, staggered entry into service or different use rates, a more complicated method to estimate the MCF is required (see Meeker and Escobar 1998, Section 16.2.2, and Nelson 2003, Chapter 3). Such more general methods are used in software packages (see Doganaksoy, Hahn, and Meeker 2006b).

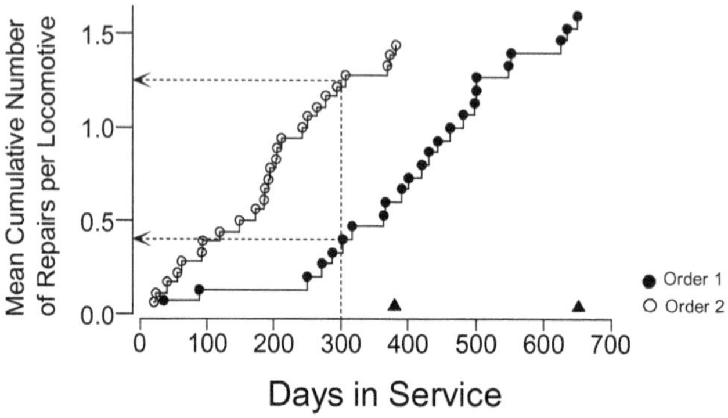

Figure 9.6 MCF estimates for braking grid repair times for two orders (the solid triangles designate the time of last repair for Orders 1 and 2, respectively).

more repairs than Order 1. For example, after approximately 300 hours the MCF for Order 2 was about 1.25, as compared to about 0.40 for Order 1. A formal analysis (Doganaksoy and Nelson 1998), moreover, showed a statistically significant difference between the MCFs of the two orders for most ages.

This difference was attributed to differences in manufacturing since the usage environments were similar for both orders. The supplier changed the type of die that cut the braking grids between the times when the locomotives in the two orders were built. This was found by Engineering to be the cause for the differences; the type of die used for the first order was to be used in future production.

The Value of Added Information Further analysis would have been possible had more detailed information been gathered. For instance, individual grid brake identifications and the positions of the repaired braking grids in the locomotive may have helped determine whether any positions were especially vulnerable. Such data also would have permitted analyses of the times to repair for the *individual* braking grids (as a nonrepairable product). More extensive records are to be maintained in the future.

9.4.4 Further Reading

Nelson (2003) and, at a more advanced level, Ascher and Feingold (1984), Meeker and Escobar (1998, Chapter 16), and Rigdon and Basu (2000) provide detailed discussions of reliability data analysis for repairable systems.

9.5 PLANNING THE FAILURE REPORTING SYSTEM

9.5.1 The Need

Evaluations of reliability from field data can be only as good as the data upon which these are based. The more we know about a unit's history, use and performance over time, the better and speedier can be the corrective action to avoid future failures. Segmentation analyses, for example, require clear identification of the subpopulation(s) to which each unit belongs.

Unfortunately, the needed data are rarely readily there just waiting for us to scoop up. Sometimes, although we are assured that the desired information is available, we find out that what has been recorded is incomplete and/or inconsistent, and unlikely to provide the foundation for a sound analysis. Often, this is because the data have been obtained for purposes other than in-depth reliability assessment; for example, for tracking warranty costs or for legal protection. There might also be other deficiencies. In developing a reliability tracking system for a refrigerator, we found that, although field failure data were routinely obtained, such data were purged after 2 years to "unclutter" the system.

It is, therefore, essential that proper plans are made *in advance* – and, perhaps, validated during beta site testing – to establish a reporting system that, to the degree possible, provides the data needed to expedite the detailed identification of all (or possibly a random sample of) field failures. Establishing such a system can be likened to taking out property or health insurance. We hope that we will not need it (and work diligently to make this so) – but it surely comes in useful if we do. Moreover, like most insurances, we incur the costs in advance, not knowing the specific benefits that might be forthcoming and when. Thus, one of our major challenges may be to convince management of the need for investing in such a system and showing its potential payoffs.

9.5.2 Data Acquisition Specifics

In broad terms, the data to be recorded is all the information that might be potentially useful in helping explain field problems and in building better products in the future. The specifics vary from one application to the next; the laptop computer hard drive, described earlier, provides one example. If possible, the data should include:

- Unique identification of each unit and its key parts, together with information that readily permits tracking manufacturing history, which in itself needs to be well maintained.
- Applicable past performance data, such as end-of-line readings.
- Information on when and where the product was used.
- The use environment, such as ambient conditions and usage rate and of any other variables that might be related to the likelihood of product failure.
- Maintenance and service information.
- Product performance (and its degradation) over time.
- Abnormalities or failure symptoms noted by the user.
- Failure times on failed units and running times on unfailed ones.
- Detailed information about failures and corrective action, including the failure mode, cause and mechanism (see Section 5.7.2). This might require autopsies on, at least, some failed units; autopsies on a sample of unfailed units are also useful for comparison.
- Detailed information on replacement parts.

For some high-volume products, it might be appropriate, instead of getting data on every unit, to do so on a random sample.

Those responsible for product repair and service are typically in the best position to obtain the required field failure information. Their prime job, however, is to make repairs as effectively and speedily as possible. Gathering information is likely to take precious added time and might seem like unnecessary bureaucracy. Also, assessment of the root cause of failures (e.g., deterioration of a connecting part, the corrosion of an internal switch, or the overheating of a wire) often requires further investigation. Agreement must therefore be reached (and carefully documented) on what information is to be taken and how to obtain and record it consistently – along with providing appropriate recognition and compensation for those collecting such information.

9.6 A NOTE ON PRODUCT SAFETY

9.6.1 An Important Concern

In Section 5.9, we stressed the importance of identifying and removing product safety problems in design. Unfortunately, such problems sometimes still slip through and, as pointed out earlier, are difficult to uncover during in-house testing due to their low frequency and also because they might be brought on by unconventional product use.

Immediate reporting and assessment of only a single safety incident in the field is often critical – even though it might not have created serious harm; for example, a small fire that was immediately extinguished. An understanding of the cause and, if possible, underlying mechanism of the safety problem is again essential since the correction frequently requires some form of redesign or materials change.

It is, in addition, often of immediate critical importance to identify those units in the field that due to their manufacturing history and/or use environment might be especially vulnerable. This information is used to determine what customers should be selected for special warnings or what units need be considered for product recall. Such evaluations again call for a combination of physical understanding and statistical analysis (such as segmentation studies).

Safety degradation information can provide important advance warning of an impending safety problem. For example, unusual overheating may be a precursor of a fire hazard and might be uncovered by proactive monitoring (see Section 9.9.4).

Good field data that help speedily identify, assess and correct product safety problems are again of paramount importance. When a problem occurs, it is unlikely that much of the needed information is readily available – *unless* this was actively planned for in advance.

9.6.2 Pharmaceutical Products

Discovery of adverse effects of a product after its release is of particular concern for pharmaceuticals and is receiving increasing attention, as a consequence of some highly publicized cases[8] that occurred despite the extensive testing prior to release described in the next chapter.

How can this happen? Some reasons are:

- The probability of occurrence of the adverse effect might be quite small, and, to attain a high probability of detection, requires a sample much larger than that which is often practical in clinical trials. Also, segments of the population that are especially vulnerable, for example, the elderly, may not be included in such studies in sufficient numbers to detect an adverse effect.
- Patients with specified ailments or abnormal laboratory measurements or ones who are taking certain other medications may be excluded from the clinical trials, but may receive the drug once marketed. In addition, physicians sometimes prescribe a drug, approved for a particular ailment for a specified group of patients (e.g., adults), to those for which it was not intended.
- The adverse effect might appear only after lengthy product use – often beyond that which is practical for approval testing.

There is, understandably, considerable pressure to release medication, especially for a serious disease, once its effectiveness in treating that disease is demonstrated. Thus, safety monitoring needs to continue after product release. Quality Progress (March 2007) reports that, according to a recent Pricewaterhouse Coopers study, pharmaceutical companies are shifting to postmarket drug safety initiatives rather than relying solely on preapproval clinical trials and the federal approval process. All of this makes it especially important to have in place a statistically sound system that tracks adverse (and beneficial) side effects and closely monitors the results.

9.7 TRACKING CUSTOMER SATISFACTION

9.7.1 The Issue

Reliability and safety are not our sole product concerns. We are, interested in *all* issues that may lead to customer dissatisfaction or a feeling of "less than delight" about our product. Purchasers of a washing machine might, for example, feel that the product is "too noisy," that it does a "poor job" cleaning the laundry, that it is inconvenient to use, or that it was delivered with chips on its surface. Or they may be dissatisfied with the service rendered (or not rendered) – for example, delivery, billing and repair. All of these add up to the "word of mouth" reputation of our product and impact future sales.

Ideally, market research initially identified potential sources of dissatisfaction and helped ensure that these are addressed during product design. Any remaining concerns need be speedily discovered, reported, and corrected. We need, in addition, be sensitive to changes over time in what customers are seeking and to new or

8. See FDA web site (www.fda.gov/opacom/Enforce.html) for a listing of U.S. alerts on drug, food, and cosmetic products.

forthcoming competitive offerings. Finally, in addition to knowing what we might be doing wrong, it is important to determine what customers think we are doing right – so that we can learn to do so all the time on all products.

Statistical concepts and tools are used to track sources of dissatisfaction experienced or perceived by customers, to quantify their impact, and to help avoid their reoccurrence. Problems may be ranked by frequency of occurrence using a Pareto chart and/or tracked over time by control charts that signal changes.[9] Such analyses, moreover, need to be part of a process with clearly defined lines of responsibility for corrective action.

9.7.2 Getting Good Data

The biggest challenge is again that of gathering meaningful information. Data about customer dissatisfaction might be obtained from complaints to the company service center or from service calls. Response cards, distributed with the product at delivery or given to the customer for mailing when a repair is performed, provide further information. Statistical estimates are often not needed in such situations, but when they are, we need be aware of biases due to the frequent self-selection of respondents.

Similar to other applications, we need to understand the root causes of customer concerns in order to address these effectively – even though customer feedback often tends to be imprecise. A further concern is with customers who do not voice their dissatisfaction and "suffer in silence." Whether or not a problem is reported may depend on its severity, the tolerance level of the user, whether or not the product is still under warranty and, in general, whether or not customers see any personal benefit in reporting their dissatisfaction.

As in our initial market research (Section 3.3.1), a well-crafted and executed survey can provide information to quantify customer satisfaction. Such a survey may include overview questions concerning the likelihood of repurchasing the same brand (and perhaps of purchasing other products from the same manufacturer).

As in other applications, in order to address customer concerns effectively, we need to understand their root causes, even though customer feedback often tends to be imprecise. A washing machine may, for example, be characterized only as "leaking water" or "sometimes not working." Thus, it is necessary to develop consistent and maximally meaningful ways to assess, categorize, and analyze customer feedback.

9.8 COMPETITIVE EVALUATIONS AND ADVERTISING CLAIMS

9.8.1 The Problem

Frequently, we need to conduct studies to compare the reliability, performance, and general customer satisfaction of our products against our competitors'. The results

9. The varying number of units in operation and their ages need be taken into consideration in constructing such charts.

may be used to provide information for strategic planning, marketing, developing new product design, product pricing, and formulating advertising claims.

9.8.2 Getting Competitive Product Data

Our ability to procure information on competitors' products may be limited. In purchasing such product on the retail market, we need to obtain as broad a sample as possible. For example, we want to avoid limiting such purchases to a single store, especially at one time, so as to decrease the chances that our samples are all from, say, the same manufacturing batch.

9.8.3 Formulating Advertising Claims

Our findings may be used to make advertising claims about our product – and how it compares with competitors'.

The specific wording of advertising claims often has statistical ramifications. A typical consumer may, for example, have difficulty differentiating between the two statements:

- Our product has (been shown statistically to have) an average life that is longer than that of any of our major competitors.
- No major competitive product has (been shown statistically to have) an average life that exceeds our product's.

Yet there is a large difference in what is needed to justify these two statements statistically. And even though the second statement may be technically correct (perhaps because of inadequate sample size), its subtlety may not be understood by many. We need, thus, be concerned how a specific claim might be perceived by the general (not statistically savvy) audience at which it is targeted. See Sidebar 9.2 for further discussion.

SIDEBAR 9.2: ADVERTISING CLAIMS: THE NEED TO GET IT RIGHT

Advertising claims, and especially ones involving comparisons with other products, are subject to close scrutiny. They need to meet the requirements of government, and sometimes industry, regulating agencies and fully comply with "truth in advertising" laws and regulations. They will be carefully examined by competitors, who – perhaps with the help of their own statistical experts – would like to show that the studies upon which the claims are based are flawed and, perhaps, take legal action.

Statisticians, working with knowledgeable engineers and lawyers, help ensure that product claims have been soundly demonstrated statistically. This requires:

- Obtaining samples that, if not random, provide as broad coverage as possible of the company's and, where applicable, competitive products.[10]

10. A possible, and perhaps unavoidable, source of bias is that such sampling is likely to be of early-built product so as to allow timely use of the results.

- Having a sufficiently large sample size.
- Conducting meaningful tests and comparative evaluations.
- Analyzing the results correctly (e.g., taking proper account of measurement errors).
- Documenting the results (including the raw data) and the analyses, and being prepared to share these.
- Desisting from using statistical jargon that may be misinterpreted by nonstatisticians.

9.9 PROACTIVE PRODUCT SERVICING

9.9.1 Goals

All products, like all humans, eventually fail and die. The approaches briefly described in this section aim to minimize – through statistically based proactive servicing (or just-in-time maintenance) – the associated cost and inconvenience – and sometimes danger – of field failures.

Proactive product servicing has two major goals. The first is to avoid *unscheduled* shutdowns, if possible. The second is to ensure speedy and inexpensive repairs, when field failures do occur – so as to minimize their deleterious impact. Performing repairs during scheduled maintenance on automobiles, aircraft engines, locomotives, and medical scanners is, for example, a great deal less disruptive and costly than coping with unexpected field failures. The emergence of long-term service agreements (see Sidebar 9.3) has provided added impetus to such efforts.

SIDEBAR 9.3: THE CHALLENGE OF LONG-TERM SERVICE AGREEMENTS

Some systems equipment manufacturers now offer Long-Term Service Agreements (LTSAs) to customers.[11] Under these, manufacturers assume the responsibility for both routine and unscheduled maintenance of their systems and for fixing product failures for a predetermined time, say 15 years. The service provider typically warrants system availability. Power generation equipment may, for example, be guaranteed to be operable 99% of the time or to have no more than two unscheduled shutdowns over a 6-month period. If the guarantee is not met, the service provider must pay a, often substantial, penalty. LTSAs represent a major extension of simple product-servicing warranties that have traditionally been sold for appliances and other products, such as the 3-year extended warranty on the laptop computer described in Section 9.3.1.

Through LTSAs, manufacturers sell to customers not just products, but a guaranteed ongoing level of service. This gives such service providers added motivation to build high-reliability products.

In addition to the statistical challenges described in this section, LTSAs require predictions of needed future servicing and the associated costs. These help management decide whether or not an LTSA should be offered (or accepted by a customer) and price such services.

11. Occasionally the LTSA provider is different from the manufacturer.

Statistical concepts and tools apply to proactive product servicing in developing (1) product maintenance schedules, (2) proactive parts replacement strategies, and (3) automated monitoring to identify impending failures.

9.9.2 Product Maintenance Scheduling

Many systems, from automobiles to aircraft engines, are serviced periodically. Typical examples are automobile oil changes, scheduled thermal barrier coating of turbine components, and the replacement of filters in air conditioners. Routine maintenance should be scheduled to provide an optimum trade-off between the cost and inconvenience of such servicing and the likely greater cost and inconvenience due to unscheduled failures that servicing might avert. We will briefly illustrate the concept by an automobile maintenance scheduling example; similar considerations apply for other systems.

Automobile manufacturers have traditionally advised purchasers how frequently to change oil and lubricate car parts. They also stipulate servicing for the cars they lease; for example, service every 6 months or every 3,000 miles, whichever comes first. This approach can be improved upon by taking into account how the car is operated; for example, by considering such factors as driving speed, number of stops and number of cold starts. The more intensive the use environment, the greater is the need for servicing. This suggests developing a system that determines the frequency of routine maintenance based on driving and cost considerations.

Dynamic maintenance scheduling, or the so-called "condition based maintenance" provides a further opportunity (see Adjengue, Yacout, and Ilk 2007). For automobiles this might involve monitoring oil degradation and other measures of deterioration over time to determine when (in miles) the next maintenance should be performed. The car operator is then advised, perhaps upon car start-up, of the next recommended servicing for various assumed driving scenarios.

For both of the preceding situations (static and dynamic), statistical and operations research methods help develop an optimum procedure.

9.9.3 Proactive Parts Replacement Strategies

Many equipment field failures occur due to the wearout of relatively inexpensive parts, causing costs due to unscheduled system shutdowns that are far greater than the cost of the part itself. To avoid this, we want to replace potentially vulnerable parts with new ones at strategically selected times during routine maintenance.

Sometimes, an impending failure can be detected by inspection of the part or by embedded instrumentation (discussed shortly). In other cases, we need to rely on the estimated statistical lifetime distribution of the part for such assessment. If the hazard function (see Section A2 of Appendix to Chapter 5) for a part increases over time,[12] the probability of failure of that part prior to the next (evenly spaced) scheduled maintenance also increases with time and proactive replacement warrants consideration. One might, for example, replace the part if the estimated probability

12. Such as for a part with a Weibull distribution for lifetime with a shape parameter exceeding 1.

of failure prior to the next routine maintenance exceeds a specified threshold, say 1 in a 1,000, or alternately if this probability is twice its initial (prior to first maintenance) value. The specific plan needs to balance the cost of prematurely replacing a part from which one might get further life against the chances and cost of the part failing in service.

The resulting statistical evaluations can again be made more powerful by taking operating environment into consideration. For example, in deciding when to replace a part in a locomotive, the analysis should consider the terrain (e.g., flat, mountainous, numerous tunnels) in which the locomotive will be operating.

9.9.4 Automated Monitoring for Impending Failures

Concept A final line of defense in avoiding or mitigating the impact of field failures is provided by new technology – remote (and often continuous) monitoring of products using sophisticated instrumentation.[13] Car owners have become acquainted with this approach through lights flashing on their dashboards – signaling a possible reliability problem (e.g., impending engine malfunction), a safety concern (e.g., door not closed properly), an environmental issue (e.g., emissions problem) or, sometimes, a false alarm (possibly due to malfunction of the instrumentation).

Forewarning of an impending system failure allows repairs to be executed in a minimally intrusive and costly manner – ideally, without users even being aware of the problem. As a minimum, it can help identify what part(s) need to be replaced and the technical service personnel best suited to make the repair speedily.

Statistical concepts and tools help ensure that the best possible data are obtained to signal an impending problem and to develop rules for making the most prudent and cost-effective decisions from such data (see Barone 2006).

Locomotive Engine Example Modern locomotive engines are equipped with numerous sensors that read such properties as oil pressure, oil temperature and water (coolant) temperature. The resulting information is used during normal operation to make automatic systems adjustments (e.g., to control fuel and oil flow), using control algorithms.

The data may, in addition, be used to identify impending failures and avert these by shutting down the system before serious damage occurs. In light of the inconvenience and lost revenue associated with shutdowns, it is imperative, however, to minimize the number of false alarms so that the engine is shut down only when absolutely necessary.

Failure Modes Consider, specifically, the occurrence of a large drop in engine oil pressure. This could be due to any one of the three reasons; these are shown in Table 9.2, together with the associated problem severity and the action that would be taken *if* it were known that this was, indeed, the reason for the oil drop. Such actions range from activating a backup sensor to shutting down the engine.

13. Such monitoring is differentiated from dynamic maintenance scheduling (Section 9.9.2) in that it signals a specific impending failure as opposed to helping schedule routine maintenance.

TABLE 9.2 Possible Reasons for Engine Oil Pressure Drop and Consequences

Possible Reason for Oil Pressure Drop	Problem Severity	Appropriate Action
Oil pressure sensor failure	Not a problem per se, but can lead to erroneous control and even engine failure.	• Activate back up sensor. • Replace sensor at next routine maintenance.
Cooling system failure	Important potential problem. Prolonged operation with a failed cooling system can lead to engine failure.	• Reduce engine load. • Activate secondary cooling systems. • Monitor all critical systems. • Repair at next routine maintenance or sooner.
Oil pump failure	Critical problem; resulting insufficient lubrication can rapidly lead to engine failure and severe damage.	• Halt all engine functions. • Repair before reactivating engine.

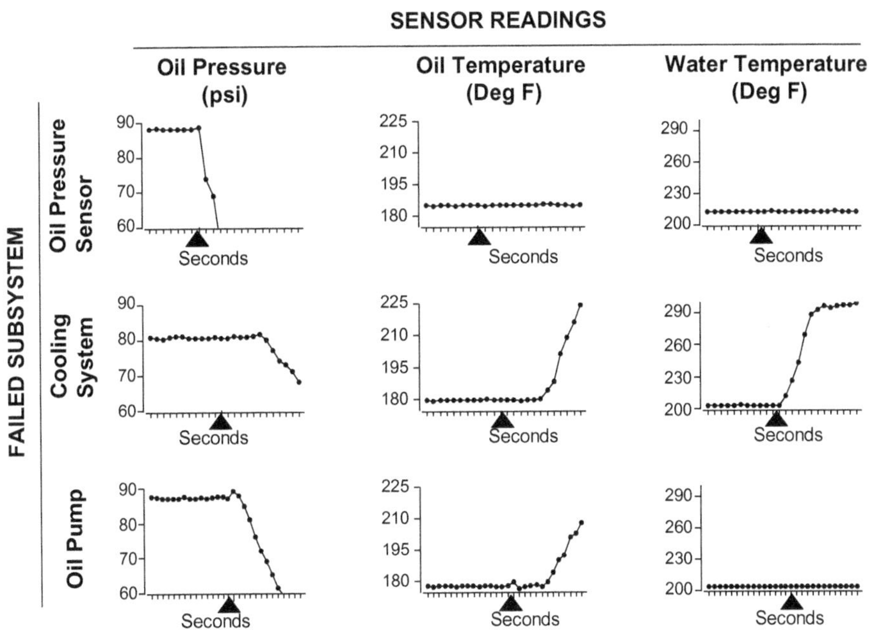

Figure 9.7 Readings on three sensors for three different problems.

Thus, when an oil drop occurs, we need to find the reason so as to take the appropriate action. Sensor data on *how* oil pressure is changing over time, as well as on oil temperature and water temperature, help us do so.

Data Acquisition An engine was induced in a lab test to fail in each of the three modes shown in Table 9.2 so as to observe the resulting sensor measurements. Figure 9.7 shows the readings on the three sensors over a 30-second time period for each mode. The solid triangle marks the point at which the problem was presumably triggered, as evidenced by an appreciable change in at least one of the three measurements.

Analysis Each of the three incidents resulted in a sudden sharp drop in oil pressure. The accompanying changes in oil temperature and water temperature, however, differ. Thus, data from the three sensors can help us distinguish between the three reasons for the oil pressure drop. In particular:

- For the faulty sensor, the drop in oil pressure was *not accompanied* by any perceptible change in either oil or water temperature.
- For the cooling system failure, a sharp rise in water temperature *preceded* the drop in oil pressure, which was also accompanied by a sharp rise in oil temperature.
- For the oil pump failure, a sharp rise in oil temperature *followed* the drop in oil pressure; water temperature remained unchanged.

Toward an Algorithm The data in Figure 9.7 were from a single engine under fixed-test conditions in the lab. Similar data were obtained on different engines operating in the field under varying environments and involving various oil pressure drop incidents. The resulting patterns were sufficiently distinct and similar to those depicted in Figure 9.7 to permit a useful algorithm to be developed for using the three sensor and some further measurements to identify the reason for the oil pressure drop. This led, in a great majority of cases, to appropriate remedial action, including the all-important decision of whether or not to shut down the system. The analyses required the use of advanced, multivariate and time series methods.

Further Comment Statistical methods play an important part in proactive product servicing. Such work, as always, is conducted in collaboration with knowledgeable engineers and subject matter experts. Other analytic methods also are also used in addressing such problems; these include machine learning, neural nets, and Bayesian belief networks (see Hastie, Tibshirani, and Friedman 2003, and Pearl 1988).

9.10 MAJOR TAKEAWAYS

- Despite the required upfront focus on designing and building reliable products, their performance in the field still needs to be scrutinized.
- Reliability data tracking is used to assess product field reliability; its goals include early identification of failures to promote action to avoid their repetition, mitigation of the harmful impact of failures of product already built, and apprising management.
- Segmentation – that is, breaking down the population into meaningful subsets, doing separate analyses on each and acting on the resulting information – is a useful tool for reliability tracking.

- MCF plots are useful in the reliability analysis of repairable products.
- Careful advance planning of the failure reporting system is a key requirement for the rapid identification and correction of field reliability problems.
- Good field data are also required to help speedily identify, assess and correct product safety problems that were not previously uncovered.
- Customer response data and well-conducted surveys can improve our understanding of customer satisfaction and dissatisfaction and act thereon.
- Competitive evaluations are conducted for various purposes, including making advertising claims. This requires great care to help ensure the statistical validity of the claim and to avoid misleading the public.
- Proactive product servicing is becoming increasingly important, especially with the emergence of LTSAs. Statistical concepts and methods are used to develop optimum product maintenance schedules and proactive parts replacement strategies, and in automated monitoring for impending failures.

DISCUSSION QUESTIONS

General

1. Why do Figures 9.2 and 9.3 lead to more meaningful assessments of reliability for the laptop computer hard drive than Figure 9.1? Also suggest some ways in which the dynamic reporting and corrective action system described in Section 9.3.1 might be utilized.

2. In the locomotive braking grid failure example (Section 9.4.3) the usage environments were said to be similar for both orders. How would you assess this claim?

3. Discuss the data gathering and statistical analysis implications for rapidly quantifying and addressing the following safety problems:
   - A fire resulting from the malfunction of an electric iron.
   - Young children getting injured while playing with a remote controlled race car.

4. You want to survey consumer satisfaction and dissatisfaction for a new brand of toothpaste. The results will be used in a possible advertising claim, as well as to help improve the development and marketing of the next generation product. Provide details on how you would perform such a survey.

5. A battery for use in a variety of consumer products (e.g., flashlights and portable radios) is believed to have a longer average life than those of two major competitors. Develop a detailed study to get statistically valid data to form the basis of an advertising claim of the product's longer life.

6. Pricing a LTSA sometimes involves estimating the long-term reliability of a product. How would you go about obtaining needed data to make such an estimate for a new product with limited field experience?

7. Propose a proactive parts replacement strategy for an automobile.

8. Suggest some further situations that could benefit from proactive product servicing. What specific approach or combination of approaches would you suggest for doing this?

Technical

1. In the laptop computer reliability tracking example (Section 9.3.1), we noted that those availing themselves of the additional 3-year extended warranty were not a random sample of all purchasers. How would you assess any resulting bias in the analysis and how might you correct for this?

2. In the laptop computer reliability tracking example, we indicated that the originally proposed management overview chart "gives useful information for estimating replacement part requirements." Elaborate.

3. In the laptop computer reliability tracking example, we indicated that making improvements suggested by Figures 9.2 and 9.3 "required further statistical analyses by product subpopulation and failure mode." Elaborate.

4. In the laptop computer reliability tracking example it was stated that "The (early warning) system ... gives information similar to that of a statistical control chart applied to field failures." How does a control chart for tracking reliability (over time of production) differ from a conventional control chart on in-process or end-of-line measurements? How would you develop such a chart?

5. We stated in comparing segmentation with multiple-failure mode analyses (Section 9.3.3) that "In both situations, the results of the individual analyses can subsequently be combined to obtain an omnibus analysis for the entire population." How would you conduct such omnibus analyses in each case?

6. In the locomotive braking grid failure example (Section 9.4.3), we suggested that if information on individual grid brake identifications and the positions of the repaired braking grids in the locomotive were available, more informative analyses would have been possible. Elaborate.

7. In a footnote in Section 9.7.1, dealing with using control charts to track customer dissatisfaction, we said that "the varying number of units in operation and their ages neèd be taken into consideration in constructing such charts." Elaborate.

8. We stated in Section 9.8.3 that in comparing our product with competitors' requires, among other things, "having a sufficiently large sample size." How would you determine what is a sufficiently large sample size for a claim of product superiority?

9. In Section 9.9.2, on product maintenance scheduling, we state that "statistical and operations research methods help develop an optimum procedure." Elaborate.

10. We stated in Section 9.9.3 that in deciding whether or not to replace a part at the time of inspection, based on its estimated time to failure distribution "the specific plan needs to balance the cost of prematurely replacing a part from which one might get further life against the chances and cost of the part failing in service." How would you perform such an evaluation?

PART III

OTHER APPLICATIONS

IN THIS part of the book we describe further applications of statistics in chapters dealing with:

- Pharmaceutical products
- Financial services
- Business processes
- Other applications

Some of these applications (the chapter on pharmaceuticals and much of the chapter on other applications) deal with manufactured products. Therefore, our discussion in the previous part of the book also applies. In addition, we stress specific issues that characterize the use of statistics in these applications.

CHAPTER 10

PHARMACEUTICAL PRODUCTS[1]

10.1 ABOUT THIS CHAPTER

Statistics and statisticians have contributed to pharmaceutical manufacturing, research and development, and marketing for decades. In the United States, the Food, Drug, and Cosmetic Act of 1938 and the Kefauver-Harris Amendments of 1962 created a favorable environment by requiring that newly introduced pharmaceutical products be proven to be both safe and effective. To achieve such proof, a program of preclinical and clinical trials needs to be designed, executed, and reported. Recognizing the possible dire consequences of doing otherwise, the pharmaceutical industry and its regulatory agencies came to appreciate that this requires controlled statistically based investigations. As a result, this industry is, today, one of the largest employers of statisticians – and, especially of ones with advanced degrees.

In a subsequent chapter, it is stated that there are few areas of application in which statisticians have a financial impact comparable to that in financial services. The work that statisticians do in pharmaceutical applications also has important payoffs. In addition, pharmaceutical statisticians – together with colleagues engaged in other areas of medical research – can claim that there are few areas in which statisticians have comparable impact on *humans*.

We begin by reviewing the drug development process and by providing an overview of the statistician's role. We then describe some of the issues statisticians face in preclinical and clinical studies, respectively (preclinical studies involve testing the compound directly and animal testing; clinical testing is targeted principally at the intended human patient population.) A detailed example of a clinical trial is provided. We then briefly consider the regulatory review and various other, mostly post-approval, activities. The chapter ends with a brief enumeration of some statistical tools used in pharmaceutical applications and with information on where to go to learn more.

1. This chapter was prepared by contributing author Robert Starbuck, Assistant Vice President, Special Projects, Wyeth Research.

The Role of Statistics in Business and Industry, by Gerald J. Hahn & Necip Doganaksoy

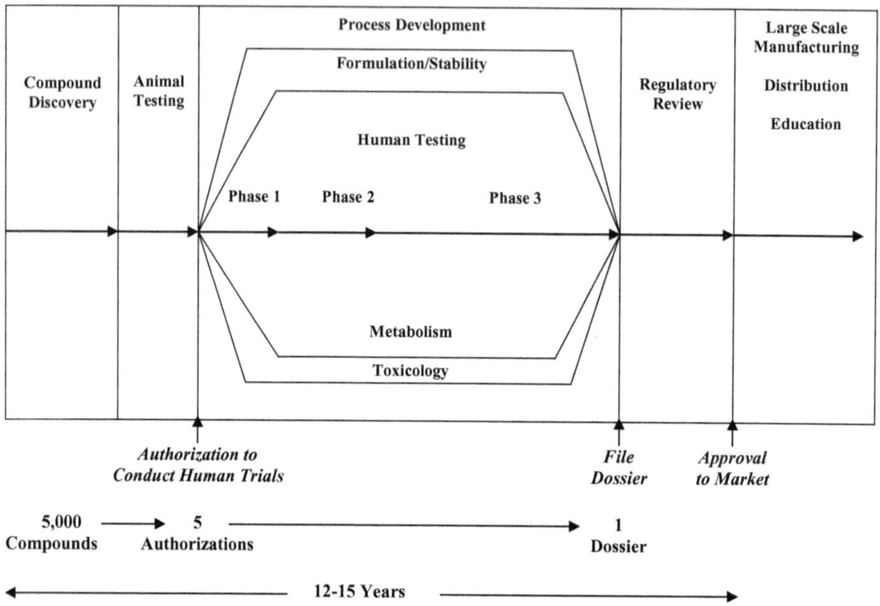

Figure 10.1 Drug development process overview.

10.2 DRUG DEVELOPMENT: PROCESS OVERVIEW

The drug development process is summarized briefly (and incompletely) in Figure 10.1.[2] The process begins with the search for and, hopefully, discovery of new compounds or biological agents; the most promising move into animal testing.

If the product continues to appear viable, clinical development involving the testing of humans begins – subject to authorization from the appropriate regulatory agency. This involves three (and sometimes four) phases and requires the results of all phases to be favorable. If the results continue to be positive, a dossier – a detailed report providing all the findings to date and generally requesting approval for putting a product on the market – is submitted to the regulatory agency. Marketing, large-scale manufacturing, and distribution of the product begin upon approval.[3] Clinical studies are, however, likely to extend beyond the filing of a dossier to assess long-term impact or for various other reasons.

This process has traditionally taken 12 to 15 years to complete. The industry is striving to shorten this time to accelerate the availability of new drugs to patients and to maximize the time that the product is actually on the market prior to patent expiration (in the United States, generally 17 years from the time the patent is granted, which typically occurs during Compound Discovery). It has been estimated

2. Adapted from a short course developed and taught by Bruce Rodda and Robert Starbuck.

3. Manufacturing may, in fact, commence prior to approval to expedite getting the product to pharmacies after approval.

that for every 5,000 compounds considered during discovery, typically only one gets as far as filing a dossier.

Our discussion focuses on the development of drugs, but applies as well to vaccines, biologicals, and other forms of therapy (e.g., radiation therapy). In addition, over-the-counter (OTC) products typically start out as prescription products, and as such, also go through at least the initial drug development phases. Thus, for some products (e.g., nonsteroidal inflammatory drugs like Advil™) enough evidence is acquired for the medical and regulatory community to believe that the general public can safely use the product without medical advice.

10.3 THE ROLE OF A PHARMACEUTICAL STATISTICIAN: AN OVERVIEW

Perhaps due to the regulatory environment, much of the statistical work in the pharmaceutical industry continues to be done by professional statisticians.

10.3.1 Application Areas: Overview

Statisticians in the pharmaceutical industry are typically engaged in one or more of the following activities:

- Support of basic research: Help identify potential new drugs from a vast library of compounds, performing in vitro (test tube) and in vivo (in the body – animal or human) studies on promising candidates, genetic profiling, etc.
- Pharmaceutical development: Determine how to make a new chemical or biological entity and explore what process steps maximize product yield and minimize cost.
- Preclinical and clinical studies: Initially, study the impact of a new drug on animals and healthy volunteers[4] to characterize what the body does to the drug, i.e., absorption, distribution, metabolism, and excretion. Then extend these studies to assess the impact of the drug on humans and patients with the targeted disease.
- Marketing and commercialization: Further scientific studies to characterize the new product, surveys of physicians and patients, preparation of publications and public presentations, etc.
- Manufacturing: Guide the production of the drug in final market formulation, usually, in large quantities (e.g., assessing uniformity of drug content, stability, and dissolution).

The great majority of statisticians in the pharmaceutical industry work on preclinical or clinical studies.

4. The use of healthy volunteers in initial studies avoids the possible confounding variability and bias that can occur in patients taking other medications.

10.3.2 Nature of Work

Pharmaceutical statisticians are engaged in the data acquisition process and in program development and maintenance. Their biggest contributions are often in the planning and analysis of studies that differ in nature from those conducted in the past; a similar approach may then be applied to future studies. They are also most likely to become involved in studies that are especially critical to the business and/or call for specialized experimentation and more complicated designs and analyses.

10.3.3 Organization

Statisticians are organized within pharmaceutical companies in a variety of ways, including large centralized groups, decentralized groups aligned by therapeutic area or geographical location, and independently operating multidisciplinary teams.

Irrespective of the organizational structure, statisticians involved in the complex process of drug development typically also belong to multidisciplinary teams that provide the collective expertise necessary to conduct a successful clinical development program. Collaborators are drawn from a wide array of scientific disciplines, and often from a variety of therapeutic areas; they may include physicians, clinical scientists, programmers, data managers, clinical writers, clinical pharmacists, pharmacologists and pharmacokineticists, health outcomes assessment professionals, and project planners.

Not all studies require a professional statistician. It is not uncommon for a biologist, or other knowledgeable professional, to carry out the planning and statistical analysis of a pharmaceutical study using methods developed or approved by a statistician. In such cases, statisticians may contribute by teaching scientists about basic statistical concepts and demonstrating how to employ relatively simple, easily accessible and robust methods. Prudent scientists, however, call on a statistician to review the resulting test plan and analysis.

10.3.4 Some Important Skills

Many statisticians working in pharmaceuticals hold Ph.D. degrees. Others have Master's degrees (see Lakshminarayanan 2007).

To be effective, statisticians must feel comfortable with relevant biological, physical and chemical concepts and terminology, and understand their impact. They must also appreciate logistical constraints imposed by laboratory practice and workflow restrictions. They need to be fast learners so as to meet the challenges presented by advances in technology and to keep abreast with fellow scientists.

Successful statisticians supporting drug studies have strong communications and interpersonal skills, interest in learning and understanding regulatory requirements, and an appreciation of the business aspects of pharmaceutical research and development. Such skills, in fact, are as important, or even more so, than their statistical analysis skills (also see Sidebar 10.1).

SIDEBAR 10.1: PHARMACEUTICAL VERSUS OTHER INDUSTRIAL STATISTICIANS: THE WORK ENVIRONMENT

Time Pressures

The time pressures on pharmaceutical statisticians are similar to those on other statisticians in business and industry. This might seem a little surprising, given the relatively long time it takes to bring a drug to market. Many decisions are, however, made along the way on a day-to-day basis.

General Acceptance of Work

Numerous laws, regulations, and industry practices, adopted in the United States and many other countries in the second half of the last century, have assured statisticians in the pharmaceutical industry their proverbial place in the sun. Such acceptance is not always found in other application areas.[5] As a consequence, pharmaceutical statisticians may spend less time "selling" their potential contributions.

Opportunities for Advancement

Perhaps, because of the mainstream nature of statisticians' work, the pharmaceutical industry provides especially promising opportunities for promotion to higher levels in the organization.

Impact of Regulatory Agencies

Regulatory agencies play an important role in the lives of pharmaceutical statisticians, and, especially, those engaged in clinical studies. These agencies have their own rules, practices, and stringent reporting requirements that can add some bureaucracy to a pharmaceutical statistician's activities.

Government and industrial regulatory agencies are also important in applications of statistics beyond pharmaceuticals – especially in issues dealing with safety or the environment. However, in many such cases, failure to meet specifications established by, say Underwriters Laboratories, leads to modifications in the product design. Thus, if an automobile fails to satisfy environmental regulations, it is likely to be redesigned. In contrast, failure of a drug to meet regulatory requirements often results in abandonment of the drug.

Public Spotlight

Statisticians engaged in pharmaceutical research, as well as in other medical applications, and their colleagues, frequently find themselves in the public spotlight. In light of the significant impact that drugs have on human life, hardly a day goes by without some pharmaceutical study receiving attention in the media.

Predictability of Work

Pharmaceutical statisticians face some exceptional challenges. There is, however, often a common thread in the work, at least, of those engaged in clinical studies. This commonality is the planning and analysis of trials to evaluate the efficacy and safety of a new drug, and obtaining information to submit to a regulatory agency. This may result in greater predictability, and perhaps less variety, in the work of many pharmaceutical statisticians, as compared to industrial statisticians – especially the ones who support entire businesses or deal with a variety of product types.

5. Government agencies that are principally dedicated to collecting statistics may be another exception.

10.4 PRECLINICAL STUDIES

If at first, the idea is not absurd, there is no hope for it.
—Albert Einstein

10.4.1 Discovery Research

Background The search for new drugs begins in discovery research. Biologists strive to identify biological targets (e.g., a receptor on a nerve cell or the surface of a body organ) upon which a small molecule, protein, or vaccine might act to prevent, arrest, or reverse a disease. When a potential target is identified, in vitro assays are developed, which provide the ability to measure the impact of a compound or vaccine on the target. For small molecules, one or more selected assays might be adapted to automated formats that permit tens of thousands to hundreds of thousands of compounds to be tested. This has led to the use of high throughput screening as a tool for identifying a workable subset of promising candidates – a topic closely linked to the concept of combinatorial chemistry, discussed in Section 3.5.1.

From these efforts, a small subset of promising compounds is identified, and medicinal chemists work to improve these candidates' desirable properties. Promising compounds or proteins then undergo further testing with in vitro, and then in vivo, assays that more closely mimic the disease process. Good screening and prediction models can greatly reduce the attrition rate of potential new drugs or biological products.

The Role of Statistics Statisticians work with scientists throughout the discovery process to help develop and refine assays, fit appropriate models to the resulting data, and design and analyze experiments that distinguish true effects from noise and experimental artifact. As in other areas, helping ensure that the right data are obtained upfront is of paramount importance.

Due to improvements in analytical methods and instrumentation and various technological advances, such as the use of DNA micro-arrays, the quantity, complexity, and types of data generated in discovery research have increased tremendously. Statistics is, consequently, being used increasingly in the reduction, summarization, standardization and automation of calculations, and in performing multivariate analyses.

10.4.2 Drug Safety Evaluation in Animals

Background Once a compound has been shown capable of producing its desired effect through either in vitro or in vivo studies or both, further experiments are conducted with animals to evaluate potential safety in humans. The required level of safety varies from one product to the next and must be balanced against potential benefits. Such studies follow government regulations, like those provided by the International Conference on Harmonization (ICH) in ICH E9 (1999).[6]

6. www.fda.gov/cder/guidance.

Experiments are typically conducted in appropriately chosen animal species, such as mice and monkeys, and evaluated by animal models to establish the maximum tolerated dose and to assess whether it seems safe to initiate testing in humans.[7] Drugs that are highly toxic in animals are unlikely to proceed to human trials. Investigations may be conducted to:

- Determine the absorption, distribution, metabolism and excretion of the drug in animals
- Develop assays for the drug and its major metabolites (if any).
- Determine the maximum dose with no toxic effects.
- Identify adverse effects that might also carry over to humans.
- Test models of activity sensitive to known effective drugs. An example is putting a mouse into the center of a box to determine how quickly it moves to the perimeter – its normal survival instinct. An effective antianxiety drug will make the mouse less anxious.
- Identify organ toxicities. These are, typically, liver and kidney, but other organs may also be impacted.
- Characterize effects on reproduction.

In the United States, such studies are typically conducted in accordance with the Good Laboratory Practices laws and enforced by the Food and Drug Administration (FDA) to help ensure the integrity and quality of nonclinical study data.

A Simple Example A toxicology study is to be conducted over a 28-day period to assess the impact on the body weight of rats of administering various doses of a proposed new drug. (Appreciable reduction in body weight gain in rats is an indicator of a possible deleterious effect in humans.) Twenty male and 20 female rats from one strain and breed were selected for the study and exposed to one of four treatments: control, low dose, mid dose, and high dose. To achieve comparable starting weights, the 40 rats were stratified by gender and body weight in groups (or strata) of four, and, within each group, randomly assigned to one of the four treatments. The rats were then given a normal diet, and weighed initially and four times subsequently at 1-week intervals.

Figure 10.2 displays the resulting mean body weight for the female rats. (The conclusions for the male rats were similar.) This chart suggests that:

- The low dosage does not appear to have an impact different from that of the control.
- The medium dosage might lead to a small reduction in average body weight (less than 5% after 28 days) as compared to the control.
- The high dosage appears to lead to an appreciable reduction in body weight (around 15% after 28 days) as compared to the control, likely disqualifying this dosage from further consideration.

7. Additional animal experiments may be needed to support clinical trials of long duration, and for selected populations, such as women of childbearing age.

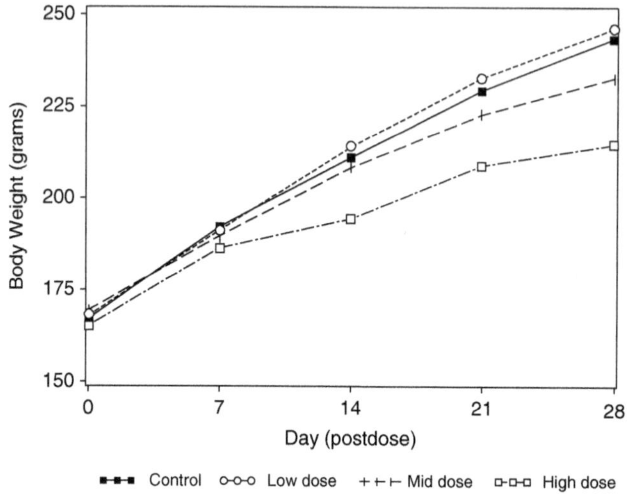

Figure 10.2 Mean body weight of rats subjected to different dosages of a drug.

These findings can be made more definitive by a statistical evaluation of the mean differences (and the associated confidence intervals) between the various dosages and the control.

10.4.3 Formulation Development

Background In the formulation development stage, scientists take compounds that have been produced in small quantities during the discovery phase and work toward transforming them into potentially marketable, high-quality pharmaceutical products. This process entails devising analytical testing procedures and manufacturing processes for commercial scale quantities. Most importantly, it means developing a stable formulation that consistently delivers a specified dosage of the active ingredient.

Analytical (chemical or biological) methods initially define the desired physical and chemical characteristics of the drug. Some important properties are the amount of active ingredient (potency), the uniformity of the active ingredient within the formulation, and the amount of active ingredient dissolved in solution over time. Studies are then performed to ensure that a product with the desired properties can be readily produced. Other studies are conducted to assess such factors as stability, potency, and physical properties.

Assessing Measurement Capability A key factor in such evaluations is the reliability of the instruments used in making the assessments. Statistical methods help analytical chemists determine the suitability of instrument settings, ensure that they meet specifications for accuracy and precision, evaluate repeatability and reproducibility, and, in general, optimize the instrument. This may typically involve running a planned experiment to quantify the variability in measurement due to such factors as different analysts preparing the samples or taking the measurements, variation among sample preparations, and instrument and reading error. The resulting data might be evaluated graphically and by an analysis of variance to assess and quantify the relative impact of each of the contributors to measurement variability (see Section 3.4).

Developing a Manufacturable Product The manufacturing process is simultaneously developed to build an effective product. This typically requires screening a large number of formulations and process variables and assessing their impact, using statistically designed experiments. These need to minimize the number of test runs, due to logistic and time constraints and the scarcity of active pharmaceutical ingredients, while still providing information about critical interactions. A sequence of experiments, similar to those in the child car seat example (Section 4.6.2) is often the best strategy.

10.4.4 Stability Assessment

Background A drug, in addition to being effective immediately after it has been produced, also needs to retain a high level of potency when kept in its container for a defined period of time. The drug's expiration date – the date after which it should no longer be used, as stamped on its container – is of particular interest. This is the elapsed time after manufacture for which the characteristics of a drug, packaged in a specified manner, can be expected to remain within established specifications with a high probability. Such specifications are typically taken as at least 90% of the dosage potency claimed on the label. An estimate of the drug expiration date is generally included in the dossier submitted for market approval to the regulatory body.

A stability study is required to determine a drug's expiration date. Statistics enters in designing the testing schedule, determining the number of samples to be analyzed, setting the drug characteristic specifications and establishing the expiration date from the resulting data. For further information on stability testing in the United States, see the FDA (1998) draft guidance on stability testing and various ICH studies.[8]

Example[9] Expiration dates for a new drug at four dosage levels were required. We describe the conduct of the study and its analysis at one dosage level.

Three batches of the product were manufactured at different times and packaged into sealed bottles of 100 tablets. Ten tablets were then selected from one bottle from each batch and ground and mixed. The potencies of each of the three mixtures were measured five times each; the resulting values and their means for each batch were recorded. This procedure was repeated at three-month intervals during the first year, at 6-month intervals during the second year, and annually thereafter, each time taking 10 additional samples from the same bottle for each batch.

8. www.fda.gov/cder/guidance/.
9. An example of a more complex stability study will be included in the ftp site for this book.

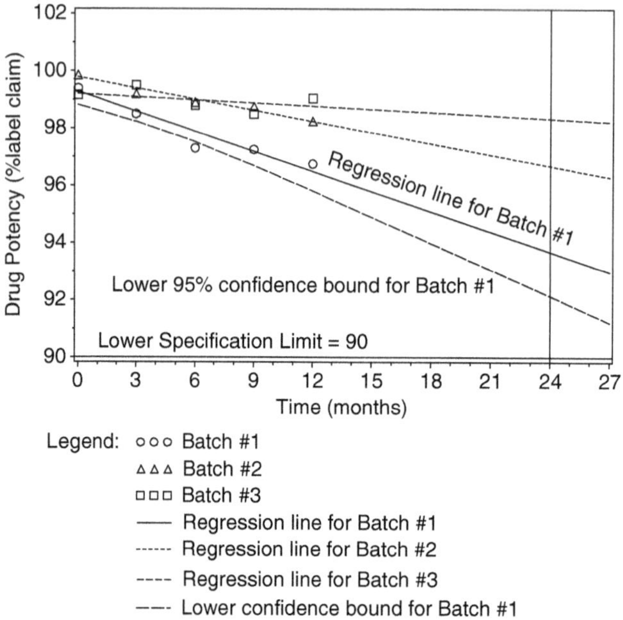

Figure 10.3 Mean drug potency at 3-month intervals during first year for three batches (and fitted regression lines) for stability study.

Figure 10.3 displays the mean potency for each batch at each readout time after one year. At that time, a preliminary estimate of the expiration date is desired. The plot of the data suggests a linear relationship between time and potency. However, the rate of degradation, as measured by the slope of the line for each batch, differed among the three batches. Separate regression lines were, therefore, fitted for each batch. It seemed reasonable from this plot, and from past experience, to extrapolate these lines to 2 years. The resulting fitted lines are shown in Figure 10.3, together with the lower 95% confidence bound on the mean for the batch that showed most degradation (batch 1).

Regulatory agencies usually grant expiration dates in yearly increments. Following standard industry practice, the lower 95% confidence bound on the mean for the worst batch was compared with the lower specification limit of 90% for potency (horizontal line in Figure 10.3). This led to a proposed expiration date of 2 years since the lower 95% confidence bound, extrapolated to that time, still exceeded 90%.

The study was to be continued for another 2 years to ensure that the results are maintained and, possibly, to determine whether a 3-year expiration date might be warranted.[10] The updated results will then be resubmitted to the regulatory agency, possibly with a request to extend the expiration date.

See Sidebar 10.2 for some further technical considerations.

10. The confidence bound will approach the fitted line, improving the chances of such demonstration, as additional data are obtained.

SIDEBAR 10.2: FORMULATION STABILITY STUDY: SOME TECHNICAL CONSIDERATIONS

The example illustrates a commonly used approach for assessing stability. This is by no means the only method for obtaining an expiration date – nor even the best from a statistical perspective. It is, however, widely used and enjoys broad acceptance by regulatory agencies. It also illustrates the type of compromise that frequently needs to be made between what might be the most desirable approach and what is doable in practice.

Matters of concern to statisticians include:

- *Number of batches.* Three batches were used in this study. To provide a better estimate of drug shelf life and inter-batch variability, it is preferable to sample more batches. This was not practical in this study due to cost and manufacturing capacity limitations, and because similar evaluations were also to be conducted at three other dosage levels. Testing a minimum of three batches has become common practice for obtaining stability data. Efforts were made to have the three batches be as representative of production as possible.
- *Frequency of testing.* Unless the product is expected to degrade very rapidly, the proposed frequency of testing seemed adequate for fitting a degradation curve.
- *Proper simulation of use conditions.* The trials should be conducted in a manner that simulates as closely as possible (or possibly accelerates) the different ways that the drug is handled, and might degrade in customer use. For example, consideration should be given to whether frequent opening or closing of the bottle, and periodic removal of tablets, especially under humid or other possible unfavorable conditions, may result in more rapid degradation of the remaining tablets.
- *Sampling plan.* The specifics of the sampling plan and the number of samples to be tested from each bottle depend upon the precision of the analytical method and the ability of the manufacturing process to produce uniform tablets. In this example, five measurements from a single grind of 10 tablets from a single bottle were judged to be satisfactory.[11]
- *Methodology.* Use of the lower confidence bound of the fitted equation for the worst batch to estimate the expiration date is a so-called quick and dirty method. More analytically rigorous approaches can be applied (e.g., taking the slope of the fitted degradation line for each batch to be a random variable). This would be especially appropriate if the study had involved sampling from more than three batches. With only three batches, the approach that was used is statistically defensible, although its exact properties are difficult to assess.
- *Confidence interval or tolerance interval?* One can argue that this evaluation calls for examining individual tablets (rather than a mix of 10 tablets) and for constructing a statistical tolerance interval, rather than a confidence interval on the mean, to estimate the expiration date. This would result in a wider interval (and possibly inability to justify a 2-year expiration date). Which type of interval to use depends upon the specific protection that the expiration date is intended to provide. The use of a confidence interval on the mean might be defended by the fact that the interval is being constructed on the worst batch. Also, this is the approach established by the regulators, and is reasonably well understood.

11. Measurement error tends to inflate the distance of the lower confidence bound from its fitted line and make it more difficult to demonstrate an acceptable expiration date.

10.5 CLINICAL STUDIES

> **There is no substitute for honest, thorough, scientific effort to get correct data (no matter how much of it clashes with preconceived ideas).**
> —Edmund C. Berkeley

10.5.1 Overview

A clinical research project involves a targeted evaluation and a series of clinical studies or trials that collect information to characterize the efficacy and safety of the new drug *in the intended patient population*. Since clinical studies involve humans, particular attention must be given to minimizing possible risks, while obtaining needed information.

The clinical development plan provides a starting point for this effort. It states the goals of the project, characterizes current and future markets and commercial competition, delineates the proposed clinical studies and their schedules, identifies go/no-go decision criteria, and provides other relevant information that will lead to submission for regulatory agency approval. Statisticians typically interact with colleagues in other disciplines in crafting and critiquing such plans.

Upon internal approval of the clinical plan and subsequent authorization by the regulatory agencies, work proceeds on planning the initial trials. For each trial, a protocol is created that provides instructions to the physician investigator(s). This document includes descriptions of the criteria for admitting subjects into the trial, the number of subjects to be studied and the rationale for that number, the nature and timing of initial data to be collected on subjects, the treatments and treatment regimens to be administered to the subjects, and the key efficacy and safety variables and how they are to be analyzed. Statisticians are intimately involved in the development of such protocols.

Clinical research typically consists of, at least, three phases: Initial human testing, exploratory studies, and large-scale confirmatory studies.[12] Each phase involves increasingly greater exposure of humans to the drug and requires success of the previous phases.

We briefly consider the goals of each phase and then illustrate some of the statistical issues by describing a specific Phase 2 study.

10.5.2 Phase 1: Initial Human Testing

The drug is introduced initially into a small sample (e.g., 50 to 100) of, most often healthy, human subjects. Some goals of Phase 1 testing are to:

- Evaluate the maximum tolerated dose (single and multiple doses).
- Characterize possible adverse effects.

12. Some projects also involve a fourth phase to obtain added specialized information – for example, to characterize the performance of the drug in pediatric patients.

- Determine doses and dose regimens likely to be successful.
- Characterize possible interactions with other drugs.
- Evaluate the pharmacokinetics (defined as "the study of the absorption, distribution and elimination of a pharmaceutical in the body as a function of dose and time (or) what the body does to the drug"; see Senn 1997).
- Evaluate the pharmacodynamic effects (defined as "the study of biochemical, physiological or therapeutic responses to a pharmaceutical in the body as a function of dose and time (or) what the drug does to the body"; see Senn 1997).

This phase typically involves intensive data gathering and analysis.

10.5.3 Phase 2: Exploratory Studies

The drug is next evaluated more extensively in humans, using typically around 200 to 300 patients with the target disease addressed by the drug to:

- Confirm proof of concept, for example, evaluate the effect of a surrogate marker. In particular, in studies of oral contraceptives, inhibition of ovulation is a surrogate marker for pregnancy prevention (which is specifically evaluated only in Phase 3) and can be determined more rapidly and with fewer consequences. Similarly, the CD4 count (a measure of the strength of the immune system) in HIV patients is a surrogate for survival, the ultimate measure of efficacy.
- Determine the dose response efficacy profile over time.
- Evaluate the safety of the drug.
- Explore the effects of the drug on different patient subsets (e.g., age, gender, and race).
- Consider various measures of efficacy and safety. For example, blood pressure (BP) is taken as systolic BP and diastolic BP, and with the patient in various body positions. One of these measurements is generally designated as the primary variable.
- Determine if gradually increasing dose at the beginning of treatment and/or gradually decreasing dose at the end of treatment are needed.
- Suggest doses and dose regimens to be used in Phase 3 studies.

Statisticians working in this area are involved in planning the study and its details, including submitting the test plan to the regulatory agency for approval, as well as the analysis of the results.

10.5.4 Phase 3: Large-Scale Confirmatory Studies

The preceding phases, if successful, are the prologue to the all-important Phase 3 studies. In this market-oriented stage of development, the drug is evaluated in a substantial number of patients (typically several thousand or more across multiple clinical centers) with the targeted disease (or other conditions requiring treatment) and for longer treatment durations than in previous phases. If successful, this stage provides the most important information to bring to the regulatory agency to obtain approval to market the drug. In addition to reinforcing the results of the earlier studies, this phase aims to:

- Obtain evidence of the efficacy and safety of the drug at doses, and dose regimens, intended for application to patients.
- Evaluate the drug by patient subsets and for special groups (e.g., children and elderly).
- Quantify further health outcomes from drug treatment (e.g., number of hospitalization days, patient well-being assessment, and need for caregiver support).
- Compare the advantages and disadvantages of the drug to other drugs on the market (e.g., efficacy, adverse effects, speed of impact, and dosage frequency).
- Determine if any treatment withdrawal effects occur (e.g., drug dependency or rebound to baseline severity of illness).

When data are obtained from multiple studies, the resulting information is frequently combined by a meta-analysis (see Sidebar 10.3).

SIDEBAR 10.3: META-ANALYSIS

Individual clinical trials may not be large enough to provide adequate sensitivity to characterize differences in patient subsets of interest, such as different age groups, or to assess presumably rare events, such as encountered in safety assessments. Meta-analysis involves *pooling data* across a number of relevant clinical trials. It permits evaluations to be conducted with increased confidence due to the larger number of patients. Pooling also provides synthesis of the collective knowledge about the drug's safety and efficacy and an overall assessment of its benefits and risks.

Issues in meta-analyses include deciding which clinical trials can be pooled and over what time periods, assigning potential weights to each source of information, and determining appropriate data analysis techniques. Hedges and Olkin (1985); D'Agostino and Weintraub (1995); and Ionnidis, Cappelleri, and Lau (1998) provide further details.

At the successful conclusion of this phase, the market formulation is finalized, and regulatory approval to market the drug is sought. Tracking of, at least, a subset of patients may, however, continue to assess long-term effects – depending on the type of disease being studied. For example, tracking is likely to be continued for a cancer study, but not for a short-term disease or condition, such as a mild upper respiratory infection.

10.6 CASE STUDY: PHASE 2 CLINICAL TRIAL

A theory is worthless without good supporting data.
—A.L. Romanoff

A Phase 2 clinical study of a new antianxiety drug was conducted by a pharmaceutical company in the early 1990s. The target population was patients with Generalized Anxiety Disorder (GAD), defined as excessive anxiety or worry about a number of events or activities occurring on a majority of days over a period of at least six months. The new drug, if proven effective, was estimated to eventually generate over $400 million in peak annual sales.

10.6.1 Primary Objectives

The study's primary objectives were to quantify the proposed new drug's efficacy in treating GAD and to identify and assess possible adverse effects.

10.6.2 Design of Trials: Some Fundamental Considerations

In clinical studies, as in many other applications, often the most important thing that statisticians do is to help plan the study, including the data acquisition process.

Doses of the New Drug The impact of the drug is to be characterized at three dosage levels (10 mg, 20 mg, and 30 mg), administered daily. The low dose was anticipated to provide limited benefit, the high dose was expected to be maximally effective, and the middle dose was anticipated to provide a suboptimal benefit.

Selection of Investigators In selecting investigators, consideration is given to various factors, including the likely number of patients eligible for the protocol in the investigator's practice, location of the practice, adequacy of facilities, and "track record" of enrolling patients and conducting clinical studies effectively.

Efficacy Measurement Drug effectiveness is to be evaluated by an experienced physician as the reduction in patient anxiety from that experienced just prior to treatment. The total score on the Hamilton Anxiety Scale (HAMA), a subjective rating method (see Sidebar 10.4), is to be used as the numeric metric of anxiety.

SIDEBAR 10.4: THE HAMA

GAD is measured by physicians using the HAMA, a multi-item rating score that assesses the patient's level of anxiety (see Hamilton 1959). This requires physicians to evaluate patients for the following 14 behaviors, rating each on a 5-point scale (none, mild, moderate, severe, and very severe): anxious mood, tension, fears, insomnia, intellectual (cognitive), depressed mood, general somatic (muscular), general somatic (sensory), cardiovascular symptoms, respiratory symptoms, gastrointestinal symptoms, genitourinary symptoms, autonomic symptoms, and behavior at interview.[13] The total of the 14 ratings is the primary response variable of the study.

Since a patient's true state is neither constant nor fully known, the investigators must make educated assessments of the 14 features, based on their observation, knowledge and experience. These are, by necessity, somewhat subjective – as compared to more quantitative assessments, such as taking a patient's blood pressure – and generally result in greater measurement error in the total score than more direct measurements. This inflates the variability of the evaluation of the drug's efficacy, and thus increases the number of patients that must be studied to achieve a desired level of statistical precision. In addition, there may be differences between physicians in rating patients' conditions.

An initial meeting was held to arrive at a common understanding of the way that physicians rate each item and to develop and document in the study protocol as uniform a procedure as possible.

Safety Measurement Drug safety is to be evaluated from assessments of the participants' body weight, blood pressure, laboratory tests and electrocardiograms, and from a physical examination to identify and document any adverse indications. Such assessments are made both immediately prior to treatment and at specified times during the study.

Comparator Group Inclusion of a comparator group against which the impact of a proposed new drug is to be compared is essential in many clinical studies. Frequently, a placebo (0 mg of new drug) is used as a comparator. One or more drugs in current use may also be included in addition to, or sometimes in place of, the placebo. In this study, a placebo was chosen as the comparator. In the subsequent discussion, a "treatment" will refer to one of the three levels of dosage *or* the placebo.

Selection of Participants The study admission (inclusion/exclusion) criteria and procedures for enrolling subjects were carefully considered and documented in the study protocol. These influence the enrollment rates, and the ability to generalize the study results and maintain the blinding of the investigator and subjects to subject-treatment assignments. Participants were recruited by clinical investigators from their patients according to the study protocol; this required patients to have an initial HAMA total score of at least 18. Advertising and the involvement of patient advocate groups are two other methods of recruiting patients.

Administration of Treatments Each patient received the same treatment throughout the trial. A possible alternative would be to change the treatment at specified times. This, however, is rarely done due to the likelihood of carryover effects.

Number of Centers A single-center study is easier to manage, but may not meet enrollment targets within the desired time frame and may produce less generalizable results than multiple centers. It is, therefore, usually advisable to employ multiple

13. Instructions for scoring these items will be shown in the ftp site of this book.

centers. This broadens the scope of the study and increases its validity. In this trial, nine study centers were used.

10.6.3 Design of Trials: Some Details

Titration and Tapering For some drugs, a targeted dose is achieved by starting at a lower dose and stepping up (titrating) the dose until the targeted level is achieved. This allows subjects to adjust to the drug more successfully and avoid adverse effects. In addition, for some drugs, stepping down (tapering) the dosage until the treatment is discontinued entirely helps subjects avoid adverse effects due to abruptly stopping treatment. For this drug, neither titration nor tapering was felt necessary.

Randomization In the vast majority of clinical trials, the assignment of treatments to patients is randomized to avoid selection bias. Sometimes sampling is stratified based, for example, on disease severity, age, or race. In this study, patients were randomly assigned to treatments so that approximately the same number at each center received each treatment.

Blinding Most trials are "double-blind," that is, neither patients nor investigators know the assignment of treatments to patients. In most cases, those sponsoring the study are also blinded. Double blinding was used in this study.

Study Duration The treatment should be long enough to allow the impact of a drug to be observed. In anesthetic and dental pain relief studies, treatment effects are measured in minutes or hours. In HIV or Alzheimer's disease studies, it may require a year or longer to observe treatment effects. In this trial, patients were to be treated for 5 weeks; this is the time that extensive empirical evidence shows is required to detect the effect of a drug in patients with GAD. Longer studies will be conducted in Phase 3 to assess possible longer-term adverse effects and to meet regulatory requirements.

Measurement Frequency HAMA assessments are to be made just prior to treatment and weekly thereafter for 5 weeks. When the onset of a drug's impact is important, observations are typically taken more frequently just after the start of treatment. Measurements are costly and inconvenient; frequency of testing needs to be justified, based, for example, upon concerns about adverse reactions.

10.6.4 Analysis Plan: Fundamentals

Deciding Upfront The statistician, in planning the study, needs to define and document the key analyses to be conducted on the resulting data and the underlying assumptions. This is needed to determine required sample sizes, to establish appropriate data collection methods, and to avoid the results biasing the analysis. This maximizes scientific credibility – even though initial plans do get changed for various reasons, including advances in statistical methodology, opinions or new guidelines from regulatory agencies, and failure of the data to meet initial assumptions. Changes in analysis methods should, like the original analysis plans, be well documented.

The analyses may involve estimation, hypothesis testing, or both and need to be clearly stated in the protocol. Frequently (as in this example), confidence intervals provide more useful information than hypothesis tests (see Borenstein 1994).

Application to Example The key analysis of the data from the study is to relate the change from baseline (i.e., prior to treatment) of the mean HAMA total score to the administered drug dosage after 5 weeks of exposure. This is to be based upon fitting a linear trend to the observed change in mean score against dosage level (0 mg, 10 mg, 20 mg, and 30 mg). An average reduction of four units in the HAMA total score for a 30-mg dosage, or of 1.33 score units per 10 mg of dosage, relative to the placebo, is considered clinically important and, if present, should have a high probability of resulting in statistical significance.

In addition, it is desired to estimate the mean difference in the change in HAMA total score after 5 weeks between each of the three dosage groups and the placebo group with an (individual analysis) two-sided 95% confidence interval not exceeding ±3 HAMA total score units.

Similar specifications were established for assessing drug safety.

10.6.5 Sample Size Determination

Basic Concept A key question, invariably, is how many patients to include in a clinical study to obtain the desired information with a specified degree of precision. Actual sample sizes will, in addition, be impacted by practical and regulatory considerations.

Samples size is traditionally determined by formulating the null hypothesis to be evaluated and the associated allowable Type I error (rejection of the null hypothesis when there is no effect) and Type II error (failure to reject the null hypothesis if a specified alternative hypothesis is, in fact, true) probabilities. In setting a Type II error probability, the magnitude of the true effect (sometimes referred to as a clinically relevant difference) for which a high probability of hypothesis rejection is desired needs to be specified. An initial estimate of the standard deviation associated with a measurement (or rate of occurrence for "success" or "failure" measurements) is also required.

A further criterion for sample size determination that is gaining increased recognition is the number of patients required to achieve a specified expected length for the confidence interval to contain the drug's estimated average impact at a specified dosage, compared to, say, a placebo.

Application to Case Study In this study, the null hypothesis is that the new drug is no better than the placebo in reducing the mean HAMA total score over a 5-week exposure period. The alternative hypothesis is that the score after 5 weeks of exposure decreases at a rate of 1.33 score units per 10-mg dosage, according to a linear relationship, with increase in drug dosage from 0 to 30 mg.

The Type I and Type II error probabilities were set to be 0.05 and 0.15, respectively. In addition, the standard deviation for an individual HAMA total score was, based upon past experience, taken to be eight units.

Using this information in a sample size formula for a linear trend test, the number of patients to study was determined to be 65 for each of the three dosage levels and for the placebo, or a total sample size of 260 (see Mendenhall 1968, p. 416). The actual number of patients totaled 290 to allow for early withdrawals (see Sidebar 10.5 for further discussion). Twenty of the originally selected patients did, indeed, withdraw from the study before receiving any treatment, resulting in actual sample sizes of 69, 68, 68, and 65 for the placebo and three dosage groups, respectively.

In addition, it was determined that with a sample size of 65 for each of the four treatment groups, one can expect to estimate the true mean difference in the 5-week change in HAMA total score for each dosage group versus the placebo group within a 95% confidence interval of ±2.75 HAMA total score units.

The actual allocation of patients to the nine treatment centers ranged from 23 to 71. Equal allocation could not be achieved because the centers could not enroll subjects at the same rate; requiring such equal enrollments would have delayed completion of the study.

SIDEBAR 10.5: A TECHNICAL NOTE

The sample size determined in the example ignores various practical and theoretical considerations. For example, the Type I and Type II error probabilities that are being controlled are based upon a single comparison: a linear trend test. If one looked at comparisons simultaneously (e.g., comparison of each of the individual dosages versus the placebo), the overall Type I error probability with regard to finding, *at least, one* significant difference is larger. Numerous methods have been developed to control Type I error when multiple comparisons are desired (see Proschan and Waclawiw 2000).

10.6.6 Data Collection

Importance It is critical to the success of a clinical study that the data needed to conduct the desired analyses are collected and recorded completely, consistently, and unambiguously. Those developing the detailed procedures for data collection must be intimately familiar with the general environment in which the data are collected, as well as the analyses to be conducted; they must also clearly document these procedures. All involved – including those building the database, monitoring the study sites, querying the data, developing the tabulations, conducting the analyses and preparing the study report – need to understand the way data are to be gathered and checked, the general structure of the database, and how the data are to be analyzed and reported.

So what specific data should be collected in a particular clinical trial? Whatever might be needed to support the study objectives, including data about variables that may be of future interest. For example, body mass index may be an important covariate. Its value is determined from a patient's weight and height, both of which need to be measured.

Other data may be required to relate an observation to a given reference point. For example, blood pressure needs to be measured, both prior to and throughout the study, if we might wish to evaluate the impact of a treatment on blood pressure.

Application to Case Study In this trial, information on physical characteristics of study participants is to be obtained at various times, in addition to their HAMA scores and other response variables. These characteristics include physical variables, such as height and weight, and diagnostic factors, such as "episode" duration and the previous use of benzodiazepines.

10.6.7 Preliminary Data Analysis

Before embarking on the analysis of the data, it is useful to obtain an initial impression of what the data look like. This may suggest other issues to keep in mind in the subsequent analysis. Figure 10.4 displays the HAMA total score for four "typical" patients.

Some of the insights that can be gleaned from this plot are:

- Patients begin the clinical trial with different initial HAMA total scores. The analysis will account for these differences by using the pretreatment score as a covariate.
- Some patients did not participate for the full 35 planned days of treatment.
- All four patients showed improvements over the course of the study. This is not uncommon; there is a natural so-called "regression to the mean" for many diseases (e.g., you may see a doctor when your cold gets bad, but the natural

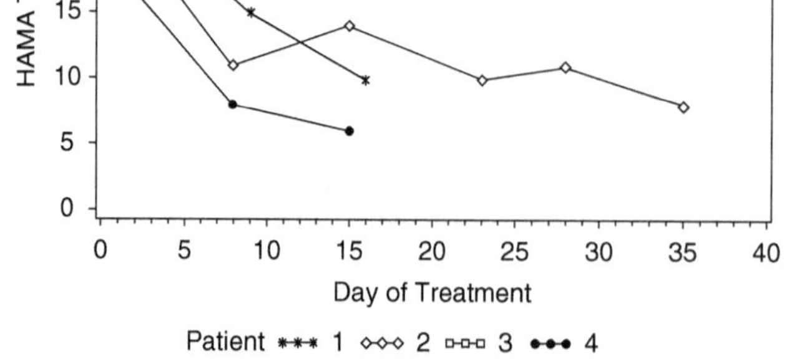

Figure 10.4 HAMA total score for four patients.

course of a cold is to improve over time; the same improvement might have occurred even if you had not been treated; see Weisberg 2005, Chapter 1). In addition, the act of participating in a clinical trial, being seen by a physician, and taking a "treatment" – even an inert one – often in itself results in some perceived improvement ("Hawthorne" effect). For this reason, a placebo was included in the study as the standard against which actual drug dosages are to be compared.

- There is not always a decrease in a patient's HAMA total score from one week to the next. This may be due to short-term variability in a patient's condition, measurement error, a natural fluctuation in the course of the disease or, most likely, a combination of factors.
- The improvement tends to level off over time. One reason for this is that the HAMA total score has a lower limit of zero. This supports the decision to end the study after 5 weeks.
- Not all patients are evaluated on the treatment day specified in the protocol. In fact, none of the four patients were evaluated exactly 7 days after the start of treatment – three were evaluated on day 8 and the other one on day 9.

10.6.8 Missing Measurements

Missing HAMA Scores HAMA assessments for some of the item scores were missing for some cases. A decision was made to use an observation if no more than two of its14 item scores were missing. Missing item scores were imputed from the mean of the scores of the nonmissing items to estimate a total score for all 14 items.

Dropouts Patients routinely drop out of clinical trials due to real or conceived adverse experiences or side effects from using the drug, or the belief that "the drug is not doing anything for me." In other cases, dropout may be for reasons that, presumably, are unrelated to the treatment – such as forgetfulness, time pressures, or lack of discipline.

It was decided that all patients who received at least 4 days of treatment and had at least one HAMA assessment after the start of treatment should be included in the analysis. In addition, the number of dropouts, the ascertained reasons for dropout (and especially whether dropout might be related to treatment), and how each dropout was handled were to be reported.

Missing Measurements and Measurement Irregularity Patients are frequently not assessed at the times specified in the protocol. This results in (1) readings taken at a day during the week that is different from the designated day, and (2) weeks being skipped altogether.

This issue was addressed as follows:

- In the summary tables and figures, the results were grouped by the week they were taken, rather than when they were scheduled. Figure 10.5 provides a graphical display of patient assessment times and their associated groupings.

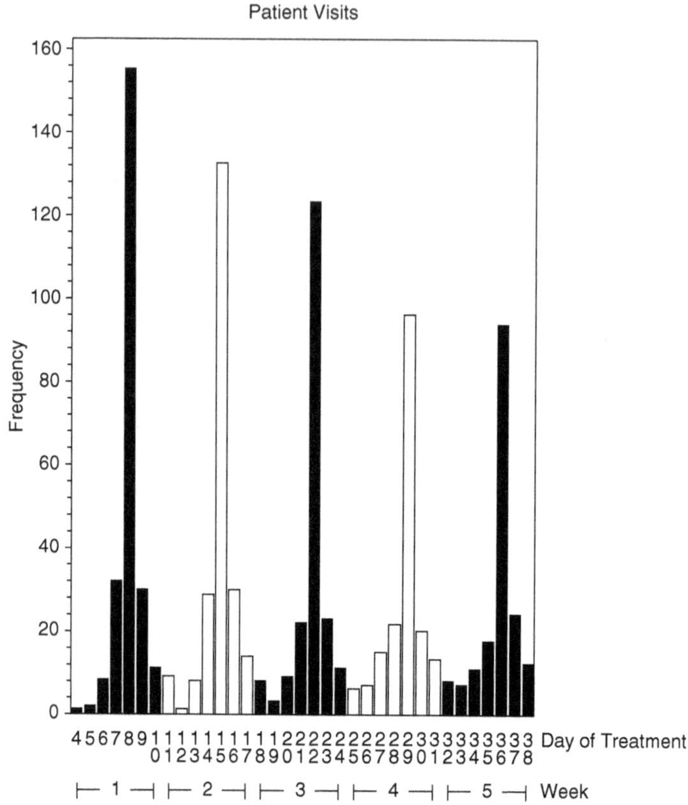

Figure 10.5 Actual patient assessment times and associated groupings.

- In regression analyses of HAMA total score versus time of treatment, the actual, rather than the scheduled, treatment days were used.
- For those patients for whom a 5-week measurement was not available, the last observed value was used. Other approaches, such as fitting a statistical curve to the available data, could also have been used.

10.6.9 Results of Study

Assumptions It is assumed that the standard deviations of the HAMA total score and of the other key response variables are the same for each treatment group. A less important assumption (due to the central limit theorem) is that the total scores are normally distributed. These assumptions need to be evaluated. If they do not appear to be met, alternative methods may have to be used. For example, if the standard deviations of the changes in HAMA total scores differ among the treatment groups, an analysis that does not assume equal standard deviations is used. The protocol and a companion statistical analysis plan specify the assumptions underlying the methodology, how these are to be evaluated, and what alternatives might be used if they are found not to apply.

Assessment of the data in this study did not suggest any of the assumptions to be questionable.

Summary Tables and Figures The following results are, after adjustment of the observations for the pretreatment HAMA total score, which was taken as a covariate. Similar results were obtained without such adjustment.

Table 10.1 gives the results after each of the 5 weeks of the study (and final 5-week projected values adjusted for dropouts and missing values) showing for each treatment:

- the sample size (n)
- the mean change from baseline of the HAMA total score

TABLE 10.1 Summary of Changes from Baseline in HAMA Total Scores (After Adjustment for Covariate) After 5 Weeks

Week	Treatment	n	Mean Change from Baseline[a]	Mean Difference Dosage Group – Placebo	Standard Error of Mean Difference	95% Confidence Interval on Mean Difference	
1	Placebo (mg)	58	−5.2				
	10	62	−5.9	−0.7	0.9	−2.4	1.1
	20	65	−5.5	−0.3	0.9	−2.0	1.5
	30	54	−6.5	−1.3	0.9	−3.1	0.6
2	Placebo (mg)	57	−6.9				
	10	55	−8.9	−2.0	1.2	−4.3	0.3
	20	58	−9.2	−2.3	1.1	−4.5	0.0
	30	48	−8.4	−1.5	1.2	−3.8	0.9
3	Placebo (mg)	50	−9.0				
	10	50	−10.6	−1.6	1.3	−4.1	0.9
	20	57	−10.0	−1.0	1.2	−3.4	1.4
	30	38	−11.1	−2.1	1.4	−4.7	0.6
4	Placebo (mg)	46	−9.3				
	10	47	−11.8	−2.5	1.5	−5.0	−0.1
	20	44	−11.8	−2.5	1.5	−5.0	0.0
	30	37	−11.9	−2.6	1.6	−5.2	0.0
5	Placebo (mg)	41	−10.3				
	10	46	−12.0	−1.7	1.3	−4.5	1.1
	20	46	−13.0	−2.7	1.3	−5.5	0.2
	30	35	−12.2	−1.9	1.4	−4.9	1.2
Final[b]	Placebo (mg)	64	−8.7				
	10	64	−10.2	−1.5	1.3	−4.1	1.1
	20	67	−10.8	−2.1	1.3	−4.7	0.5
	30	55	−10.0	−1.3	1.4	−4.0	1.4

[a]Adjusted for covariate (baseline HAMA total score).
[b]Projected 5-week results, adjusted for dropouts and missing data.

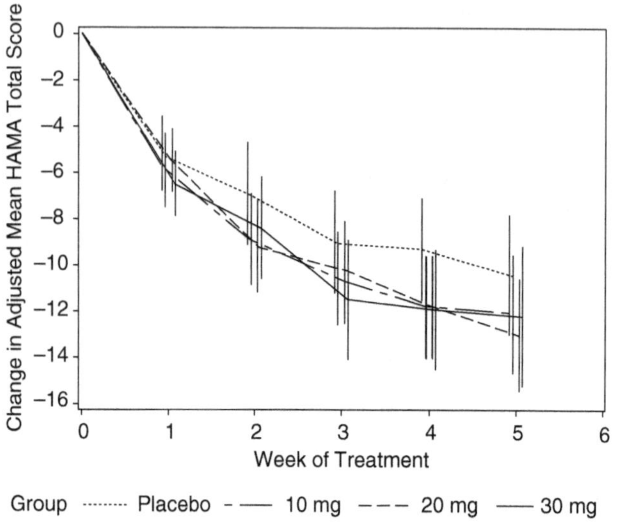

Figure 10.6 Mean change from baseline in HAMA total score (after adjustment for covariate) for four treatments with 95% confidence intervals.

- the mean difference of the change from baseline in the HAMA total score, relative to the placebo
- the standard error of the preceding mean difference
- a two-sided 95% confidence interval on the preceding mean difference

Figure 10.6 is a plot of the weekly adjusted mean changes from baseline of the HAMA total score, with two-sided 95% confidence intervals, for each treatment.

Findings The following major conclusions (using principally the final projected values) were drawn from the results shown in Table 10.1 and Figure 10.6 (and various other analyses):

- The mean HAMA total scores for each of the three dosage groups was significantly reduced from its baseline value over time – but so also were those for the placebo. The mean changes in the final HAMA total score from baseline ranged from 10.0 to 10.8 among the three dosage groups. The placebo group, however, also showed a mean change of 8.7. Most of the changes, took place during the first three weeks for each of the treatments.
- The mean differences (i.e., final HAMA total score – baseline HAMA total score) between an individual dosage group and the placebo group was *statistically insignificant* at a 5% significance level, for each of the three dosages. Equivalently, each of the two-sided 95% confidence intervals on the differences in the mean change for dosage group versus placebo overlapped 0.
- The best final result was obtained for the 20-mg dosage group (with a mean improvement of 2.1 relative to placebo), rather than, as had been expected, for the 30-mg dosage (with a mean improvement of 1.3). The data, thus,

refuted the initial assumption of a linear relationship between dosage and improvement. (A statistical test of the slope of a hypothesized linear relationship between dosage and HAMA total score resulted in a p value of 0.30 – providing no evidence of a linear relationship.) Any "lift" that the drug might have provided appeared to have come mainly from going from no dosage to a 10-mg dosage.

- The *lower 95% confidence bounds* on the difference between each of the dosage group mean changes in HAMA total score and that for the placebo group overlapped –4.0 for each of the dosage groups (but only barely for the 10-mg and 30-mg dosages). Thus, one cannot rule out the *slight possibility* that the true impact of the new drug does, indeed, meet the stated goal of an average reduction of four units in the HAMA total score.[14] A larger sample is required to reduce the lengths of these confidence intervals and to get more precise assessments.
- There is a great deal of variability in the results among individual patients in the changes of the HAMA total scores (analysis not shown here). This suggests the possibility that the drug might be effective for a particular *subgroup* – for example, patients with especially high starting HAMA total scores or elderly patients. Additional analyses need to be conducted to assess such conjectures.

10.6.10 Reports

The findings of the study were summarized in a formal report, prepared for the regulatory agency.

Written reports tend to play a more important role in pharmaceutical studies than in many other applications due to the seriousness of the consequences and the requirements of regulatory agencies. Thus, the trend to replace written reports with less formal PowerPoint presentations has not been as marked in the pharmaceutical industry as elsewhere.

Reports are generally directed at one or a combination of the following audiences:

- Regulatory agencies in the United States or abroad
- The professional community (often via papers in refereed journals or presentations at technical meetings)
- The general public and potential customers (sometimes in the form of press releases)

There are important differentiations even within these areas. For example, the "professional community" can be fellow researchers in a specialized area or the community of practicing physicians. Clearly, the nature of the presentation will vary depending upon the target audience.

14. The upper 95% confidence bounds also suggested the possibility that the drug could actually be slightly less effective than the placebo since these were all positive.

For clinical studies, a report is generated for each clinical trial. Statisticians, together with many others, contribute to such reports (often resulting in a long list of coauthors). The details of the statistical analyses are frequently reported in an appendix.

10.6.11 The Bottom Line

Animal testing had, in this study, provided some encouragement that the proposed drug might meet its goal of reducing the mean HAMA total score in humans by at least four units relative to a placebo in a 5-week period. Phase 1 testing, involving a small number of humans that did not have the disease, did not suggest any major stumbling blocks. Therefore, proceeding to Phase 2 testing seemed warranted.

The results from Phase 2 showed that it was not very likely that the drug would provide the desired mean improvement. In addition, in subsequent analyses, no subpopulation emerged for which the drug provided promise of being especially effective. There was, moreover, some evidence of adverse side effects at the highest dosage.

As a result of these findings, the clinical development program was terminated. This was, of course, disappointing. However, it avoided the much greater cost and human disappointment that would likely have been incurred from Phase 3 clinical trials. Thus, the Phase 2 trial did its intended job and resulted in a not unusual outcome. A closeout report was filed with the FDA. This was much briefer than the dossier that would have been prepared if the outcome had been successful.

We note that although a variety of highly sophisticated statistical analyses were conducted (most of which are not shown here), the major results could be summarized by a simple tabulation and some incisive graphical displays. This was a consequence of the careful planning that went into the test program.

10.7 REGULATORY REVIEW

Successful Phase 3 testing culminates with the preparation of a dossier, providing the collective evidence from the clinical trials and requesting market approval, for submission to regulatory agencies in the countries in which the drug is to be sold.

The extensiveness of the reviews of such dossiers by regulating agencies varies from very thorough in the United States to less intensive in some countries. Members of the regulatory agency typically raise further questions and company personnel respond to these either in personal appearances or by written responses. Additional clinical trials may also be conducted in response to requests or to support commercialization.

Statisticians become involved at this stage in explaining the studies and their analyses. They also participate in the planning and analysis of any additional trials.

10.8 SOME OTHER APPLICATIONS

10.8.1 Process Improvement

The large human risks and potential gains associated with the development and introduction of a new drug call for a highly disciplined approach. Similar challenges have led to the use of Six Sigma concepts in other businesses. In the pharmaceutical industry, regulatory agencies help ensure a structured and data-oriented approach. Nunnally and McConnell (2007) consider Six Sigma for pharmaceuticals.

There are also further important opportunities for process improvement similar to those discussed in other chapters. An important example is reduction of the lengthy time span from conception of a product to bringing it to market, while still fully satisfying regulatory requirements.

10.8.2 Manufacturing

Once a drug has been approved for general use, the manufacturer is anxious to proceed with high-volume production as rapidly as possible. After successful scale-up, the manufacturer needs to ensure that the product continues to be built according to specifications.

The statistical and other issues raised at these points are similar to those described in earlier chapters. These are especially important in pharmaceutical applications because of the serious consequences of a potentially faulty product.

10.8.3 Field Performance Tracking

In light of the criticality of pharmaceutical products, their performance in the field is carefully tracked, per our discussion in Section 9.6.2. Better records are, therefore, generally maintained for pharmaceuticals than for many other products – although these may not always be readily accessible due to privacy issues. The resulting evaluations – often based upon a large number of total observations – might identify subgroups that may especially benefit from a drug (see Sidebar 10.6).

SIDEBAR 10.6: THE PHARMACEUTICAL STATISTICIAN AS A DATA DETECTIVE

Iressa™ is a drug approved for treating lung cancer. Based upon the results of Phase 3 clinical trials and experience in the field, it was found to be highly effective for about one-seventh of the patients that were treated with it, and ineffective for the remainder.

Those responsible for designing the clinical trial and the field database kept good records on the background of patients who received the drug. These records were searched for what might differentiate the subjects for whom the drug was successful from those for whom it was not. These analyses suggested that the drug was more effective in women than men, in nonsmokers than smokers and in Japanese than non-Japanese patients – and they estimated the success odds associated with different groups. This suggested groups that *might* be targeted for this expensive (then $1,900 per month) drug.

These results provided the starting point for research by other scientists. By exploring the biological makeup of the most susceptible groups, it was found, according to a *New York Times* article (April 20, 2004), that "The people who improve sharply have a genetic mutation in their tumors that makes their disease highly vulnerable to the drug." Confirmation of this result will allow doctors to determine better who should and should not receive the drug. Moreover, it introduces opportunities for potential proactive measures for individuals that have the identified genetic mutation, and also for research on the possible impact of the drug on individuals with other diseases.

This example illustrates one way that pharmaceutical statisticians contribute to interdisciplinary research teams. In addition, it reiterates the importance of advance planning to obtain potentially useful data. Such analyses also need proceed cautiously. With many such assessments, some groups are going to appear significant just due to chance. Thus, any conclusions need to be validated by biological explanations, as in this example, and by further empirical validation with independent data.

Manufacturers also collect, summarize, and interpret spontaneously reported adverse events to characterize further the safety profile of the drug. Such postmarketing surveillance has an epidemiological focus. For example, adverse event data collected through patient self-reports and through national health registries are routinely subjected to data mining.

10.8.4 Other Activities

Additional activities in bringing a new drug to market in which statisticians might become involved include:

- Performing post-approval clinical trials and continuing previously started programs to assess long-term impact, as part of a larger, more comprehensive risk management program
- Education of the medical and patient communities, and further publication of results
- Product extensions (e.g., a once-a-day to replace a three-times-a-day formulation)
- Life cycle management (e.g., evaluating the impact of reducing the dosage, or changing the treatment regimen)

10.8.5 Opportunities Beyond the Pharmaceutical Industry

Regulatory agencies also provide important opportunities for statisticians. The challenges include developing statistically based rules for drug approval, assessing the adequacy of the clinical trials proposed for a new drug, and then evaluating the validity of the analyses conducted by the manufacturer – culminating in a decision on whether to move on to the next phase of testing or even approval of the drug for a specified market.

There are also significant statistical applications in health-related areas beyond the pharmaceutical industry, such as medicine and hospital administration (see Section 13.6.1). A further related area, epidemiological studies, is discussed briefly in Section 13.6.2.

10.9 SOME STATISTICAL TOOLS USED IN PHARMACEUTICAL APPLICATIONS

Those engaged in pharmaceutical research use many of the same tools for data analysis as those in other application areas, as suggested by the examples. Commonly used statistical methods in discovery research include linear and nonlinear regression and fixed and mixed effects linear models. Partial least squares, cluster analysis, class prediction (sometimes including regression trees), and visualization of high-dimensional data are gaining popularity. In drug safety evaluation, the statistical techniques used range from simple parametric and nonparametric tests and the construction of statistical intervals to mixed models for repeated measure studies, and from categorical and survival-type models to models for genomic data. In stability studies, frequently used experimental designs include fractional factorial and incomplete block plans.

New methods continue to be developed and applied. According to Senn (1997, p. 23), topics especially important to those involved in drug development, introduced in the 20 years starting in 1972, are generalized linear models, proportional hazards models, the log-rank test, ordered categorical data and Gibbs sampling. Senn adds that "the 1990's have seen considerable work on meta-analysis, dose-finding, crossover trials, equivalence studies, surrogate endpoints, repeated measures, survival analysis, multiple endpoints, sequential analysis, compliance and missing data (to mention but a few areas of particular relevance to drug development)."

These powerful methods, together with improved computational capabilities, have allowed pharmaceutical statisticians to make increasingly greater contributions. At the same time, as is evident from the examples in this chapter, the new tools should not detract from the fundamentals, such as providing expert guidance in the study planning process and providing insightful, graphical displays of the resulting data.

10.10 SOURCES OF FURTHER INFORMATION

Senn (1997) discusses many of the issues that arise in the planning and analysis of regulatory clinical trials. Guyton and Hall (2005), a guide to physiology, also include important information for statisticians who plan to go into medical or pharmaceutical research. Armitage and Colton (2005) and Chow (2003) provide further technical information.[15]

15. The ftp site for this book will include a listing of additional relevant articles in specialized areas.

The statistical methods employed in pharmaceutical research continue to evolve rapidly. For updates, see the following journals:

- *Biometrics* (International Biometric Society)
- *Clinical Trials: Journal of the Society for Clinical Trials* (Society for Clinical Trials)
- *Controlled Clinical Trials* (Elsevier Science)
- *Drug Information Journal* (Drug Information Association)
- *Statistics in Medicine* (John Wiley & Sons)
- *Statistics in Pharmaceutical Research* (American Statistical Association)

10.11 MAJOR TAKEAWAYS

- The process of discovering and developing new pharmaceutical products is lengthy and complex. The recognized and profound value of the scientific method in pharmaceutical research and the consequent need for a statistical approach has resulted in the pharmaceutical industry becoming one of the largest employers of professional statisticians; they typically become members of study teams for new drugs.
- Pharmaceutical products need to be proven safe and effective by statistically valid preclinical and clinical studies.
- Preclinical studies include discovery research, drug safety evaluation in animals, formulation development, and stability assessment.
- Clinical studies are targeted to meet the requirements of regulatory agencies whose approval is needed to market the product. They consist of, at least, three phases: initial human testing, exploratory studies, and large-scale confirmatory studies, each presenting important statistical issues in study planning and analysis.
- Other important statistical applications arise in process improvement, manufacturing, field performance tracking, and also in health-related activities beyond pharmaceuticals.

DISCUSSION QUESTIONS

General

1. What are some traits that seem especially important for successful statisticians working on pharmaceutical applications?
2. How might you improve the study to evaluate stability described in Section 10.4.4?
3. In the example in Section 10.4.4, "efforts were made to have the three batches be as representative of production as possible." How might this be done?

4. In the clinical trial (GAD) study in Section 10.6, patients were evaluated prior to treatment and weekly for 5 weeks after the start of treatment. What are some other evaluation schedules that might be used, and what are their pros and cons?

5. In the GAD study and similar clinical trials, what are some alternatives for grouping observations by time on therapy?

6. The participants in a clinical trial are generally selected from volunteers. How might this bias the results and how might you try to minimize such bias?

7. Suggest added examples of important opportunities for process improvement in the pharmaceutical business.

8. How might you go about working toward a "reduction of the lengthy time span from conception of a product to bringing it to market, while still fully satisfying regulatory requirements" (Section 10.8.1).

9. Research and present further developments that may have occurred in the Iressa™ study (Section 10.8.3) since the publication of the April 2004 *New York Times* article.

10. Suggest some further examples of statisticians acting as "data detectives."

Technical

1. In the example in Section 10.4.4, a suggested more rigorous method of analysis is to take "the slope of the fitted degradation line for each batch to be a random variable." Elaborate on how you might use this approach in estimating the drug's expiration date.

2. Why might one want to use a statistical tolerance interval rather than a confidence interval in the example in Section 10.4.4? Elaborate on the comment "which type of interval to use depends upon the specific protection that the expiration date is intended to provide." Contrast the two types of intervals and compare them to a prediction interval on a future observation.

3. We say (in Section 10.6.3) that in some clinical trials, sampling "is stratified based, for example, on disease severity, age, or race." What are the advantages of stratified sampling in such studies?

4. In the GAD study, it was determined (Section 10.6.5) that a sample size of 65 for each of the four treatment groups is required to achieve the desired Type I and Type II error probabilities. Verify this result.

5. It was also claimed in the GAD study that a sample size of 65 would result in a 95% confidence interval of expected length ±2.75 HAMA total score units on the 5-week difference in the mean scores for a particular dosage versus the placebo. Verify this result. How might you assess the degree to which the actual confidence interval attained will vary around its expected length?

6. In the GAD study, "missing item scores were imputed from the mean of the scores of the nonmissing items to estimate a total score for all 14 items" (Section 10.6.8). How might this be done? What are some potential problems? What are other possible ways of addressing missing scores?

7. In the GAD study, what other statistical approaches might be used to compare the effect of a drug to that of a placebo? How do these compare with regard to the underlying

assumptions, handling of missing data, adjustment for variation in baseline GAD, and determination of sample size?

8. In the analysis of the GAD study (Section 10.6.9), what further treatment group comparisons might be of interest? What approaches can be used to control the experiment-wise Type I error? What impact would these approaches have on the power of the experiment and on the required sample size to achieve the same power?

9. In the GAD study, all patients who received at least 4 days of treatment and had at least one HAMA assessment after the start of treatment were included in the analysis, even though some or all of the subsequent data on such patients might be missing. What are some alternative approaches for handling such missing data? What are the pros and cons of each approach?

10. In the GAD study, a more refined method of handling dropouts in the analysis might be to give less weight to observations with less exposure time. How might you construct such a weighting factor? Why might it make sense?

11. What specific approaches might have been used in the Iressa™ study to identify subject groups for which this drug might be especially effective and also to reduce spurious results?

CHAPTER 11

FINANCIAL SERVICES[1]

11.1 ABOUT THIS CHAPTER

Go where the big bucks are.
—Lynne Hare

Financial Services is an exceptionally broad application area of statistics. This field is especially important in light of the financial consequences that accompany decisions and their impact on national and global economies. Financing consumer and commercial activities is a key ingredient of stable economic growth and expansion.

Statistics is a necessary ingredient to the effective management of *millions* of loans and *billions* of financial transactions in the United States. Credit exposure in the billions of dollars is managed in the United States each year by the application of statistical models. There are few other areas in which statisticians have comparable financial impact.

The use of statistics in financial services applications has, moreover, seen appreciable growth in recent years due to the availability of large amounts of data, improved computing environments, and advances in methods for building and deploying statistical models to aid decision-making.

It is impossible, in a single chapter, to do justice to the many applications of statistics to financial services. Instead, we focus on those areas in which we have been most involved, notably the estimation of the future performance of loans. The problems faced in these applications are representative of those encountered by the entire financial services industry, and often call for similar technical approaches.

The chapter is organized as follows:

- An introduction to financial services and its uncertainties
- A description of credit scoring for consumer loan approval
- An example of building a credit-scoring model
- Discussion of some challenges in model development and implementation, principally in the context of the credit scoring example
- A broader look at modeling in the credit management process

1. This chapter was prepared by contributing authors William Alexander, Director of Quantitative Research, Wachovia Securities and Tim Keyes, Manager, Strategic Portfolio Decisioning, Corporate Financial Services, General Electric Company.

The Role of Statistics in Business and Industry, by Gerald J. Hahn & Necip Doganaksoy

- Specific application areas
- Some comments about statisticians in financial services
- Suggestions for further reading

11.2 FINANCIAL SERVICES AND ITS UNCERTAINTIES

The key to optimal decision-making is the quantification of uncertainty.
—C.R. Rao

11.2.1 The Industry

The financial services industry involves a broad range of activities including consumer and business lending, savings and checking accounts, insurance, investments, mergers and acquisitions, trading of financial instruments, and funding via the capital markets. We are all customers of this industry through our dealings with banks, finance and insurance companies, and the finance arms of suppliers of goods and materials.

Table 11.1, showing the debt outstanding for various sectors of industry in the United States, demonstrates the enormity of the financial services industry; the value of the loans owed to commercial banks rivals the entire national debt.

11.2.2 Business Uncertainties

Many – if not all – transactions in the financial services industry can be characterized as the trading of cash flows at one point in time with other cash flows in the future. The timing and amount of the future cash flows is not known with certainty. Accurate quantification of these uncertainties is the key to success.

Consider, for example, a mortgage loan made by a bank to an individual. From the bank's perspective, there is an immediate, certain cash outflow in return for a series of uncertain cash inflows (payments) over time. The payments are uncertain in both their amount and timing. First, not all customers repay their mortgages. Some may default, requiring the bank to foreclose on the collateral property, often without

TABLE 11.1 2007 U.S. Debt by Sector

Sector Total	September 2007 Total Debt ($ Billions)
Bank Credit, Commercial Banks[a]	8,952
Consumer Debt[a]	2,482
Revolving Credit[a]	920
U.S. Public Debt Outstanding[b]	9,008

[a]Federal Reserve Economic Data (FRED II), research, stlouisfed.org
[b]U.S. Department of the Treasury, Bureau of the Public Debt, www.publicdebt.treas.gov

recovering the full principal. Second, most mortgages prepay – that is, the full amount owed is paid back early. Prepayment changes the timing of cash flows and results in a reduction in income since the bank forgoes future interest. Thus, the bank's long-term profitability depends on the accuracy of its assessment of uncertainty. Good predictions enable banks to originate loans that form a profitable portfolio and to do so with well-understood risk.

The importance of gaining a better understanding of uncertainty can hardly be overestimated. There are many examples of companies suffering serious losses, and even becoming insolvent, because of bad forecasts.

11.2.3 Statistics in Financial Services

A key role of statistics in financial services is to help quantify uncertainty. This typically calls for statistical modeling. In the mortgage loan example, such models can be used to estimate the expected prepayment and loss rates.

In addition to helping quantify uncertainty, statistical modeling can also expedite processing and reduce costs, especially when there are many, often relatively small, opportunities to be evaluated. A further advantage of statistical models is their objectivity in light of possible legal requirements and challenges.

Obstacles Statistical modeling also presents challenges. Data are required to build such models. When entering a new business or launching a new product, there may be little or no data. Or there may be little data available on the credit behavior for important segments of the population – such as recent immigrants or young adults. In such cases, it may not be possible to build useful statistical models. Models based solely on assumptions or heuristics may be developed, but they cannot be tested without data.

For some processes, the past is not a good predictor of the future. This observation brings to light a distinction between many financial services and other industrial applications. Most engineering, manufacturing and pharmaceutical applications hinge heavily on understanding an underlying physical system. Though perhaps complex or difficult to measure, these systems have well-understood structure and often have components, based on chemistry or physics, that are largely deterministic. In contrast, financial services analytics requires forecasting human behavior. Such forecasts inherently have a high degree of variability.

Alternatives Given these obstacles, varying degree of reliance has been placed on statistical modeling in the financial industry. In the context of loan approvals, for example, three scenarios have emerged:

- Decisions on all loans, except the few that fail to meet an initial screen (e.g., unusually large loans, or applications from consumers with little credit history), are made directly from a statistically based model with little or no human intervention. The process may include a right of appeal by rejected applicants. This scenario is especially prevalent when there are a large number of applicants and each loan is a small fraction of the overall portfolio. The consumer credit card industry uses this approach.

- The statistical model is again applied to all applications that meet specified guidelines. This can lead to approval or rejection of a loan application. But it can also result in a third possibility for the more difficult cases – referral of the loan for manual review.
- High-level managers make the approval decision with input from knowledgeable underwriters. Statistical models are not used automatically in making decisions, but may be part of the process. Large commercial loans and portfolio acquisitions are typically handled in this manner. There are several reasons for this. First, the consequences of an individual decision are so great that a high-level manager needs to bear the responsibility. Second, the relevant data may be sparse, making it difficult to build a model with sufficient statistical precision to operate satisfactorily on its own. Third, each case has its own idiosyncrasies that cannot be readily quantified or generalized.

11.3 CREDIT SCORING FOR CONSUMER LOAN APPROVAL

At one time, all consumers applying for credit were personally interviewed. Based upon the interview and supporting information, a credit manager decided whether credit should be granted and under what terms. There were guidelines for making decisions, but these could be readily overridden. It was the manager's responsibility to balance the trade-offs between various risks: Is only two months in a new job offset by a higher paycheck? Is it worse to have been 60 days late on a credit card or 30 days late on a mortgage? The process had a large element of subjectivity that caused decisions to be suboptimal, inconsistent and, perhaps, based on factors unrelated to the likelihood of repayment.

Modern computing and statistics permit the systematic collection, retention and analysis of loan data. These have led to the widespread development and use of databased credit-scoring models. Such models are built using data from past loans to relate risk characteristics known when a loan approval decision is made to subsequent payment performance. In statistical parlance, the risk characteristics are the explanatory variables and payment performance is the response variable. Such models produce a "credit score" for each loan, measuring the risk of nonpayment. Going forward, credit scores are calculated for new loan applicants based upon their risk characteristics and used to decide whether or not to approve the application.

Common predictive characteristics include the applicant's payment performance on previous loans (usually obtained from credit bureaus), information about the applicant (e.g., income, time at residence, and job category) and characteristics of the proposed loan (e.g., LTV: the ratio of Loan amount To Value of the asset being financed). Examples of loan payment performance include whether the borrower has been delinquent, or defaulted, and the amount of any monetary loss. The simplest models classify loans as "good" or "bad" based on their payment performance.

For a general treatise on credit scoring, see Lewis (1992).

11.4 CREDIT SCORING EXAMPLE

11.4.1 Problem Statement

The following simple example illustrates the basic approach. The next section deals with some added complexities that need to be considered in practice.

Information is available on loans originated during two consecutive years and their performance under the existing system of subjective loan approval. During Year 1, there were 14,123 loan applications. Of these, 11,721 were approved, yielding an approval rate of 83%; 11,116 of these approved loans ended up being booked – a "pull-through" rate of 95%. The payment performance of the booked loans was then tracked over the 12-month period after their origination. Any loan that became 90 days delinquent (i.e., missed three consecutive monthly payments) was classified as "bad"; all others were classified as "good." Using this criterion, 10.1% of the approved loans were bad – a rate deemed unacceptable by management. As a result, one of the approval criteria, the required LTV, was lowered from 95% in Year 1 to 80% in Year 2.

In Year 2, there were 13,543 loan applications. Of these, 9,874 were approved, using the revised acceptance criteria, an approval rate of 73%. Of these applications, 9,095 were booked, a pull-through rate of 92%. These resulted in 8.0% bads during their first 12 months. The drop from 10.1% in Year 1 was claimed to be a consequence of the stricter approval rules. (There were, of course, also other factors that varied from Year 1 to Year 2.)

Management, however, was now concerned with the sharp drop (18%) in booked loans from Year 1 to Year 2. A new analysis was commissioned to determine opportunities to expand volume for the least added risk. This involved building a credit score model using the 11,116 booked loans in Year 1 (the so-called training or model-building set) and assessing the performance of this model on the 9,095 Year 2 loans (the holdout or validation sample).

11.4.2 Predictors of Loan Performance

Three variables predictive of loan performance are:

- The loan's LTV: Loan applications with LTVs greater than 95% in Year 1 and greater than 80% in Year 2 were rejected. Applications with LTVs less than 10% were also rejected in both years because such loans are unprofitable.
- Rent/Own: This is an "indicator" variable signaling whether or not the applicant owns or rents their current home. This variable is coded as 0 = owns and 1 = rents.
- D60: The number of times any of the applicant's *other* credit accounts has been 60 or more days overdue in the 6 months prior to the application. This variable takes on the values 0, 1, 2, or 3 for 0, 1, 2, or 3 60-day past due accounts, respectively. Credit policy in both years prohibited making loans to applicants with four or more D60's.

TABLE 11.2 Characteristics of Booked Loans

Variable	Metric	Year 1	Year 2
Payment Performance	Proportion Bad	10.1%	8.0%
LTV	Loan to Value		
	Minimum	10%	10%
	Lower Quartile	65%	64%
	Median	72%	71%
	Mean	70%	68%
	Upper Quartile	76%	75%
	Maximum	95%	80%
Rent/Own	Proportion Rents	21%	19%
D60	Number of 60-day Overdue Accounts		
	0	75%	73%
	1	12%	12%
	2	8%	9%
	3	5%	6%

TABLE 11.3 Logistic Regression Estimates for Credit Score Model fitted to Year 1 Data

Parameter	Estimated Value	Standard Error	t
Intercept	11.140	0.361	
LTV	–0.110	0.005	–24.0
Rent/own	–1.615	0.072	–22.5
D60	–0.559	0.034	–16.5

The three variables represent the data types commonly encountered: continuous measurements (LTV), counts (D60), and ordinal (Own/Rent).

Table 11.2 gives some descriptive statistics for the loans booked in each of the 2 years.

11.4.3 Model Building

The response variable (Y) is created by coding each good loan as $Y = 1$ and each bad loan as $Y = 0$. Since Y is binary, logistic regression provides a natural modeling technique (see Hosmer and Lemeshow 2000). Let p denote the probability that a loan is "good"; then the logistic regression estimates the (natural) log of the odds ratio, ln $(p/(1 - p))$, of a loan being good, as a function of the three explanatory variables.

Table 11.3 gives parameter estimates for the logistic regression model, fitted from the Year 1 data, together with their associated standard errors and t values. The latter show each of the three explanatory variables to be highly significant statistically.[2]

2. This is as one would expect, especially in light of the large sample size.

TABLE 11.4 Credit Score Model

Characteristic	Points	Condition
Base	800	
LTV	−5.22	For every LTV percentage point from 10 to 95
Rent/own	−76.6	If rent
	0	If none
D60	−26.5	If one
	−53.1	If two
	−79.6	If three

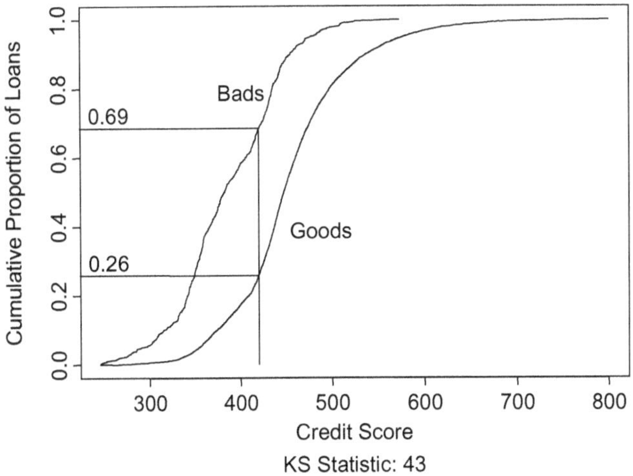

Figure 11.1 Cumulative distributions of credit scores for Year 2 good and bad loans.

To estimate the probability, p, the model output Y^* is transformed by $e(Y^*)/(1 + e(Y^*))$. It is, in addition, common practice to rescale the log odds ratio to yield values on a more convenient and intuitive scale – in this case to provide a "credit score" that ranges from 200 (worst) to 800 (best).[3,4]

The resulting credit score model is shown in Table 11.4.

11.4.4 Model Assessment

The ability of the credit score model to separate good loans from bad ones is assessed in Figure 11.1 by applying the model to the Year 2 data. This figure, sometimes

3. The rescaling is a linear transformation of the log-odds scale; in the example, credit score = 323.49 + 47.46 × log odds value.

4. The so-called "subprime" loans (a big issue in 2008) are ones that, for example, are issued to individuals with low credit scores on this transformed scale (or cannot be scored due to lack of credit history, recent employment changes, etc.).

referred to as a Kolmogorov–Smirnov (KS) plot, shows the cumulative distribution of credit scores for the good and bad loans separately.

Figure 11.1 shows that bad loans tend to have lower scores than good one. For instance, about 58% of bad loans, but only about 17% of good loans, have a score less than 400. The greater the separation of the two curves, the better the model discriminates bad loans from good loans. The maximum vertical distance between the two curves is known as the KS statistic. A KS statistic of 100% means that the model provides perfect separation. In this example, the KS statistic is 43%, attained at a score of 420. This is considered to be very good for this type of application.

11.4.5 Model Use

Basis of Comparison From Table 11.4, we note that reducing LTV by 15 percentage points (from 95% to 80%) increases the credit score by 78.3 points (i.e., -5.22×-15), all else being equal. In comparison, rejecting applications with three D60s increases the score by 79.6 points and declining renters increases the score by 76.6 points. The true utility of the score, however, is *not* achieved by evaluating variables individually, but by performing a *composite* risk assessment.

The two plots in Figure 11.2, using Year 1 data, evaluate the consequences of controlling credit risk using LTV alone and using the credit score, respectively. The ordinate of each plot is the proportion of bad loans in the portfolio that is eliminated if the corresponding LTV or credit score cutoff is used to differentiate unacceptable from acceptable applications. (The credit score cutoff rejects applicants with a score below the cutoff, whereas the LTV cutoff rejects applicants with an LTV exceeding the cutoff.) The horizontal scale at the top of each graph is the percentage of loans booked in Year 1 that would be rejected if the corresponding cutoff were employed.

Use of LTV to Limit Risk We note from Figure 11.2a that the strategy that was actually used in Year 2 (lowering the maximum acceptable LTV to 80) was

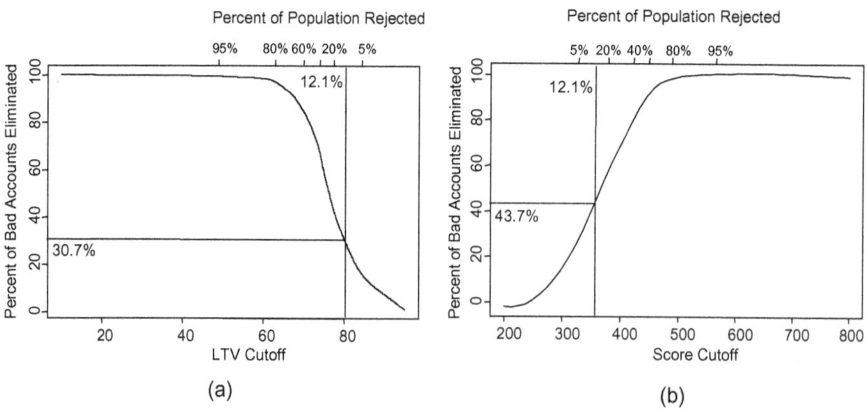

Figure 11.2 Comparison of loan approval strategies using (a) LTV cutoff and (b) credit score cutoff based on Year 1 data.

estimated, based on Year 1 data, to eliminate 30.7% of the bad loans at the expense of rejecting 12.1% of loans that would have been accepted under the prior guidelines. It turned out that the resulting estimate of 8% bad loans (i.e., 10.1% × (1 – 0.307)/(1 – 0.121)) was right on – there were, in fact, 8.0% bad loans in Year 2.

Use of Credit Score as an Acceptance Criterion Consider now the alternative of using a credit score cutoff to reduce the percentage of bad loans, as shown in Figure 11.2b. We note that a credit score of 357 would be expected to reject the same percentage of applicants (12.1%) as did the policy of rejecting LTVs over 80. Such a strategy, however, would be expected to eliminate 43.7% of the bad loans, for an expected portfolio bad rate of 6.5% = 10.1% × (1 – 0.437)/(1 – 0.121). Using the credit score cutoff, rather than LTV alone, is thus estimated to eliminate 43.7% as opposed to 30.7% of bad loans and reduce the portfolio "bad rate" from 8.0% to 6.5% for the same drop in business volume.

Similarly, we can see from Figure 11.2b that if we target eliminating 30.7% of bad loans (as under the earlier strategy), we would use a score cutoff of 338. With this cutoff, the volume drop would have been estimated to be only 7.0% rather than 12.1%.

11.4.6 Conclusion and Comments

The example illustrates that the most effective way to control risk is not by controlling risk characteristics individually, but by using a composite credit score. The final business decision requires balancing losses and volume. Increasing the cutoff score not only reduces the risk, but also results in more loans being rejected.

Implementing a credit score is not an invitation to open the "floodgates" to accepting credit score applications. In this example, the applications on which the model was built were constrained by existing credit policies. Thus, no applications with four or more D60s or LTVs higher than 95% were approved.

Finally, we note that this is another example of an analytic study (Section 1.4.3). We are using the results from Year 1 to extrapolate to Year 2, under the assumption that the underlying model does not change (see Sections 11.5.1 and 11.5.6 for further discussion).

11.5 MODEL DEVELOPMENT AND IMPLEMENTATION: SOME IMPORTANT NITTY-GRITTIES

The preceding simple example provides a flavor of the process involved in the development of a credit score model – and financial modeling, in general. The actual process is far more complex. In the example, practical and technical issues such as the following need to be addressed:

- The loans to use in building the model
- The appropriate measure of loan performance
- Selection of explanatory variables

- Acquisition of, and quality checks on, the data
- Selection of an appropriate modeling technique
- Construction of the model and its functional form
- Evaluation of the ability of the model to differentiate between good and bad loans in the future
- Determination of the model's operating region, or the range of risk characteristics for which it can be applied
- Selection of a cutoff score and its consequences
- Integration of the model into, or modification of, existing credit policy
- Legal and regulatory considerations
- Documentation of the model and its use
- Monitoring the actual use of the model

Some of these issues are discussed further in this section, principally in the context of the credit scoring example. Similar considerations apply in other areas of lending and in many other financial services applications.

11.5.1 Data Selection

The theme "it's all about the data" has particular relevance for financial model building applications.

It is not uncommon for the lion's share of the time on such projects to be spent assembling and refining the data prior to performing any analyses. There is no substitute for the deliberate, thoughtful and detailed work required to do this. The data acquisition process may seem plodding, time-consuming, and not nearly as much fun as model building. However, it is frequently the single most important part of the entire modeling operation. It is especially important to consider the process that generated the data and holistically transforming process knowledge into appropriate data acquisition methods.

The bottom line is this:

The Choice of the Statistical Technique Is Much Less Important than Data and Process Considerations

A disciplined process to assessing and gaining an understanding of the data is especially important. One such approach, called SIPOC, is described briefly in Sidebar 11.1.

SIDEBAR 11.1: THE SIPOC APPROACH FOR DATA ASSESSMENT

SIPOC, a tool frequently used in Six Sigma applications, is helpful in understanding historical data, especially in financial services applications. It provides a mechanism for differentiating "nice to have" versus "need to have" data, and to help ensure that the assembled data are used in a meaningful way.

SIPOC calls for a process to go from the supplier of the data (who provides the inputs) to the customer (who uses the resulting outputs) through the following steps:

- S (Supplier) – Assess the quality and expertise of the supplier of the data, and the purpose(s) for which the data were obtained.
- I (Input) – Obtain crisp operational definitions of all data elements.
- P (Process) – Specify how the data are to be used in a manner that is transparent to those who depend on the results.
- O (Output) – Ensure that the end result of data processing is the procurement of information that is meaningful to the end-user.
- C (Customer) – Validate that the output from the data processing is used downstream in other process steps and understand each element of the process.

See Hoerl and Snee (2001, Chapter 3) for further details.

Population Selection In determining the population to be used for model building, it is imperative to consider how this population might differ from the population of future applicants. The goal is to make the two populations as similar as possible. The following are some questions that need be addressed as illustrated by the preceding example:

- *Has the business environment changed?*
  To what degree does the general business climate that existed when the data to be used for model building were collected differ from that which new loans will experience? For instance, a downturn in housing prices can worsen payment performance and more intensive collection of delinquent loans can improve it.
- *Has the product or access channel changed?*
  Changes in the product (e.g., terms of the loan) or access channel (i.e., the vehicle for applying for a loan) can materially impact performance. Lenders have found, for example, that otherwise presumably identical loans originated through the Internet tend to result in a higher default rate than traditionally originated loans. A model developed from data that does not include Internet loans applied to a situation that includes such loans is likely to underestimate the expected percentage of bad loans.
- *Has the applicant population changed?*
  New criteria for loan authorization by a competitor can change the characteristics of the applicant population. Competitors' improved ability to identify and solicit lower-risk applicants may, for example, result in some better-paying applicants migrating away from our product.
- *How to handle the change in accepted population?*
  The population of accepted loans based upon implementation of the new model is inherently different from that used in developing the model. In particular, some applications that would have been accepted previously are now declined and others that would previously have been declined are now accepted.

The consequences of rejecting loans that were previously approved can be evaluated from the available data. Assessing how previously declined applications would have performed had they been accepted, or so-called "reject inferencing," is more difficult (see Sidebar 11.2).

We need also be concerned with possible changes in the guidelines for granting loans beyond the application of the model itself. Thus, in the example, applications with four or more D60s or with LTVs higher than 95% were to be automatically rejected in the past. Future use of the model assumes that this continues to be the case. Such policy might, moreover, have been flaunted on occasion in the past by human analysts. Such cases need be removed from the data used for model development.

SIDEBAR 11.2: REJECT INFERENCING

Ash and Meester (2002) review a variety of ways of addressing "reject inferencing." They advocate adding data on previous rejects to the database by following how these fared over the same time period with other creditors. They also compare various approaches and their limitations, and provide further references. Overall, there seems to be no completely satisfactory way of handling the reject inferencing problem. Reject inferencing is one example of the substantive statistical problems that arise in statistical modeling in financial services applications (see also Heckman 1976).

The preceding questions are more easily asked than answered. In selecting data for model building, one is constantly faced with the conflict between using only the most recent data versus including older data. In the example, newer loans will be more representative of the current lending environment and processes. Inclusion of older loans, in addition to increasing the database size, may provide longer payment history and reflect broader economic conditions. The resulting trade-offs require an understanding of the intricacies of the data and of the underlying business environment. A possible compromise might be to weight more recent observations more heavily in the statistical analysis. Also, various approaches might be implemented and compared. Once a model has been selected, its usefulness needs to be carefully evaluated and constantly monitored for its continued applicability (see Sections 11.5.3 and 11.5.5).

Explanatory Variable Selection The goal in selecting explanatory variables is to find readily available characteristics that are most successful in predicting payment performance.

An explanatory variable should also make sense. There are, moreover, important legal restrictions on the type of information that may be used in making loan approval decisions (see Sidebar 11.3). Variables that are expected to have direct impact on ability and willingness to pay – such as income, existing debts and past credit performance – are usually most appropriate.

SIDEBAR 11.3: SOME LEGAL RESPONSIBILITIES

Financial service businesses in the United States are highly regulated. Laws governing the origination, servicing and collection of consumer debt, and the maintenance of consumer credit histories include the Fair Lending Act, the Equal Credit Opportunity Act, the Fair Credit Reporting Act, the Fair and Accurate Credit Transactions Act, the Fair Debt Collection Practices Act, the Gramm–Leach–Bliley Act, the Truth in Lending Act, and the Real Estate Settlement Procedures Act.

Not all characteristics that might be predictive can be included in a model. For example:

- Use of gender, marital status, age, race and various others personal attributes is prohibited.
- Geography cannot be considered in the credit decision at too fine a level. For instance, the 5-digit U.S. zip code associated with the applicant's address cannot be used; using fewer digits is acceptable.

In addition, there are direct and indirect costs associated with obtaining data on explanatory variables. Direct costs include those of purchasing data (such as from credit bureaus), conducting appraisals, and checking the veracity of applicant-provided information such as employment status. Requiring applicants to provide detailed background information is an example of an indirect cost that can delay loan approval while awaiting such data and even result in canceling of the application.

The final assessment of the usefulness of an explanatory variable rests in the model development and evaluation process. Many variables will be omitted because they appear to add no predictive value or because their added predictive value does not justify the costs of data collection.

11.5.2 Model Fitting

The actual model fitting process – the aspect on which most books in statistics focus – estimates the relationship between the explanatory variables and the performance variable(s). Many tools are available to do this; linear regression, logistic regression, survival analysis, and tree-based methods seem to be the most popular for credit score development. Other methods include generalized additive models, Markov chains, time series analysis, and neural nets.

The specific method selected depends on the details of the problem, the personal preferences and experience of the analyst, and the level of acceptance a particular methodology enjoys in the organization. Rosenberg and Gleit (1994), and Hand and Henley (1997) provide general discussions on model selection in credit management.

Technical concerns that need be considered in model development include:

- Explanatory variables may interact. For instance, the effect of increasing LTV may differ for applicants with good and not-so-good credit history.

Interactions are typically handled by adding interaction terms to the model or by building separate models for different segments (subsets) of the population.

- There may be a large number of explanatory variables and these may be highly correlated, perhaps associated/related in a complex manner. The resulting collinearity needs to be considered both in building the model and in its subsequent use (see Belsley, Kuh, and Welch 1980, Chapter 3).
- The validity of the assumption underlying the use of the method and especially their robustness to the application at hand need to be examined.

11.5.3 Model Evaluation

Traditional Methods Statistical texts provide methods for assessing the usefulness of a fitted model. For the well-known case of regression analysis, numerical measures include the standard error of the fitted equation, the coefficient of determination, and the standard errors of the estimated parameters. Residual plots and partial regression leverage plots are two graphical approaches. All these methods provide best-case evaluations of the performance of the model for future prediction since they are based on the data used to fit the model. "Leave-one-out at a time" diagnostics, such as PRESS, DFBETAS, and DFFITS, strive to address this by refitting the model omitting individual observations and then predicting the values of these observations (see Belsley, Kuh, and Welch 1980, Section 2.1).

Holdout Samples Another approach to model evaluation, illustrated in the earlier example, is to use a holdout sample. Holdout samples generally provide a more realistic assessment of future performance than traditional methods. This is because the relationships may drift over time or be otherwise unstable. Or the model may have so many parameters that it is over-fit, resulting in severe statistical instability. The key question of how to select a holdout sample is considered in Sidebar 11.4.

The fitted model can then be evaluated by plotting the predicted values against the actual values in the holdout sample and summarizing the results by such measures as the mean absolute prediction error or the root mean squared prediction error. A comparison of the diagnostics from the training set with those from the holdout sample provides a further assessment of model stability.

SIDEBAR 11.4: SELECTING A HOLDOUT SAMPLE

A holdout sample may be selected by randomly assigning each observation to the training or holdout data set. Another option is to use the observations prior to some date as the training set and the subsequent observations as the holdout sample. This simulates implementing the model for decision-making at the time the holdout sample begins.

A holdout sample is a particularly attractive option for consumer loan approval modeling because there is usually ample data. When the data are sparse a possible approach for holdout sampling is segmentation into k time-based subsets. Each subset is then held out in turn; the model is built from the remaining observations and then evaluated on the subset that was held out. The "leave-one-out at a time" approach can be considered an extreme case of a holdout sample.

Another approach is to make assessments with multiple holdout samples of varying sizes. For example, evaluations can be conducted using the last 10%, 20%, 30%, 40%, and so on, of the data as a holdout sample. Such a process may provide insight on how long a model applies into the future.

One further idea used by some practitioners is to divide the data into three distinct parts, possibly time-based. The first part is used for building alternative candidate models. These are then tested and compared on the second part; that is, the earlier models are applied to the data and an assessment of each model's predictive quality is made. The data may also be used to rebuild the best fitting model and to address any observed shortcomings. The fitted model previously identified as best fitting is then evaluated on the third part.

Final Model Fit After all holdout evaluations have been conducted, it is generally advisable to obtain a revised fit of the model. This may include all the data or, depending upon the results of a holdout evaluation, just the most recent data. Another possibility is to use all the data, but give more weight to recent observations.

11.5.4 Model Implementation for Decision-Making

Moving a statistical model from a development to a decision-making environment is a major task. Like the model building process itself, close collaboration between the model builders, management, finance, and information technology is required.

Before it is implemented, management needs to know the performance that can be expected from a credit-scoring model. A financial analysis, in fact, ties in closely with the model evaluation process. If the model is to be used directly for decision-making, a cutoff score needs to be selected. The aim is to balance the conflicting goals of volume growth and profitability. Development of a cutoff score and a financial assessment associated with use of a model may be done together. To arrive at an appropriate cutoff score, the analyst translates the consequences of using alternative cutoff scores into financial terms.

11.5.5 Model Monitoring and Updating

Once a model is implemented, its ongoing predictive performance needs to be carefully monitored. All models eventually become obsolete; the purpose of monitoring is to signal when a model's decision-making capabilities degrade to an unacceptable level.

The monitoring process involves comparing actual to predicted performance. This is complicated by the fact that it takes time for performance data to accumulate. For example, if a model's performance during its first 2 years is to be evaluated, it will be that long before the first loans will have aged enough to assess the model.

Since a lengthy wait is usually unacceptable, other measures of model performance that are available more quickly need to be considered. For example, reducing the time window by using a less severe definition of "delinquency" (e.g., 60 days, rather than 90 days overdue) will yield a richer sample of "bads" sooner.

Control chart concepts might be used to monitor the model's ongoing predictive performance, as measured by, say, PRESS residuals (for continuous response variables), or error rates (in assessing go/no-go approval decisions); these signal model degradation.

When changes are detected, a search for the underlying causes needs to be undertaken. The introduction of new products, changes in servicing and collection strategies, modifications in the definitions of the variables used to calculate the credit score, changes in factors that are not captured in the model, and fluctuations in the economy are some possible causes. Such investigation might suggest possible actions that should be taken, such as changing the terms of loans (e.g. interest rate charged) or altering the model cutoff score. Next, a model improvement/rebuild effort is usually undertaken.

Monitoring and updating is a continuous process. The cycle of model building, monitoring, and rebuilding is never-ending. We expect that adaptive or dynamic model updating will be increasingly popular. This involves continuously updating the model as new information about its performance is obtained.

11.5.6 Some Further Statistical Considerations

All models are wrong; some are useful.
—George Box

As already pointed out, the credit score development example in particular, and, for that matter most statistical applications in financial services, involve analytic studies. One rarely samples at random from the population about which decisions/inferences need to be made, as is the case in enumerative studies. Instead, it is assumed that the population from which the sample is drawn is sufficiently similar to the future population to provide a framework from which useful decisions can be made. Thus, in the credit scoring example, the results from Year 1 are used to extrapolate to Year 2, under the assumption that changes do not have a serious effect on predictions.

We make relatively little use of formal statistical inference (e.g., confidence intervals and hypothesis tests) in this chapter because of the analytic nature of the underlying studies. This is because such inferences can be misleading (e.g., statistical intervals too narrow) because they *do not* reflect the added uncertainty arising from the analytic nature of the study (as well as that due to the violation of specific statistical assumptions, such as independence of observations).

Much of statistical modeling in financial applications can be thought of as "driving using the rear-view mirror". This does not mean that statistical modeling is not useful. It *is* a warning to practitioners to constantly keep a watchful eye on the assumptions that are being made. It is also critically important that we clearly explain these assumptions and limitations to all stakeholders.

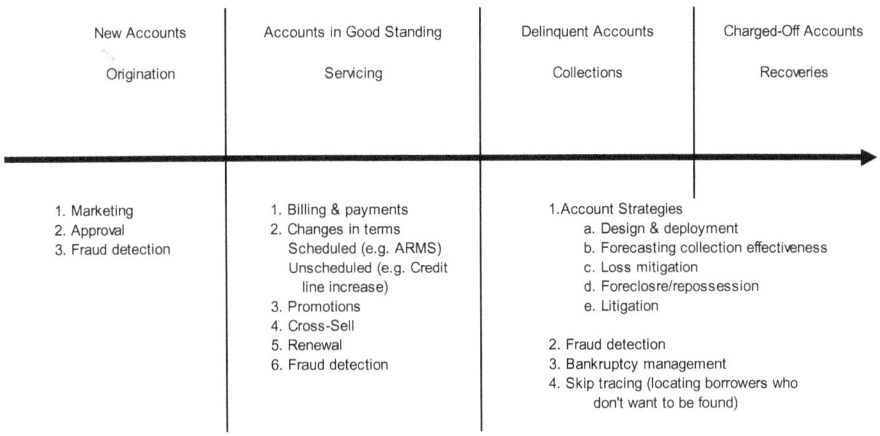

Figure 11.3 Decision points in the credit management process.

11.6 MODELING IN CREDIT MANAGEMENT: A BROADER LOOK

11.6.1 The Big Picture

Figure 11.3 identifies the decision points in a cradle-to-grave view of credit management. While all loans pass through origination, only those experiencing payment problems go to collections or recovery. Statistical models have been usefully employed at each point in the process to guide decision-making.

The level of complexity and sophistication of models used in practice varies widely. Modeling to support decision-making in credit management ranges from simple models focusing on a single event, as in Figure 11.3, to integrated cash flow models estimating the risk-adjusted profitability of the entire credit life cycle (see forthcoming discussion). The assessment of the performance of a set of loans over their entire life cycles is complex and often calls for a simulation approach.

11.6.2 Measurements of Loan Performance

More precise evaluations require more advanced measures of loan performance than those used in the consumer loan example. We consider some alternatives in increasing order of complexity.

Default Probability The simplest situation classifies default on a loan as a binary event. In the earlier example, a loan is classified as either good or bad, where bad denotes default (or perhaps high delinquency). The good/bad classification may be confined to a fixed time window after loan origination. A statistical model is built to estimate the probability of default. Profitability is managed by declining applications with an unacceptably high estimated probability of default.

Expected Loss Two loans with the same default probability can result in very different losses. For example, consider two first-lien mortgages on properties initially valued at $100,000. The first loan is for $95,000 and the second is for $80,000. Suppose that each of these loans default after payments that reduce their principal balances by $3,000, leaving balances of $92,000 and $77,000. Foreclosure proceedings are initiated, and the house is taken and sold. The sale of each house results in net proceeds of $80,000 after legal fees, repairs, sales commissions, maintenance, and carrying costs. The Loss Given Default (LGD) in these two cases is $12,000 and $0, respectively.

To take losses into consideration, we introduce the concept of the Expected Loss (EL) for a loan. This is defined as:

$$\text{EL} = \text{Probability of Default} \times \text{LGD}$$

Estimating the EL is more informative than estimating the default probability alone; it is most often modeled by developing separate models for default probability and LGD.

Expected Profitability EL is of key interest, but it still does not measure the bottom line – profitability. Thus, we may wish to model Expected Profitability.

Expected Profitability includes expected revenues (e.g., interest and fees), costs (to fund, originate, service, and collect the loan) and the impact of prepayment as well as expected loss. Expected Profitability is usually not modeled directly. Instead, each of its elements is modeled separately and then combined. The resulting estimate includes information on factors that may be nonstatistical, such as costs.

Risk-Adjusted Profitability The most sophisticated profitability models consider the variability in profitability as well as its expected value. Companies may not be willing to pursue an increase in Expected Profitability if it comes at the price of increased risk.[5]

They may, instead, choose to take less expected profit to achieve a lower probability of falling "too much" below expectations. Thus, variance is an important normalizing factor in comparing potential transactions with appreciably different risks.

A common methodology for such assessments is known as RAROC (Risk Adjusted Return on Capital; see Zalik, Walter, Kelling, and James 1996). The resulting models require estimating variability and are exceedingly complex, especially since variability must be measured at the portfolio, not at the individual transaction, level. As a result, the correlations in default probability between loans within the portfolio must be modeled. Banks often have specialized departments charged with building, maintaining and deploying such models.

5. Risk managers should not forget that truly enormous losses tend to be associated with very rare events – such as four hurricanes in one season in Florida – and not be lulled into complacency just because, for example, expected losses are manageable (see Taleb 2007).

11.7 SPECIFIC APPLICATION AREAS

11.7.1 Mortgage Loans

Characteristics of Mortgage Loans Mortgage loans are, in many ways, similar to other consumer loans. However, they also differ in some key ways:

- The size of the loan is usually larger.
- Payment is over a longer period of time with fewer defaults occurring in the first year or two. As a result, recent originations may not have a sufficient number of bad loans to permit a model to be developed.
- House values change over time.
- Loans are frequently paid off early (because of refinancing or sale of the property).
- Many loans feature important changes in terms during their lifetimes. For example, a so-called "5/1 Adjustable Rate Mortgage (ARM)" changes from a fixed rate loan to an adjustable rate mortgage after five years. There may be little data about the effect of the change on prepayments and defaults.

Issues in Model Building The special characteristics of mortgages lead to some issues that need to be considered in developing models to drive or guide the approval process.

The potential lack of an adequate database providing information on long-term delinquencies can cause difficulties. Requiring more than 2 or 3 years is likely to reduce appreciably the number of loans in the database. Modeling techniques that utilize the maximum amount of information in the data are, therefore, especially desirable. For example, survival analysis techniques might be applied (see Alexander, Grimshaw, McQueen, and Slade 2002). Since these model a time-to-event (e.g., default), even very young loans can be included in the data used for modeling.

Frequently, mortgage loan approval decisions are made using the second alternative described in Section 11.2.3. A model is developed and applied with two cutoff scores. Relatively straightforward applications whose score exceed the upper threshold are accepted, and the ones with scores below the lower threshold are rejected. Applications with intermediate scores are referred to a human underwriter – together with the output of the scoring model. As more experience is gained in using the model, the cutoff scores may be moved closer to each other, resulting in an increasing proportion of automated decisions.

11.7.2 Commercial Loans

Comparison with Consumer Loans Evaluating the credit-worthiness of a business and that of an individual has many similarities. Both require forecasting future events and usually involve evaluating cash flows available to repay the debt, past payment performance, and the value of any collateral.

There are, however, some key differences. First, commercial enterprises have more complicated financial transactions and financial statements. Individual commercial transactions also tend to vary more from one another and, as a result, the decision-making process is less routine. Second, among the best corporate customers, there are very few defaults to model. There were only 55 defaults globally of firms rated "Investment Grade," according to Standard & Poor's (2004), from 1981 to 2006. Third, commercial loan portfolios are generally made up of fewer, larger loans resulting in a more acute concentration of risk. A single default might cost a commercial lender hundreds of millions of dollars.

Role of Modeling The factors that characterize commercial loans tend to favor a manual decision-making process; statistical models tend to be more in the background than in consumer credit management. Such models can, however, provide human decision-makers with an important added, more objective, perspective.

The trend is toward greater use of statistical models, especially for relatively straightforward and repetitive situations, such as small business loans. Some larger banks and rating agencies have also developed databased risk models for mid-market commercial loans. As a consequence, companies engaged in commercial lending are hiring statisticians, decision scientists, and data miners. (See Haimowitz and Keyes 2002 for further discussion of risk analysis for commercial loans.)

11.7.3 Loan Portfolios

Background Portfolios of existing loans are frequently bought and sold on the so-called "secondary market." It is not unusual for lenders to sell loans at any point during their life cycle – from just-originated to post charge-off. Both the current portfolio owner and the potential purchaser must estimate the value of the portfolio. Doing so requires forecasting the cash flow the portfolio will generate, as well as the associated management costs.

The global financial crises of the late twentieth century spurred the development of an international secondary market for commercial loans. Lenders needed to improve their financial condition by selling nonperforming loans. This freed up capital to cover potential future losses at the cost of some immediate losses (which were sometimes reimbursed, at least partially, by global institutions such as the World Bank or the International Monetary Fund).

There are very active secondary markets for consumer loans in the United States. Many consumer finance companies buy or sell loans as a key element of their business strategy. Firms sell loans to gain liquidity, to remove subperforming or nonperforming accounts from their balance sheets or because servicing or collections is not their core competency. Other firms buy loans to leverage their area of expertise (say, foreclosure) because they do not want to become involved directly in loan origination, or as investments.

Valuation Process The value of a portfolio may be estimated as the sum of the estimated values of the accounts that comprise the portfolio. Alternately, a portfolio's value may be estimated from data on its aggregate performance or that of similar portfolios. Which approach to use depends on the available data, time, and resources. In some cases, both approaches might be used and the results compared.

For commercial loans, such evaluations are normally conducted by underwriters. Statistical modeling may play a limited role, if any. When many accounts need to be assessed within a portfolio, time constraints may not permit all to be examined manually. In such cases, an underwriter might examine only a random sample of loans, possibly stratified to give higher representation to the larger or more complex ones. The sample results can be used to obtain a statistical estimate of the total value of the portfolio.

Alternately, the sample may be used do develop a model that mimics the valuation process used by underwriters. This model is used to estimate the value of the accounts that had not been evaluated manually, generally resulting in a more precise estimate of total value than obtained by merely projecting the sample estimate. (See Sidebar 11.5 for further details on such mimicking models.)

SIDEBAR 11.5: MIMICKING MODELS

In what we have called mimicking models – in contrast to performance-based models – the response variable used to build the model is the yes/no approval decision made by a human underwriter. The explanatory variables are again characteristics of the loan and of the applicant, and may be the same as those that would be used in performance-based models. When the model is applied, the predicted value is the estimated probability that a human underwriter would approve the loan.

Mimicking models have the obvious drawback of not being fit against actual payment performance and generally do not provide new insights into the drivers of loan performance. They are, therefore, often *not* a good substitute for payment performance-based models. They tend, instead, to reveal the key criteria used by underwriters.

When little or no data exist to build a performance-based model, mimicking models may be preferable to a completely subjective approach, and help to systematize, automate, and speed up the process. Also, since there is no elapsed time required to measure loan performance, mimicking models can be built on the most recent loan originations.

Finally, mimicking models can also be based upon applications that were processed by the most expert and skilled underwriters. In this sense, they can be considered to be a form of artificial intelligence or expert system, helping bring the experience of the most knowledgeable underwriters to every lending decision, and providing a teaching tool for new underwriters.

11.7.4 Leases

Leasing of a variety of products (e.g., automobiles, office equipment, and tools) is another important form of lending. The lessor grants the lessee the use of the asset in exchange for a series of payments. The asset is to be returned to the lessor at the end of the lease, but the lessee often has the option to buy the asset at that time.

The pricing of a leased item depends, to a large degree, on the difference between the value of the item at the time of the lease's initiation and its predicted remaining (residual) value at the end of the lease. For example, suppose a new car with initial value of \$30,000 is leased for 2 years and its residual value at that time is \$20,000, the lease may then be regarded as a \$10,000 loan.

Since there is uncertainty associated with the residual value, an integral component of the leasing business is the need to forecast residual values, often using regression models. Success hinges critically on this forecast. Lessors who estimate too high may suffer severe losses when the over-valued assets are returned. On the other hand, those who estimate too low might price themselves out of the market.

11.7.5 Collections

Every champion was once a challenger.
—Ad at Saratoga Horse Raceway

Background The collection of overdue loans is a critical task in credit management. Statistical methods help develop strategies to maximize net collections (money collected less the cost to collect).

Collection departments typically select from a variety of collection strategies. Some alternatives in increasing order of severity and cost include: doing nothing, sending a reminder letter, leaving an automated telephone message, or having a collection agent or attorney call the borrower. Should nonpayment continue, a customer may be offered the opportunity to become current over a period of time, the terms of the loan may be changed (e.g., the maturity date extended), the loan may be settled by partial payment or the lender may seek a legal judgment against the borrower. If the loan is secured (i.e., property has been pledged as collateral), foreclosure and repossession are options. The lender may ultimately declare the loan uncollectible and charge it off. This does not relieve the borrower of the obligation to pay. The loan may be sent to a recoveries (post-charge-off) unit to pursue further.

Determining the best possible collection method (i.e., the one with the highest expected return) requires decision-making under uncertainty. The best strategy for collection is likely to depend on the history of the loan, other loan characteristics, and the financial situation of the borrower.

Consumer Loan Example Consider again the loan example discussed earlier. The collections department employs four strategies to collect on delinquent accounts: do nothing, call the borrower with a taped message, have an agent call, or send a strongly worded letter under legal letterhead. To decide which strategy to employ for each account, the portfolio is segmented based on three variables:

- the credit score of the loan, categorized as high, medium, or low;
- delinquency state (≤2, 3, or 4+ months overdue); and
- balance due (≤\$750 vs. greater).

This results in $3 \times 3 \times 2 = 18$ segments for each of which an optimum strategy needs to be chosen.

Each of the four strategies was applied to a random sample of accounts in each of the 18 segments. The response variable is the net average collection on each loan.

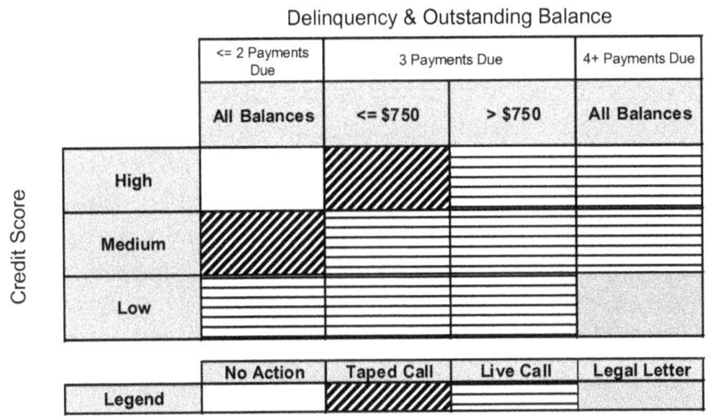

Figure 11.4 Best collection method by risk category.

The winning strategy among the four alternatives is then selected for each segment. Figure 11.4 summarizes the results (the original 18 segments were reduced to 12).

A more complex approach would be to also consider *past* collection strategies. For instance, having a live agent call at the 3-due stage might be more effective for accounts not called at the 2-due stage. Resource constraints may add further complications; there may, for example, be capacity limits on the number of calls agents can make per day.

Another, and somewhat simpler, popular approach is to conduct a "champion-challenger" study. This compares the present collection strategy (champion) with an alternative strategy (challenger). The current strategy might, for example, be the one shown in Figure 11.4. The challengers might be two alternatives, one more aggressive and the other less aggressive than the current strategy. Over time, a fraction of the overdue loans in each segment are selected randomly and assigned to one of the two challengers, rather than the champion, strategy. Net payment performance of the three strategies is closely monitored and compared. If a challenger is found to be more effective than the champion, it is adopted as the new champion. The process then continues with new challengers. (See Makuch, Dodge, Ecker, Granfors, and Hahn 1992 for further details.)

11.7.6 Insurance

Background Insurance gives financial protection against risk. Insurance policies are offered by a wide variety of financial services organizations, sometimes in collaboration with insurance companies. Familiar types of insurance include accident insurance, health insurance, life insurance, property insurance, and automobile insurance.

There is a basic difference from a risk analysis perspective between loans and insurance. For loans, the lender usually provides an initial payment, product or service to the borrower with the expectation of repayment later. Default is a major risk. For most insurance policies, the customer makes payments in advance and/or periodically, in return for compensation if and when some future event or expense actually occurs. Default is generally not an issue. Instead, the key concern to many insurance providers is whether the future event(s) will take place and, if so, when and at what cost. In most cases, a probabilistic evaluation is required. Just as in lending, the estimate is generally based on statistical projections from past data. The results are used to decide whether to provide insurance, the price and the terms.

There are significant differences in the type of risks among various types of insurance. For some products, such as personal accident, flood and trip insurance, the protection is against a rare, but high-cost, event. The key element of uncertainty is whether the event will occur and the magnitude of the resulting liability to the insurer if it does. In contrast, the occurrence of the event and the associated payout are not in question for life insurance. The chief element of uncertainty is its timing. In yet other cases, such as health insurance, we are interested in both the timing and the magnitude of multiple, but not necessarily independent, events.

Actuarial Science An entire profession – actuarial science – has grown up to address the uncertainties associated with granting insurance and with the development of decision systems to address them. Actuaries help to set rates and terms of insurance policies and are trained in probability and statistics. Discussion of the differentiation between actuaries and statisticians goes back a long time. Almost a century ago, the then president of the Casualty Actuarial and Statistical Society addressed the subject saying "despite their common interests, actuaries tend to be a separate community from statisticians, with their own professional organizations and their own formal certification process" (see Craig 1918).

Technical Challenges There is often an enormous volume of historical information on the occurrence of events such as fires, thefts, accidents, disability, loss of income, or death. However, there are exceptions. For example, data on floods may be sparse and, sometimes, outdated. In some cases, the data may have been collected in a relatively consistent manner by a government agency. Frequently, however, the available data are inexact, incomplete, or not fully relevant. As in other fields, a major challenge is to help build a sound system for data gathering.

The future may not mirror the past, and trends over time may occur. For example, life insurance is impacted by increasing longevity. The likelihood of fire might decrease as fire codes tighten. The cost of medical services increases as a result of new technology, which may further increase life spans and medical costs. Construction may affect the probability of a flood. Actuaries might apply time series or other trending analyses to forecast the occurrence of future events and integrate these with statistical models similar to those discussed earlier in this chapter. In addition, recently developed statistical approaches, such as classification and regression trees (see Section 11.7.7) and neural networks (see Berry and Linoff 1997) are being used more extensively, especially for predicting high-risk or rarely occurring events.

11.7.7 Fraud Detection

Fraudulent transactions and other illegal activities are an important concern in many financial service applications and range from credit card theft to inflated automobile accident repair claims. Financial services organizations exert much effort in detecting fraud and protecting themselves against it.

Technical Approaches Modern statistical tools, as well as various other closely related approaches, such as case-based reasoning (see Leake 1996), and neural networks are used to build models to help identify unusual behavior that may signal fraud (see Berry and Linoff 1997). Potential explanatory variables include changes in the type of products purchased, the number and amount of charges, and the mode and location of purchases. Demographic inconsistencies, such as the purchase of a motorcycle by a 90-year-old woman, might provide clues. Of course, we cannot be sure that a questionable event is evidence of fraud – the lady might be purchasing the motorcycle for her grandchild. Fraud detection is another application area in which it is desired to maximize the identification of true "culprits" (minimize the Type II statistical error), while minimizing the number of "innocents" that are incorrectly identified as culprits (Type I error).

As in other situations, the ultimate goal of the model-building effort is to arrive at a decision, such as whether or not to approve a credit card transaction or an insurance claim. Thus, fraud detection lends itself to the use of classification and regression tree analyses (see Breiman, Friedman, Olshen, and Stone 1984).[6]

Auto Insurance Claim Example This example (courtesy ANGOSS Software Corporation, www.angoss.com, Copyright 1994–2004) deals with developing a process to identify fraudulent auto insurance repair claims. A random sample of 15,420 past claims was scrutinized by expert investigators, resulting in 923 cases of fraudulent behavior. Data were available on 27 variables that might be predictors of fraudulent claims. The original analysis was conducted on a training set and verified on a time-based holdout sample with stable results; the two data sets were combined to obtain the final model, discussed here.

A classification and regression tree analysis was performed on the data. This systematically examined the effectiveness of each of the 27 variables to identify fraudulent activities, drilling down to isolate subsets of the portfolio with especially high or low expected rates of fraud.

The results are shown in Figure 11.5. The numbers in the second column of each box is the number of not fraudulent (No), fraudulent (Yes), and Total claims. The third column gives these numbers as percentages of the claims in that box (for No and Yes) and the percentage that the claims in the box are of all claims (for Total). The shading is the proportion "Yes."

We observe from these results that:

- The bulk of fraud occurs in claims in which (a) the policyholder is at fault; (b) the policy was purchased from an external agent; and (c) the policy is not liability-only. In Figure 11.5, this group consists of the left two boxes in the bottom row. It shows that 92% of the fraudulent claims (847 out of 923) occur among 45% (6,870 out of 15,420) of the total claims.

6. Alternative approaches have also been developed (see Haimowitz and Keyes 2002; Rosenberg and Gleit 1994; and Hastie Tibshirani and Friedman 2003).

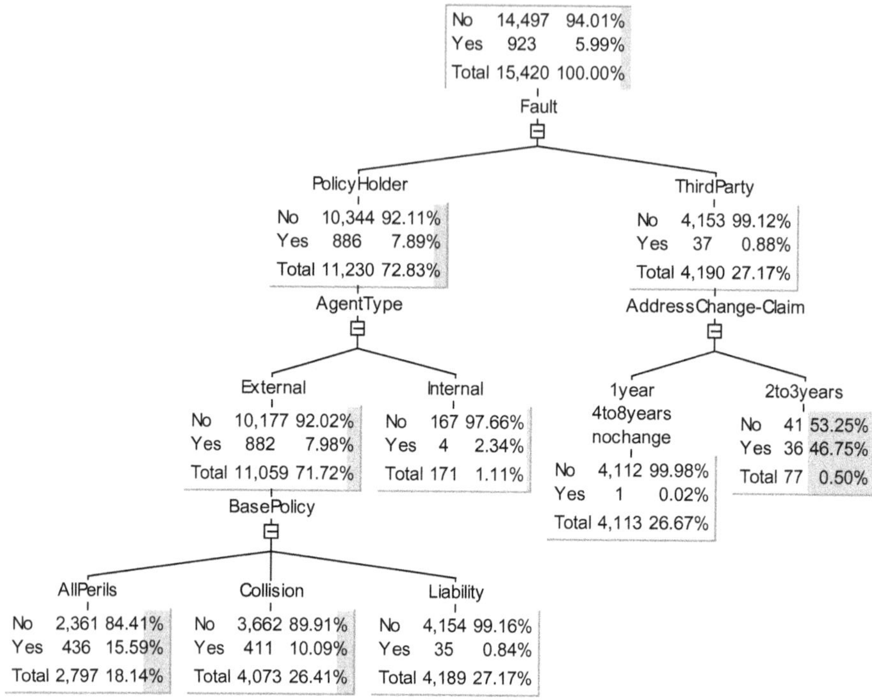

Figure 11.5 Classification and regression tree to isolate fraudulent auto insurance claims.

- The group with the highest concentration of fraud consists of those accounts in which the policyholder is not at fault, and for which there was an address change 2 to 3 years ago. There are 36 fraudulent claims out of a total of 77 claims in this group – a 47% fraud rate. There is no immediate intuitive explanation why address changes in the 2–3 year time frame for third-party-at-fault claims is predictive of fraud and an investigation into this needs to be conducted.

The results of the preceding analyses can be used to focus attention on those cases that are identified as being especially vulnerable to fraud. The company might, for example, conduct audits on segments identified with especially high fraud rates.

Further Comments The sharper the differentiation that can be achieved by the analysis, the more useful the results will be. Classification and regression trees can lead to spurious findings that are inconsistent with process or product understanding or are difficult to interpret. It is important to review the results with subject matter experts and to validate the model, preferably with independent data.

11.7.8 Some Other Areas of Application

The following are some further financial service applications of statistics:

- Database marketing to interest existing customers in other products: For example, a bank may use statistical and data mining tools to identify current holders of savings accounts who are likely potential customers for credit cards.

- Securitization of bonds (i.e., the issue of bonds that are repaid using the cash flow from an identified pool of assets (often loans)): From 2004 to mid-2007, the asset-backed commercial paper market rose from $700B to $1T, with daily volume in the $60B range.[7] Bonds are backed by various types of loans, such as credit card, home equity, auto, equipment, and student loans. Pools of loans are sold to a legal entity known as a special purpose vehicle (SPV). The SPV, in turn, sells bonds to investors, such as insurance companies, pension funds and hedge funds. The proceeds are used to purchase the loans. Payments from the loans are then used to repay the bonds. Since the pool of loans used to repay the bonds is fixed, understanding the loan pool's cash flow performance is critical. Securitization typically requires a statistical forecast of loan performance (both loss and prepayment) for the entire life of the pool.
- Financial derivatives (broadly defined as a financial transaction whose value depends on the current and future value of an asset): There has been an explosion in the variety and volume of financial derivates traded, including interest rate derivatives, credit derivatives, and options. The value of these instruments varies according to the future path of markets. These paths, and their uncertainty, can be modeled statistically.
- Market risk management: The risk to financial systems arising from exposure to financial markets became clear in the fall of 1998 with the collapse of hedge funds and the subsequent market dislocation. Work to quantify and mitigate such risks uses, among other approaches, statistically based methods, such as Value-at-Risk (VaR) modeling (see Markowitz 1952 and Group of 30 1993).

11.8 STATISTICIANS IN FINANCIAL SERVICES

11.8.1 Educational Background

Professionals engaged in statistical analysis in financial services have a wide range of backgrounds. We have seen such work conducted by people who hold degrees in finance, business administration, economics, physics, computer science, and even chemistry, in addition to statistics, mathematics and operations research. There is not a single educational path to a successful career in applying quantitative methods to financial services. Statistics departments in universities certainly do not hold a monopoly.

With advances in technology, quantitative analysis is becoming a core competency for much of the work in financial services. This bodes well for those with training in this area, either as their major field or as a minor.

7. See www.securitization.net.

11.8.2 Career Paths

Those with statistical backgrounds may be part of a group dedicated to analytics or they may be located within an operating unit of the business. Initially, much of the work may be in building and implementing models, desirably, under the guidance of an experienced professional. Knowledge of statistical software and computer savvy is essential.

Analysts might subsequently advance to becoming project leaders. This involves problem definition, data acquisition, selecting an appropriate model building approach, and implementation – as well as managing resources, preparing budgets, and presenting results to management.

The next step often may be to manage a quantitative analysis group. From there, the road diverges in a number of directions. Some move up the risk management chain within the business; others move to another corporate function, such as finance.

11.9 FURTHER READING

Not surprisingly, there is a vast literature on financial services, with varying emphasis on economics, accounting, econometrics, computations and statistics, and various other areas. The following are a few books with a strong statistical flavor.

Bluhm, Overbeck, and Wagner (2002) provide an introduction to credit risk and portfolio management. This book covers modeling theory and methods. Topics discussed include credit risk in securitization, trading portfolios, and credit derivatives. Popular products for assessing credit risks in commercial portfolios, such as Moody's KMV, are considered.

Ong (1999) discusses estimation of the variance of portfolio loss for risk-adjusted profitability calculations and elements of default estimation. Topics covered include portfolio loss forecasting (e.g., LGD), unexpected loss, default correlation, estimating portfolio loss distributions, and risk-adjusted performance measures.

Raynes and Rutledge (2003) cover a broad range from the basics of structured securities, and how they are rated and analyzed, to more complex analyses of collateral.

11.10 MAJOR TAKEAWAYS

- The long-term profitability of financial services businesses depends on their ability to accurately assess uncertainty. Statistics plays a key, and increasingly important, role.
- One of the most important applications is in constructing credit-scoring models for loans. Credit-scoring models are built using data from past loans to relate risk characteristics known when a loan approval decision is made to subsequent payment performance. Important steps in building such models include data selection, model fitting, model evaluation, model implementation for

decision-making, and model monitoring and updating. Developing credit-scoring models raises many thorny technical issues, for example, assessing how future loan applicants might differ in their performance from those used in model building.

- Specific opportunities for applying statistics deal with mortgage loans, commercial loans, loan portfolios, leases, collections, insurance, fraud detection, and numerous other areas.
- Professionals engaged in statistical analysis in financial services have a wide range of backgrounds and varied career path opportunities.

DISCUSSION QUESTIONS

General

1. We state in Section 11.2.3 that in some applications in which high-level managers make the decision "statistical models ... may be part of the process." Elaborate how this might be done and suggest some examples.

2. In discussing model fitting in Section 11.5.2, we comment that "there may be a large number of explanatory variables and these may be highly correlated, perhaps associated/related in a complex manner. The resulting collinearity needs to be considered both in building the model and in its subsequent use." What are some closely correlated variables that might be encountered in loan approval modeling?

3. The discussion in this chapter indicated that most statistical models built for financial service applications involve analytic, rather than enumerative, studies. Discuss this comment in the context of the consumer credit scoring example, and the resulting limitations in the inferences that can be drawn. Suggest some possible steps that you might take to make models more inclusive.

4. Consider the various measures of loan performance described in Section 11.6.2 and suggest situations for which each might be applicable.

5. We differentiated between payment performance models and mimicking models in Section 11.7.3. Compare these two approaches, and their pros and cons for loan approval. Indicate when you might use one approach over the other, and how you might integrate the two.

6. Comment further on the issues and challenges that can be addressed by statisticians in the other areas of application discussed briefly in Section 11.7.8.

7. What attributes, skills, and training seem especially useful for statisticians who wish to work in financial services?

Technical

1. In the credit scoring example, LTV is a continuous measurement and entered the model linearly. A common alternative used frequently in financial services applications is to bucket (or discretize) continuous explanatory variables. For instance, LTV could be divided into four ranges: LTV $< 50\%$, LTV between 50% and 60%, LTV between 60% and 80%, and LTV over 80%. From this, three indicator variables are defined and included in the model. Discuss the pros and cons of this approach and possible alternatives.

2. In a footnote to the credit scoring example in Section 11.4.3, we stated that the fact that each of the three explanatory variables were highly significant statistically is "as one would expect, especially in light of the large sample size." Elaborate on this comment and what it implies with regard to statistical significance testing.

3. Assume that in the credit scoring example, the analyst mistakenly included loans with more than three D60s that had been approved despite company policy. Explain the potential implications of this error.

4. Explain why a maximum separation between the two curves in Figure 11.1 means that the model can separate the bad loans from the good loans perfectly.

5. Discuss the applicability (and the pros and cons) of the different methods for model fitting mentioned in Section 11.5.2 and others that you might suggest. Give some examples of situations for which a particular method would be especially appropriate

6. If a fitted model is to be used to make predictions, it is usually advantageous that the explanatory variables are correlated in a manner similar to the model-build set. Explain why this is so.

7. In both Sections 11.5.1 and 11.5.3, we suggest giving more weight to recent observations. Why might this be advisable? How might you go about determining appropriate weights?

8. In the discussion of model monitoring in Section 11.5.5, it is stated that: "The monitoring process involves comparing actual to predicted performance" and suggests that control charts may be useful in doing this. Discuss how the data obtained for such control charts differ from what one might obtain in the standard use of control charts for monitoring a production process. How would you go about addressing the consequences?

9. We also state in Section 11.5.5 that "we expect that adaptive or dynamic model updating will be increasingly popular." Elaborate on this approach. Discuss its pros and cons and how you might go about implementing it.

10. In the discussion of the broader aspects of credit management in Section 11.6.1, it was stated that "The assessment of the performance of a set of loans over their entire life cycles is complex and often calls for a simulation approach." How might such a simulation be conducted?

11. In introducing classification and regression trees in Section 11.7.7, we state in a footnote that "alternative approaches (to those proposed by Breiman, Friedman, Olshen, and Stone 1984) have also been developed." Research such alternatives and compare their pros and cons versus the approach used. Suggest any improvements, in the context of the example that might make the approach used even more effective.

CHAPTER 12

BUSINESS PROCESSES

12.1 ABOUT THIS CHAPTER

This chapter deals with business processes—an area in which statistics is increasingly employed with much success. Our emphasis will be on improving an existing process, but our comments are equally relevant for the development of a new process or the major redesign of an existing one. Much of the discussion also applies to the operation of service businesses or organizations, such as airlines, repair shops, building contractors, restaurants, banks, hospitals, and schools.

We begin the chapter with a brief overview of business processes, including a discussion of the need for their improvement. We provide a detailed example, illustrating the steps involved in processing a loan application, and then discuss these steps more generally. This is followed by another example, dealing with a credit card company searching for "skipped" delinquent customers. We comment next on some applicable technical concepts and methods, and conclude with a review of the use of operations research and management science tools in the improvement of business processes.

12.2 BUSINESS PROCESSES

12.2.1 What are Business Processes?

In general, a business process is a series of interrelated activities that convert inputs into outputs to achieve a defined business goal. All work that takes place in a business can be thought of as a process. Some examples of business processes are[1]:

- Filling customer orders
- Employee recruiting, orientation and retention
- Filing a patent
- Accounting and financial reporting
- Customer service

1. The design, development and manufacture of products can also be considered to be business processes. We focus in this chapter on more transaction-oriented applications, such as those listed.

The Role of Statistics in Business and Industry, by Gerald J. Hahn & Necip Doganaksoy

Efforts to improve business processes are sometimes referred to as initiatives to improve service quality, transactional quality, or commercial quality.

12.2.2 Increased Interest in Business Processes

The growth of the services sector, and especially the emergence of such process-oriented activities as high-tech servicing, facility management, and remote diagnostics (Section 9.9), has been a major force in the growing interest in business processes. There is also increased recognition of the fact that the processes that a company utilizes to conduct its business can provide a strategic competitive advantage. Various innovative business models that have emerged on the Internet (e.g., photo processing, travel planning, auctioning, publishing, and retailing) have revolutionized entire industries. Also, business processes have benefited significantly from advances in information technology and have enabled rapid automation of complex systems and processes (see Davenport and Short 1990, and Ramaswamy 1996).

12.2.3 The Need for Business Process Improvement

Servicing Customers The performance of processes is key to business success. The following typical situations test their capability and robustness:

- Airline operations: Influx of travelers at ticket counters at peak times of day or year or, less predictably, due to flight cancellations.
- Customer service center: Unexpectedly large number of employees failing to report to work.
- Food service: Shortage of high-grade meat supply from qualified sources.
- TV broadcasting: Format incompatibility of a tape containing a commercial message received from an advertising agency—discovered only shortly before its scheduled airing time.

The imaginative acts of a few dedicated employees sometimes help save the day. The desired solution, however, is to build into the system design the capability of avoiding problems, or, as a minimum, uncovering and addressing them early on. In addition, for processes, as for products, we strive to achieve robustness to change. This requires an understanding of the drivers of variability and putting in place the necessary means to handle the unexpected.

The failure of a business process, unlike that of many products or services, may not always be immediately apparent to the service provider.[2] Customers often experience problems in the form of delayed or incorrect shipments, multiple referrals to different departmental units, or inadequately addressed complaints. These eventually prompt even the most loyal customers to take their business elsewhere. Improving business processes increases customer satisfaction, reduces internal costs, and improves productivity.

2. A notable exception that has received much attention is the unsuccessful merger of major companies. Inability to integrate processes is frequently cited as the major source of such problems.

Operational Improvement In addition to addressing customer concerns, study of business processes often reveals clues to inefficiencies and areas for improvement. Upstream problems (e.g., inaccurate sales orders, incomplete instructions, and schedule changes) frequently propagate further difficulties downstream and generate waste or extra work, often in the form of rework. Such nonvalue adding work due to process inefficiencies was characterized in Section 8.6.2 as the "hidden factory."

Need for On-Target Consistency Business processes, like products, need exhibit consistency around a target value. Manufacturers want raw materials to be delivered on time—and not sometimes early (taking up space and possibly deteriorating) and late at other times (but "on target" on the average with minimal variability). Similarly, paying bills on the due date can increase profitability (by avoiding lost interest due to early payment and loss of on-time discount due to late payment).

Some Examples Some typical business process improvement projects are:

- Improving a billing process to collect revenues faster.
- Helping a call center increase accuracy and timeliness in addressing customer issues.
- Improving an order fulfillment process to reduce cycle time and errors in product shipments.
- Reducing advertisement placement errors in TV broadcasting (e.g., back-to-back airing of commercials of competing brands).

A company-driven quality improvement and/or cost reduction initiative often provides impetus to business process improvement.

12.2.4 Some Characteristics of Poorly Designed Business Processes

The traits of a poorly designed business process resemble those of poorly manufactured products. In particular:

- The performance goals are ill defined. Goals such as increase customer satisfaction, provide world-class service or reduce downtime are noble but vague and subjective. Performance goals should be clearly defined in meaningful quantifiable terms and measured against appropriate benchmarks.
- Inadequate assessment of the capability of the process prior to start-up, resulting in defects in the system being discovered during operations, customer dissatisfaction and expense in fixing the problem.
- Jumping to problem solutions without a thorough assessment. A typical example is the implementation of a new software system to address a broken business process only to discover that the problem persists because it is inherent in the process.

12.2.5 A Team Effort

Business processes almost invariably involve multiple functions within an organization. A team effort is, therefore, essential. The team should include the broad array of activities involved with the process, and might include industrial engineers, MBAs, human resource specialists, information technologists, financial analysts, accountants, and supply-chain specialists.

Some of these individuals might, with appropriate training, take on the added role of statistical leader and act as mentors to team members in gathering data and performing statistical analyses. In other cases, a statistician might be a team member.

12.2.6 The Role of Statistics

> **The operation of any system generates information on how it can be improved.**
> —George Box

Statistical tools have traditionally been used to address problems involving business processes; for example,

- Time series analysis for sales forecasting and inventory management
- Sample surveys to study customer satisfaction (and dissatisfaction)
- Statistical sampling for financial auditing
- Multivariate analysis in market research

These generally involved the use of a specific statistical method to address a specific problem in a specific functional area such as finance, marketing, accounting, or plant operations. Ultimate success, however, requires successful performance during the *entire* business cycle from identifying potential customers to billing—including all the steps in between. Thus, in this chapter, we focus principally on the role of statistics in business process improvement from a broader perspective and at an operational level.

Shewhart and Deming were early proponents of management using statistics to improve overall business performance and the importance of understanding and reducing the sources of variability through what is now called "statistical thinking" (Section 1.6).

Example A custom manufacturing business is studying its time-to-delivery performance at two manufacturing sites to understand the causes for delays and to improve on-schedule delivery.

Table 12.1 provides the following information on delivery delays at the two sites for a particular time period:

- Delay Time: Number of (business) days between the promised and actual day of delivery (negative for early deliveries).
- Lead Time: Number of (business) days between the receipt of the order and the promised delivery date of the shipment.

TABLE 12.1 Delay Times, Lead Times, and Order Sizes at Two Sites

Site 1			Site 2		
Delay Time (Days)	Lead Time (Days)	Order Size	Delay Time (Days)	Lead Time (Days)	Order Size
6	22	23	3	3	20
3	15	15	−4	14	16
1	9	14	1	11	19
8	40	26	−1	43	31
−1	5	16	4	10	18
12	35	31	−1	22	17
8	34	29	3	36	33
0	3	15	−4	2	22
4	19	29	4	28	17
8	29	27	−1	31	29
5	6	17	3	27	28
8	42	22			

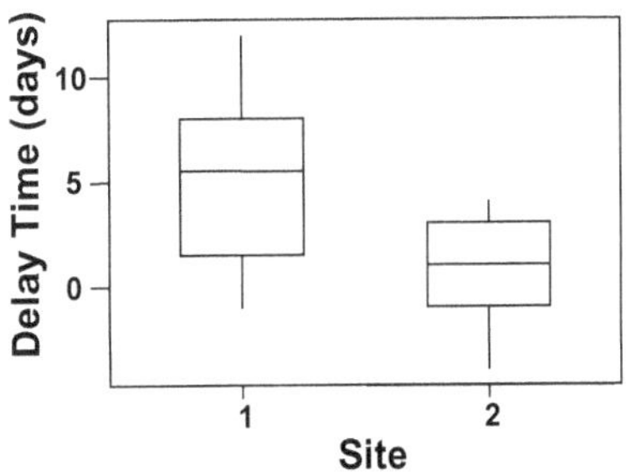

Figure 12.1 Box plots of delay times at two sites.

- Order Size: Number of manufacturing lots in the order. (Large orders take longer to manufacture.)

Initial examination, using the box plots shown in Figure 12.1, shows that Site 1 tends to have longer delay times and more variability than Site 2. However, lead time and order size also need to be considered.

A "bubble scatter plot" of delay time against lead time is shown in Figure 12.2. In this plot, the sizes of the circles are proportional to order size; filled versus unfilled circles differentiate the sites. This plot provides a succinct summary of the associations between delay time and the other three variables (order size, lead time, and site).

Figure 12.2 reveals that:

- The delay times for orders from Site 1 tend, as previously observed, to be larger and more variable than those for Site 2 (unfilled circles tend to fall above filled circles).

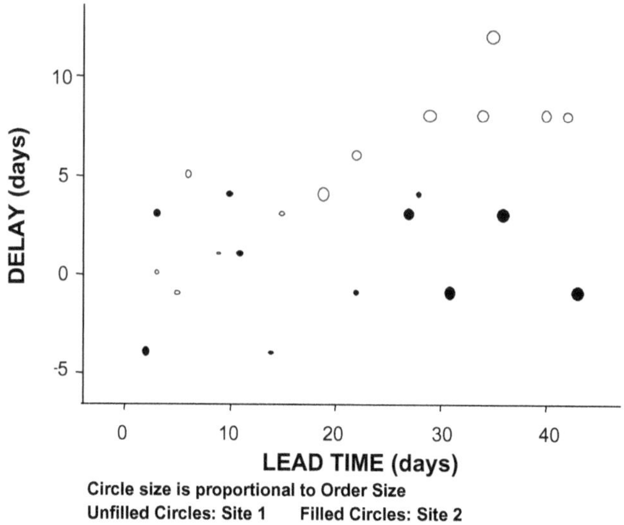

Figure 12.2 Bubble scatter plot of delay time versus lead time (differentiating site and order size).

- The lead times and sizes of orders placed at the two sites appear to be comparable in magnitude (as determined, respectively, by comparing the range of data on the *x*-axis for the two sites and comparing the sizes of the filled and unfilled circles).
- At both sites, large orders tend to have longer lead times than small orders (circle size increases from left to right on the chart for both the filled and unfilled circles).
- For Site 1 (unfilled circles), delay time increases with lead time and with size of order (as measured by size of circle). In contrast, for Site 2 (filled circles), delay time does not seem to be related to lead time or order size.
- For short lead times (e.g., 15 days or less), both suppliers seem to have comparable delay times. For longer lead times, Site 1 had appreciably longer delays than Site 2.

The plots suggest that important reductions in delay time might be gained by focusing on the long lead time/high size orders at Site 1. These were studied closely to gain clues of possible underlying causes (e.g., production scheduling, plant productivity, and product complexity) and paths for improvement.

More formal statistical analyses (e.g., multiple regression) may also be conducted and might provide further insights.

12.2.7 Systematic Approaches

Deming and Shewhart advocated the PDSA cycle (Section 2.2.5) as a systematic approach for process improvement (see Langley, Nolan, Nolan, Norman, and Provost 1996). More recently, efforts that focus on management and organizational aspects to drive business improvement have been introduced under such banners as total quality management (TQM), business process reengineering (BPR), and Six Sigma for Business Processes (see Sidebar 12.1). These—and especially PDSA and Six Sigma—involve the use of statistics to varying degrees.

SIDEBAR 12.1: SIX SIGMA FOR BUSINESS PROCESS IMPROVEMENT

Companies came to realize that Six Sigma had an important role to play in areas beyond manufacturing, including the improvement of business processes. GE, with its heavy involvement in financial services and television—as well as the numerous commercial transactions in its manufacturing businesses—was a leader in this effort. Then GE CEO Jack Welch was, in fact, referring principally to business process improvement when he declared that "variation is evil in any customer-touching process" (GE Annual Report 1998).

Other companies have followed suit. Articles in Six Sigma Forum Magazine (SSFM) detail applications at American Express (November 2001), Bank of America (February 2004), TRW (November 2004), Starwood Hotels and Resorts (November 2005), and Varian Medical Systems (May 2005); an article in Quality Progress (May 2007) discusses Six Sigma at Cigna, a healthcare benefits management company. Other SSFM articles describe noncorporate applications at North-Shore Long Island Jewish Health System (November 2004), dealing with error-free lab operations, and the Red Cross Hospital of Beverwijk in the Netherlands (February 2005), dealing with such issues as reducing invoice errors and the length of patient stays. (See Snee and Hoerl 2005 for further examples.)

12.3 EXAMPLE: IMPROVING THE LOAN DECISION PROCESS FOR A BUSINESS EQUIPMENT FINANCING COMPANY

The example illustrates the steps involved in business process improvement. To illustrate the approach, the problem is presented in a more straightforward and "linear" fashion than was actually the case. Many of the steps were, in fact, "trial and error" and there were some false starts.

12.3.1 Understand the Business Background and Problem

Consider a company that finances purchases of business equipment. In Section 11.4, we described statistical approaches to help decide who should receive loans. We now look at the overall process for handling loan applications—from receipt of the application to customer notification.

Management, in an effort to improve the quality of the services provided customers, commissioned a survey of recent loan applicants to determine what they regarded as most important in the services received. This showed the time required to arrive at a loan approval decision to be by far the most important source of dissatisfaction.[3]

3. The study likely underestimates the problem because it missed the additional potential good customers who did not apply for a loan due to the company's poor approval time reputation.

Applicants were generally told that a decision would be forthcoming within a week of filing their application. This was in line with traditional industry practice. Some competitors, however, did consistently better. Reaching decisions faster (without seriously compromising risk) would please existing customers and bring in new ones.

Management appointed a cross-functional team, including a statistician, to address the problem. Goals of making 97% of all decisions within a week and attaining a median time of, at most, 48 hours to reach a decision were established. There was general agreement – supported by examination of the delay times for some recent applications – that these goals were currently not being met.

12.3.2 Define the Process

The team started by exploring the process for reviewing loan applications. At a high level, the processing of a typical loan application follows the steps illustrated in the flow diagram shown in Figure 12.3. The comments marked in the figure suggest potential causes of process variability.

Figure 12.4 is a deployment (or cross-functional) flowchart that shows who does what at each step. It also provided some clues to potential trouble spots. For example, the return of packages deemed incomplete by loan analysts to sales managers suggests rework in the process.

12.3.3 Gather Data and Validate the Measurement System

The critical importance of gathering the needed data and obtaining reliable product measurements was emphasized in earlier chapters. This is equally important for business process improvement; it turned out to be the team's most time-consuming challenge.

Existing System The company had, in place, an automated system for easy access to information on each loan application. This met the needs of daily operations and was used by management to get snapshot summaries—such as breakdowns of loan applications by industry and size.

The system, however, was inadequate for a process improvement study. For example, a key need was to obtain accurate measurements of the time to loan decision—that is, the time from receiving the application to the time the customer received notification. Application dates, however, were recorded inconsistently. Sometimes, this was taken as the date on which the application was free of missing information, rather than when it was received.

Measuring the sources of performance variability, moreover, required gathering much relevant data about the process. Good records on various processing steps, such as the time to do various tasks, were not available. In addition, information was purged as a loan application moved through the process and the times between process steps were not always available. Cleaning the data would have been complicated and for some applications impossible.

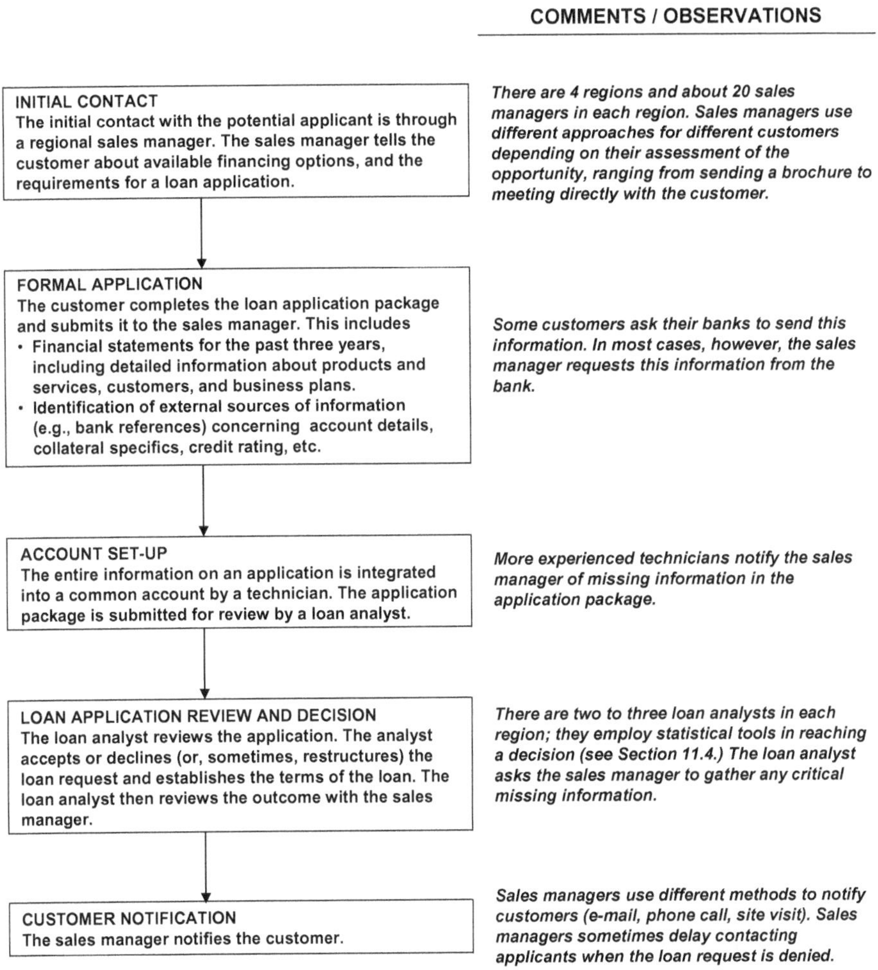

Figure 12.3 High-level process flow diagram for loan application review process.

A New Approach In light of the inadequacies of the existing data, the project team focused on getting better data on forthcoming applications. A new system was developed for capturing the needed information, based on an understanding of the inadequacies of the existing system. This was used to closely follow the applications received over the forthcoming 3 months on 743 new applications.

12.3.4 Quantify Process Performance

The elapsed days (actually measured in hours) from receipt of an application to customer notification for the 743 applications is shown by a histogram in Figure 12.5. This indicates that almost 10% of the applications took over a week to process—far short of the goal of 3% or less. The median time of 2.8 days was close to the goal of 2 days—but further reduction was desirable.

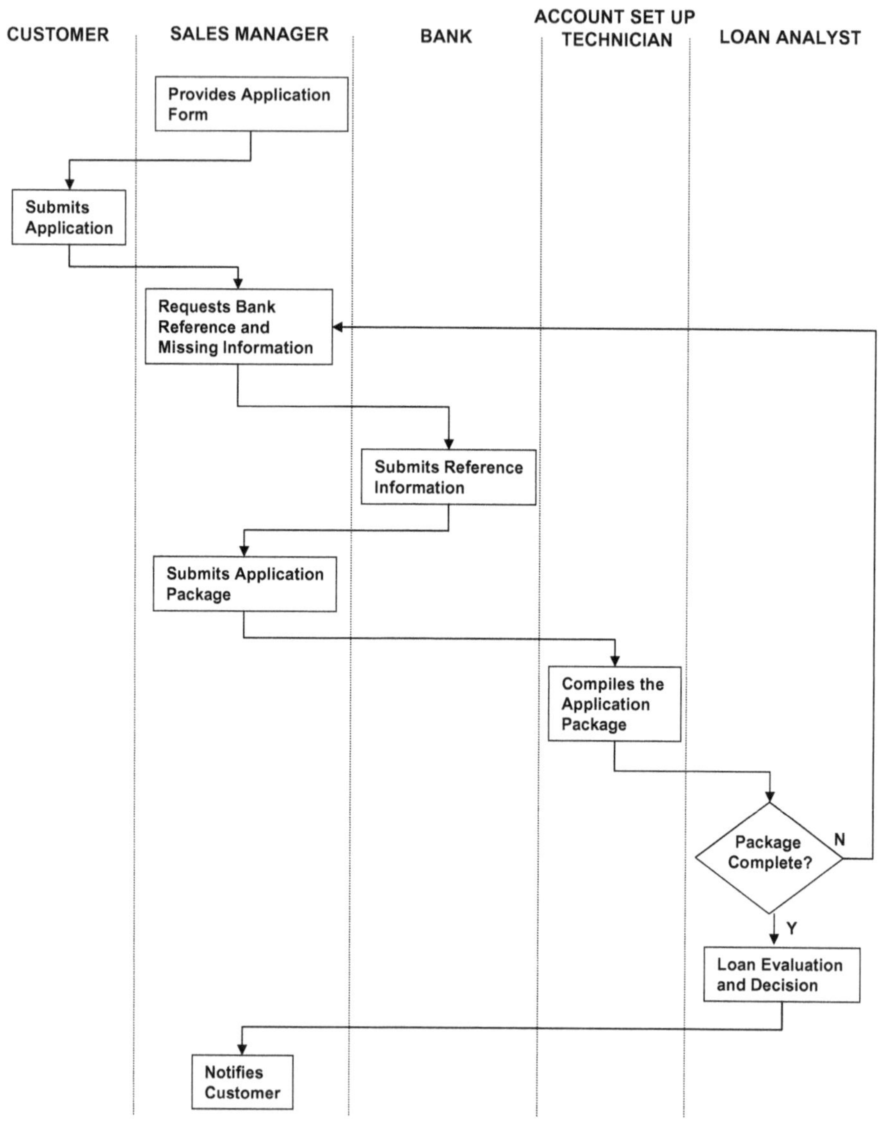

Figure 12.4 Deployment flowchart of the loan decision process.

12.3.5 Identify Potential Sources of Performance Variability

Initial discussions indicated considerable differences in opinion between sales managers and loan analysts as to the reasons for the delays in processing applications, and what needed to be done about them. Some typical explanations were:

- "Some banks are more responsive than others."
- "The completeness of the information depends heavily on the sales manager working with a customer."

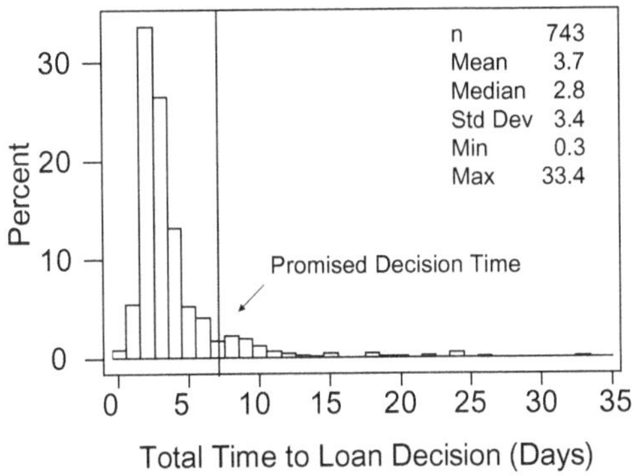

Figure 12.5 Days from receipt of application to notification by customer of decision for 743 applications.

- "The experience of the loan analyst is a key factor."
- "Big loan requests take longer to assess."
- "Large customers respond more slowly to requests for information."

Loan analysts felt that sales managers, in their desire to speed things up, frequently provided incomplete application packages that needed to be sent back to get missing information. Sales managers, in turn, considered the loan analysts unreasonably demanding. Conflicts between different functional units often come to the surface in such evaluations (see Sidebar 12.2).

SIDEBAR 12.2: DEFLECTING ORGANIZATIONAL BOUNDARIES

Businesses have traditionally been departmentalized by functional specialty (e.g., finance, engineering, manufacturing, and human resources).[4] Different organizations tend to pursue their own perceived objectives and lose sight of the overall business goals—especially when the performance of each function is differently and narrowly measured. In the example, sales managers were evaluated mainly by their ability to attract new customers; loan analysts were assessed by their skill in keeping default risk low. This can result in "turf wars" among functional groups, resulting in failure to communicate key information and inability to act cohesively.

Taking a process view helps bring out issues created by organizational boundaries and is a first step to their resolution. In addition, focusing on root causes in the process leads to a broader and more customer-centric view.

Based on these inputs and their own observations, the team developed a fishbone diagram, shown in Figure 12.6, identifying possible factors that impact loan review time. Some factors (e.g., procedure for informing customers about application requirements) could be the direct causes of variability in the decision process.

4. Similar divisions occur in other operations such as government agencies, military units, and universities.

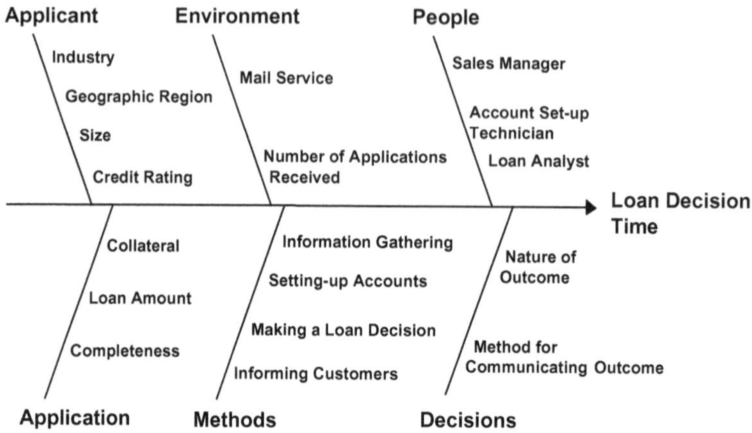

Figure 12.6 Fishbone diagram of possible factors that impact loan decision times.

Others (e.g., sales manager) might provide important clues to identifying underlying causes.

12.3.6 Measure the Sources of Performance Variability

The next step is to *quantify* the dominant contributor(s) to process variability — that is, to assess the importance of the factors suggested in Figure 12.6.

Data Gathering The new data-gathering system allowed the time required to complete each process step to be clearly and consistently recorded for the 743 new applications. In addition, data on each of the variables that may be related to the delay time were obtained. This included information on the applicant, the purpose of the loan, and identification of those (i.e., sales manager, account setup technician, and loan analyst) processing the application.

High-Level Breakdown A variety of box plots segmenting the time to make a loan decision by such variables as sales manager, region, loan analyst, loan request complexity, requested loan amount, and size of collateral were prepared. Of these, only the plot by region, shown in Figure 12.7, was revealing. This indicated that although differences between regions were small relative to the variation within regions, the longest delays tended to occur in Region 1. There was no immediate explanation for this.

Drilling Down by Process Step Additional box plots were developed to assess the contributions of individual process steps to the total delay time (see Figure 12.8). These show:

- The time (required by sales managers) to gather information
- The time (required by set-up technicians) to set up the account

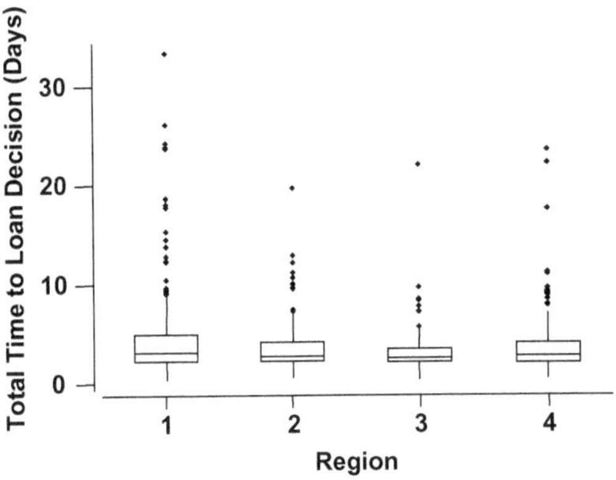

	1	2	3	4
n	250	182	175	136
Mean	4.4	3.5	2.9	3.8
Median	3.1	2.7	2.5	2.7
Minimum	0.3	0.5	0.3	0.4
Maximum	33.4	19.7	22	23.5

Figure 12.7 Time to loan decision by region.

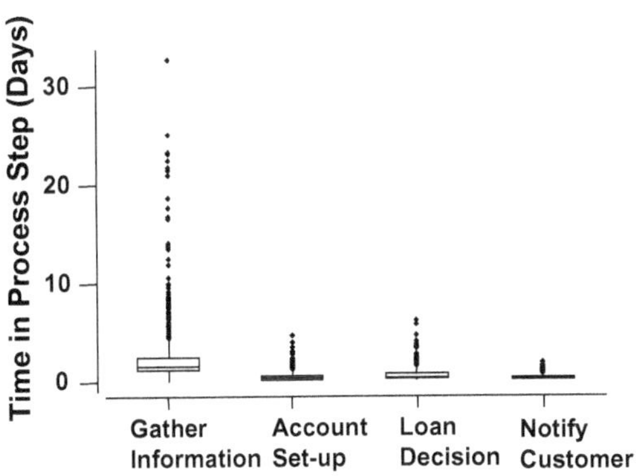

	Gather Information	Account Set-up	Loan Decision	Notify Customer
n	743	743	743	743
Mean	2.5	0.4	0.6	0.2
Median	1.5	0.3	0.3	0.2
Min	0	0.1	0.1	0.1
Max	32.8	4.6	6.1	1.8

Figure 12.8 Times required for different tasks in loan decision process.

- The time (required by loan analysts) to make a loan decision
- The time (required by sales managers) to notify customers of the loan decision

Figure 12.8 clearly showed that the information-gathering phase was the dominant contributor both to the magnitude and the variability in total cycle time. But why?

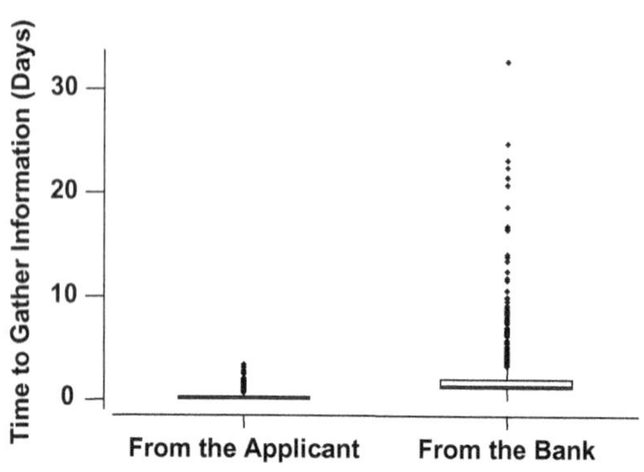

	From the Applicant	From the Bank
n	743	743
Mean	0.2	2.3
Median	0	1.3
Min	0	0
Max	3.3	32.8

Figure 12.9 Time to gather information from applicant and bank.

Drilling Down the Gather Information Step The review of a loan application was based on information from two prime sources: the applicant and the applicant's bank. These are examined separately in Figure 12.9.

Most applicants submitted all the required information with their initial applications, thus creating no delay. In the remaining cases, sales managers, working with applicants, compiled the missing information in less than three days. So this did not seem to be the problem.

The second part of the figure showed that gathering bank reference information was a key bottleneck; it could take 20 days or more. But what was causing this delay?

Drilling Down Time for Gathering Bank Reference Various plots of the time to obtain the bank reference, broken down by different variables, were made. Of these, Figure 12.10, a box plot by region and sales manager, was of most interest. This showed that some sales managers gathered the required information consistently within three or fewer days. For others, both the mean and spread were much higher. The sales managers with the longest delays tended to be in Region 1, but there were exceptions, most notably sales manager 57 in Region 3. These results called for yet further explanation.

A multiple regression analysis was performed to relate explanatory variables, such as years of service of the sales manager, loan size, bank identifier, and customer's industry to delay time. This provided no further insightful results.

Seeking an Explanation The project team held a "brainstorming" session with a small group of sales managers from all four regions to identify the reasons underlying the differences shown in Figure 12.10. Sales managers described how they conscientiously followed company guidelines, but had their own way of interfacing

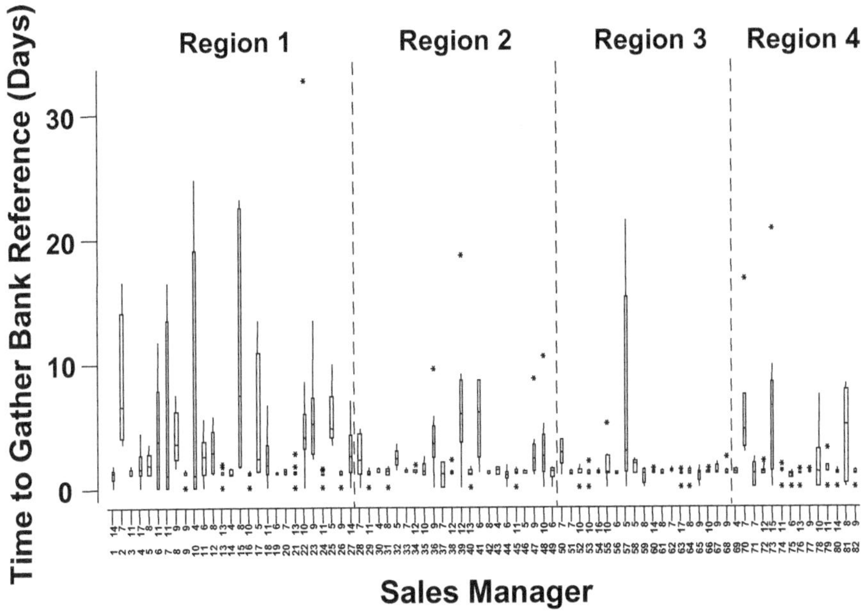

Note: The horizontal axis shows number of deals handled (first line) and sales manager (second line);
*designates extreme values.

Figure 12.10 Box plots of time to gather bank reference by sales manager and region.

with customers. These differences did not have any noticeable impact on the time to gather information.

The discussions eventually revealed that banks required a signed form by the loan applicant before releasing information. Some banks accepted the customer's signature on the loan application as sufficient. Others required a specific authorized release by the customer before proceeding; this was often time-consuming. Experienced sales managers, anticipating the delay, asked applicants to sign a release form when submitting their applications. This seemed to lead to appreciably lower average times and less variability in data gathering.

The company did not have a system by which such "best practices" could be readily identified and shared. Most sales managers who did not require advance permission came to the company through acquisitions and were concentrated in Region 1. In addition, salesperson 57, a new employee in Region 3, had not yet learned about the advantage of requesting advance permission.

Thus, though the problem reflected itself in differences among sales managers, the root cause appeared to be the absence of a definitive policy requiring advance completion of bank contact release forms.

12.3.7 Improve the Process and Validate Impact

Gathering the bank reference appeared to be the largest single source of delay in the loan decision times. The proposed fix was simple – require *all* sales managers to ask customers to sign a bank reference release with their applications.

Other ways to reduce delays (and achieve consistency) were also identified and implemented. These included:

- Adapting consistent definitions of process start and end points
- Standardizing the application form
- Using the Internet more extensively to facilitate exchange of information both internally and with customers and banks

The key questions then became "had the team correctly identified the underlying causes of the problem" and, if so, "will the proposed solutions work?" It could be, for example, that sales managers who required prerelease were also more conscientious in general or that banks that required written authorization before providing reference information also differed in other ways. And, are customers willing to provide the required sign-offs? To find out, the new procedures were "piloted" in Regions 1 and 3.

The results showed that the vast majority of applicants and banks accepted advance sign-off. This and the other changes did, indeed, seem to reduce delay times appreciably and make them more consistent.[5] Also, there was no evidence of any deleterious effects. This led to implementation of the new procedures in all regions.

12.3.8 Monitor the Process

Tracking the process over a further 3 months showed that:

- The median cycle time to process applications was reduced from 2.8 days to 1.6 days, thus beating the goal of 2 days.
- The standard deviation for the time for processing loan requests was reduced from 3.4 days to 1.4 days.
- The proportion of loan requests that were processed within a week was increased from 90.2% to 98.1%—beating the goal of 97%.

These improvements resulted in both increased customer satisfaction and cost reductions.

To help ensure that the gains are maintained, a system of control charts was developed to track the total time to process applications, the time required to gather the required information, and other key information. In addition, management was provided monthly updates of processing times. The monitoring process was also used to identify opportunities for further improvement.

12.3.9 The Role of Statistical Thinking

This example required mainly elementary, and mostly graphical, statistical methods. The main contribution of statistics was conceptual, rather than through advanced methods. Each of the three basic elements (Section 1.6) of statistical thinking was utilized:

5. The emphasis on reducing delay time also got all involved more focused on making this happen.

- A high-level process map was developed to understand the basic elements of the loan decision process.
- Several potential drivers of process cycle time variability were identified through process studies and expert inputs.
- Data on key process characteristics were obtained and used to identify the major contributors to cycle time delay. This involved an intensive drilling-down process to help identify the major sources of variation and in-depth discussions with the major players.

12.4 STEPS FOR PROCESS IMPROVEMENT

A penny saved is a penny earned.
—Benjamin Franklin

12.4.1 Understand the Business Background and Problem

It is important, early on, to achieve a sharp definition of the problem closely linked to business objectives and customer expectations. In the example, a survey made it clear that reduction of the loan decision cycle time was a key element of customer dissatisfaction.

A good understanding of the general business background is also required. This calls for involvement and full immersion in the problem, most desirably as a member of the project team.

12.4.2 Define the Process

Various tools, such as process mapping and flowcharting, help define a process (see Andersen 2007 and Tague 2004). These show the flow of both physical entities and of information. It is often helpful to start with the high level process steps and then add the details. For the loan decision example, a high-level process flow diagram and a deployment flowchart were used.

The actual operation of a process frequently differs from its documented description (and this in itself can be a contributor to variability). It is, therefore, important to supplement the process diagrams with personal observations, based upon discussion with those knowledgeable with the process, as illustrated by Figures 12.3 and 12.4.

12.4.3 Gather Data and Validate the Measurement System

Measure what is measurable, and make measurable what is not.
—Galileo Galilei

Well-targeted data gathering and the integrity of the measurement system are essential to success in business process improvement. Frequently, performance is assessed by, say, courtesy of service or expertise of staff. Such variables are often more

TABLE 12.2 Results of Study to Evaluate Consistency of Loan Analysts' Assessment of Application Package Completeness

Application	Expert Team	Loan Analyst					
		1	2	3	4	5	6
1	C	C	C	C	C	C	C
2	C	C	C	C	C	(I)	C
3	I	I	I	I	I	I	I
4	I	I	I	I	I	I	I
5	C	C	C	C	C	C	C
6	I	I	I	(C)	I	I	I
7	C	C	C	C	C	C	C
8	I	I	(C)	I	I	I	I

Circled item indicates disagreement with experts.
C = Complete I = Incomplete

difficult to quantify than measurements on manufactured products and might, for example, be rated by customers on a (ordinal) scale from 1 to 5. Such scales need to be well defined to achieve consistent results.[6]

A Further Example In the loan decision example, it was desired to understand the degree to which analysts varied in their assessment of whether or not a loan application was complete. To make this evaluation, expert analysts classified eight typical loan applications as complete or incomplete. Then six "representative" loan analysts (at least one from each of the four regions) made similar assessments. The results are shown in Table 12.2, highlighting the cases for which the analyst's assessment differed from the experts.

It was decided from this study that analyst inconsistency is not a major issue. The three cases, which resulted in differences were, however, carefully reviewed. More formal analyses of data from such studies (involving go/no-go assessments) are described by Windsor (2003).

12.4.4 Quantify Process Performance

Time to provide a product or service and defect level (e.g., percent of bills sent to customers that are inaccurate and/or incomplete) are typical performance characteristics in process improvement projects. These need to be closely linked to customer expectations. In the loan decision example, almost 10% of applications took over a week to process.

As in other applications, there are frequently multiple, and often conflicting, performance characteristics (e.g., the need to make timely loan decisions versus controlling risk) that need to be balanced.

6. Similar issues are often encountered in the food industry (Section 13.2.3).

12.4.5 Identify Potential Sources of Performance Variability

The tools used to define a process are also often used to identify potential sources of performance variability. For the loan decision example, a fishbone diagram was employed.

12.4.6 Measure the Sources of Performance Variability

Once the potential sources of process variability have been identified, the next task is to quantify their relative contributions to total variability. This, as in the example, is frequently a stepwise process, rather than a one-shot activity.

Data Gathering Improvement studies of existing processes typically start with an examination of the available data. This may identify areas that warrant closer examination and that lead to explanations of process problems. A system that provides the data to conduct such studies effectively is best included in the initial process development. This was not the case in the loan decision example; such a system was put in place with the study of the 743 new applications.

Data Analysis A frequent starting point in the analysis of the resulting data is to segment the observations (by potential explanatory variables) and to compare graphically the distributions (for continuous variables) or defect levels (for attribute variables) of the performance characteristic(s) for different segments. This was done in the example, most notably through Figure 12.10. The results suggested further assessments (i.e., "drilling down" the data) that eventually led to an explanation of many of the lengthy delay times.

Control charts (Section 6.7)—on both the process performance variables and on potential explanatory variables—are another frequently used tool to help identify potential sources of performance variability and to identify special causes that occur over time.

Thus, one can often gain much insight into a process (and perhaps dispel some misconceptions) by using appropriate graphical and simple statistical methods, such as the basic quality tools introduced in Section 2.2.5. As applications become more complex, more sophisticated analyses might be needed. Some examples are:

- Quantifying the individual components of variability (using analysis of variance) when the performance characteristic is measured on a continuous scale.
- Using specialized methods for the frequently encountered cases when performance, involves counts (e.g., number of incidences), proportions or percentages (e.g., percentage of issues resolved the first time), or ordinal data (e.g., degree of customer satisfaction). Maximum likelihood and generalized linear models are often used to model and analyze such data (see Myers, Montgomery, and Vining 2002).
- Employing data mining and related tools for situations with large volumes of data. For example, classification and regression trees (Section 11.7.7) may help identify key variables or informative subsets.

Special Studies Business process improvement projects often address questions that resemble those for products; for example, identify critical factors, trouble-shoot processes, or optimize operations. As for products, statistically designed experiments are often an effective way to obtain information for better understanding (see Sidebar 12.3).

SIDEBAR 12.3: EXPERIMENTAL DESIGN FOR BUSINESS PROCESS IMPROVEMENT

Performing a DOE on a business process often poses some significant challenges due to the difficulties in establishing a controlled environment. These are addressed in the investigations cited in Table 12.3 and in 13 case studies described by Ledolter and Swersey (2007).

DOE is also used in the skip-tracing problem described in Section 12.5.

12.4.7 Improve the Process and Validate Impact

Once the major drivers of variability have been identified, the next step is to improve the process by eliminating these or, at least, mitigating their impact.

The exact nature of the process changes depends on the specifics of the situation. For example:

- Variability in performance is mainly due to isolated (or special cause) problems, for example, inconsistently incorrect decisions by one or more credit

TABLE 12.3 Examples of Designed Experiments for Business Process Improvement

Application Area	Performance Variables	Experimental Variables	Reference
Information systems help desk	Number of calls to service resolution, response time, call length.	Type of training provided to service personnel, servicing tools (e.g., phone, web-based), accessibility to instructions, issue elevation protocol.	Anderson (2004)
Billing	Collection time, billing errors, losses.	Customer size, customer location, product type, billing method, follow-up, type of billing service (e.g., contract, in-house).	Anderson (2004)
Sales	Successful closure rate, cycle time.	Initial sales method (e.g., formal presentation, meeting with customer), number of sales people, follow-up method, sales brochure.	Anderson (2004)
Home mortgage lending	Mortgage approval times, default risk.	Credit history, mortgage size, customer region.	Hoerl and Snee (2001)

analysts. In such situations, the problem is addressed by removing the source, such as retraining or reassigning deficient analysts and improving the training of future analysts.

- Need for process improvement or redesign. In the loan decision example, the delay in obtaining sign-offs for bank reference information was addressed by requesting advance sign-offs from applicants.
- The existing system is beyond repair and needs to be replaced.

Some specific approaches for business process improvement are indicated in Sidebar 12.4. These also involve, to varying degrees, some of the other stages described in this Section.

SIDEBAR 12.4: SOME APPROACHES FOR BUSINESS PROCESS IMPROVEMENT

Process Management

Process management concepts provide basic guidelines for streamlining and simplifying operations (see Chapter 6 of Juran and Godfrey 1999 and Langley, Nolan, Nolan, Norman, and Provost 1996 for further information).

Lean Six Sigma

Lean Six Sigma (Section 8.6.2) focuses on speed, elimination of waste, standardization, and flexibility. The basic principles and tools of this approach, originally developed for manufacturing, are now being applied to the improvement of business processes (see George 2003).

Poka-Yoke

Also originally developed for manufacturing applications, Poka-Yoke or mistake-proofing, advocates common sense solutions to prevent mistakes, focusing on upstream activity (see Shingo 1986). Building automated rules into information systems to prevent the use of incomplete documents in downstream processing is a typical example. (See Evans and Lindsay 2007, Chapter 13 for further information.)

Best Practices

Additional ideas for process improvement may be gathered from studying similar processes (e.g., benchmarking) and staying on top of advances in enabling technologies. Advances in information technology allow the integration of complex work processes and can lead to radically new approaches.

An important final step before implementation is validating the proposed improvement, ensuring that it does not create any new problems, and obtaining an initial assessment of its impact. In the loan decision example, this involved piloting the proposed new methods in Regions 1 and 3.

12.4.8 Monitor the Process

Once a change has been successfully implemented, it is important to verify that the resulting gains continue to be realized over time, and to seek further possible improvements. Control charts again provide useful tools for monitoring process inputs and outputs. This was the approach used in the loan decision example.

12.5 ANOTHER EXAMPLE: IMPROVING THE SKIP TRACING PROCESS

> **He can run, but he can't hide.**
> —Joe Louis

This further example expands on an application originally presented by Hahn, Doganaksoy, and Hoerl (2000).

12.5.1 Understand the Business Background and Problem

In the consumer credit card business, losses from unpaid accounts is a huge expense, and one reason why credit card interest rates tend to be significantly higher than those on other loans. Businesses have developed various ways of getting delinquent customers to pay up on overdue accounts (Section 11.7.5). These techniques, however, are of little help if the customer has "skipped town" and cannot be located. This leads to the need for skip tracing—that is, obtaining valid addresses and/or phone numbers to reach delinquent customers.

12.5.2 Quantify Process Performance

The loss associated with skipped accounts can be readily quantified. So can the potential net gain, once these accounts are located and contacted.[7] Currently, about 58% of skipped accounts can be traced successfully. Even a 1% increase in this number can lead to a significant decrease in lost revenue.

12.5.3 Define the Process and Opportunities for Improvement

Figure 12.11 provides a high-level description of the skip-tracing process.

The main focus of this project was on the database search. There is, undoubtedly, some "optimal" combination that maximizes net collections. But how can we determine what it is?

12.5.4 Validate the Measurement System and Gather Data

Measurement Consistency A key initial task was to ensure a consistent identification of a skipped account. To qualify, an account must result in inability to locate its owner after the initial search shown in Figure 12.11. Typical characterizations

7. The actual gain depends upon how many of the located accounts make payment and on collection costs.

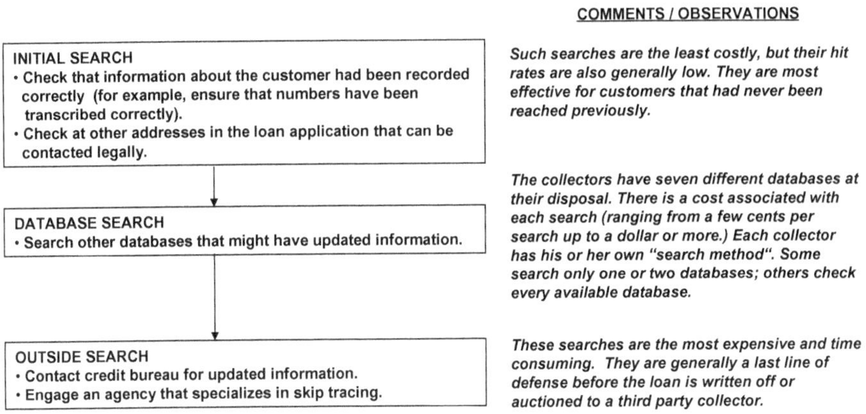

Figure 12.11 Process flow diagram of the skip-tracing process.

were no phone number given, phone not in service or disconnected, and wrong or restricted number. An account could be incorrectly labeled as skipped if, for example, a recent reported change in contact information had not been entered into the database.

A random sample of 100 accounts was independently studied by an expert locator. Each was deemed to have been correctly identified as "skipped."

Historical Data Past data could not be readily analyzed to find an optimal strategy. For example, collectors typically used the cheapest database first, and turned to more expensive ones only after these had failed. Therefore, the hit rates for the expensive databases were likely lowered because they had been used on the more difficult accounts. Also, collectors did not consistently record the approaches they had employed.

Planned Investigation Instead, a designed experiment was conducted to evaluate the impact of using each of the seven available databases, and to identify interactions among them. A 2^7 full factorial plan, calling for each of the 128 possible combinations of the seven databases, was employed. A random sample of 100 recently skipped accounts was traced using each of the 128 strategies.

The response variables were the "hit rate" or effectiveness of the tracing process (i.e., the percentage of accounts successfully located) and the cost for so doing.

12.5.5 Findings and their Implementation

Table 12.4 shows the results from seven of the resulting 128 experiments.

The complete data and their analysis shed much light on effective skip tracing. First, the least expensive database was nearly as effective as the most expensive one. Second, contrary to expectations, there was less overlap in the effectiveness of

TABLE 12.4 Results from Seven Runs from the Full Factorial Experiment for the Skip-Tracing Process

Run No.	Database							Effectiveness (%)	Total Cost ($)
	1	2	3	4	5	6	7		
1	–	–	–	–	–	–	–	0	–
2	+	–	–	–	–	–	–	38[1]	4.46
3	–	+	–	–	–	–	–	11[2]	1.53
4	+	+	–	–	–	–	–	42[3]	5.99
5	–	–	+	–	–	–	–	36	1.27
6	+	–	+	–	–	–	–	55	5.73
7	–	+	+	–	–	–	–	39	2.80
⋮								⋮	

[1]Database 1 alone found correct addresses for 38 of the 100 accounts
[2]Database 2 alone found correct addresses for 11 of the 100 accounts
[3]Databases 1 and 2 together found correct addresses for 42 of the 100 accounts
+ Database Used
– Database Not Used

individual databases than originally thought. Thus, customers that could not be found with one or two databases might still be located by using additional databases.[8]

The results of the experiment, taking into consideration both effectiveness and cost, led to an "optimal" search method that employed up to five of the seven databases in a specified sequence (i.e., using one database after another until either the skip is located or all five databases have been used).[9]

12.5.6 Improve the Process and Validate Impact

The optimum procedure was taught to all collectors. The skip account identification effectiveness subsequently increased from 58% to 63%.[10] The first year net financial benefit, from collecting additional cash, and avoiding writing off bad debt, after taking collection costs into consideration, was almost $3 million dollars—a great return for the level of effort required for the project.

A further benefit of the evaluation was that it provided a quantifiable basis for deciding on the required number of collectors.

12.6 SOME FURTHER CONCEPTS AND METHODS

12.6.1 Control versus Uncontrollable Variables

It is useful, as for products, to distinguish between controllable and uncontrollable, or noise, variables that impact business processes. Staffing levels, training,

8. This showed up as a lack of interactions in the statistical analysis.
9. The actual application also considered account balance.
10. Even though invariably, other factors changed also, the improvement seemed to be mainly attributable to the change in the collection process.

compensation, pricing, and supplier selection are examples of controllable variables. Product demand, actions of competitors, personnel availability, and variability in delivery times for raw materials are (principally) uncontrollable variables.

The goal again is to make the process as robust as possible to the variability introduced by uncontrollable variables through appropriate setting of the control variables. Some uncontrollable variables may be, at least partially, predictable, and, therefore, more readily addressed. For example, the demand for most products and services has time-of-day and seasonal fluctuations that can be estimated from past data. Some of the adverse consequences of such fluctuations can, therefore, be controlled by adjusting staffing levels appropriately.[11] In other cases, we need to mitigate the unfavorable effect of uncontrollable variables by appropriate actions on controllable factors, such as by employee training and by working closely with suppliers and identifying potential alternate suppliers.

12.6.2 Discrete Event Simulation

Discrete event simulation (Section 8.6.3) permits the assessment of a business process before its actual implementation. It lets us compare alternative strategies and eliminate the unattractive ones without incurring the expense of trying them out. It may also be employed to evaluate the impact of changes in the environment and to conduct "what if" evaluations, such as comparing alternative staffing strategies.

Discrete event simulation is especially useful when:

- There are interdependencies in the system, due, for example, to shared resources
- There is appreciable variability
- Analytical methods are unavailable or difficult to use due to problem complexity

12.6.3 Failure Mode and Effects Analysis (FMEA)

An FMEA (Section 5.7.4) can anticipate and address potential failure modes of a business process. Table 12.5 displays a segment of an FMEA for a new life insurance underwriting process, showing two potential causes of failure in assigning an application to its appropriate risk class. This analysis helped the team prioritize its efforts.

12.7 OPERATIONS RESEARCH AND MANAGEMENT SCIENCE

Business process improvement frequently calls for using the tools of operations research and management science (OR/MS) introduced in Section 1.7.3.

11. Note resemblance to feedback and feedforward control.

TABLE 12.5 Segment of an FMEA for a Life Insurance Underwriting Process

Process Step	Potential Failure Mode	Potential Effect(s) of Failure	Severity	Potential Cause(s) of Failure	Occurrence	Current Controls	Detectability	RPN	Recommended Actions	Responsibility and Target Completion Date
Assign case to a risk class	Over or under estimation of risk	Assignment to wrong risk class	7	Incomplete or incorrect information about applicant or application	7	Checklist	6	$7 \times 7 \times 6 = 294$	Evaluate and improve input data quality assurance.	Gerry End of Q3
				Rules for risk classification out-of-date	3	Ad hoc	8	$7 \times 3 \times 8 = 168$	Define conditions likely to lead to a change in risk classification rules; update rules.	Necip End of Q2

Definitions:

Severity	1 = low	10 = high
Occurrence	1 = rare	10 = often
Detectability	1 = difficult	10 = easy

RPN (Risk Priority Number) = Severity × Occurrence (Frequency) × Detectability

12.7.1 Applications to Business Process Improvement

Some typical situations for business process improvement (in addition to those indicated earlier) addressed by OR/MS methods are:

- Placement of TV ads for a product in (differently priced) time slots to maximize viewing by likely purchasers, subject to meeting a specified advertising budget.
- The so-called "traveling salesman problem," routing a salesperson so as to minimize the total time required traveling between customers in different locations. This has led to numerous transportation and logistical applications, such as routing truck deliveries and scheduling house calls for home service providers.
- Scheduling airplanes and pilots to maximize utilization of both, subject to meeting maintenance requirements for the planes and in-flight hour and home-base requirements for the pilots.
- Staffing a customer service call desk most efficiently, while maintaining or exceeding a desired level of customer responsiveness.
- Deciding where to locate product distribution centers so as to minimize costs, while meeting customer demands for delivery timeliness.

12.7.2 OR/MS Tools

The tools most frequently associated with OR/MS are mathematical programming[12] methods. These involve the use of mathematical models to optimize a process, subject to constraints. Linear programming (LP), the simplest form of mathematical programming. is used to optimize a linear "objective function" subject to constraints that can be expressed as linear inequalities (see Sidebar 12.5 for a simplified example).

Other OR/MS methods stem from probability theory, such as Markov chains and queuing theory. The latter involves the mathematical analysis of systems with waiting lines and is used, for example, by service operations to determine the resources needed to operate most efficiently (see Hall 1991).

Some additional OR/MS methods are dynamic programming, game theory, decision analysis, and inventory theory. Discrete event simulation may also be considered to be an OR/MS tool.

SIDEBAR 12.5: SIMPLE LINEAR PROGRAMMING EXAMPLE

Problem Statement

A manufacturer wishes to use excess capacity on three manufacturing lines over a 1-week period to build two added products. The problem is to determine how much of each product to build to maximize profit assuming

12. The early development of mathematical programming preceded computer programming, thus pre-empting the name.

- The processing time available for allocation over the 1-week period on the three lines is 8, 18, and 6 hours, respectively.
- Product 1 requires 0.1 hours of processing on line 1 and 0.3 hours of processing on line 2.
- Product 2 requires 0.2 hours of processing on line 2 and 0.1 hours of processing on line 3.
- The per unit contribution to profit margin is $200 for Product 1 and $500 for Product 2.

Mathematical Formulation

The "decision variables" x_1 and x_2 denote the number of units of Products 1 and 2, respectively, to produce over the 1-week period. The function to be maximized—called the "objective function"—is the total contribution to profit, that is:

$$\text{Maximize } 200x_1 + 500x_2$$

subject to the following capacity constraints for each of the three lines:

$$0.1x_1 \le 8$$

$$0.3x_1 + 0.2x_2 \le 18$$

$$0.1x_2 \le 6$$

Furthermore we require $x_1, x_2 \ge 0$.

This is an LP problem since it calls for optimizing a linear objective function, subject to meeting linear constraints.

Solution

Applying the LP algorithm yields the optimal solution $x_1^* = 20$ and $x_2^* = 60$ units (using the superscript * to designate optimal values), that is, build 20 and 60 units of the two products, respectively.

12.7.3 Applying Linear Programming to Production Planning

Goal Tyagi, Kalish, Akbay, and Munshaw (2004) use LP to develop a multiyear production plan to let a global thermoplastic resin manufacturer optimally allocate plant manufacturing capacities to different products to meet predicted demand.

The resulting plan is used to make long-term high-level production planning decisions, such as the desired capacity at each plant and supplier/logistic agreements. Poor decisions could lead to underutilized plant capacity, unmet customer demand, overproduction, or excessive transportation costs. Actual production decisions are made over shorter time horizons and may also employ mathematical programming.

Process Overview Production involves a two-stage process: polymer production and resin finishing. Figure 12.12 displays the key elements of the underlying supply chain.

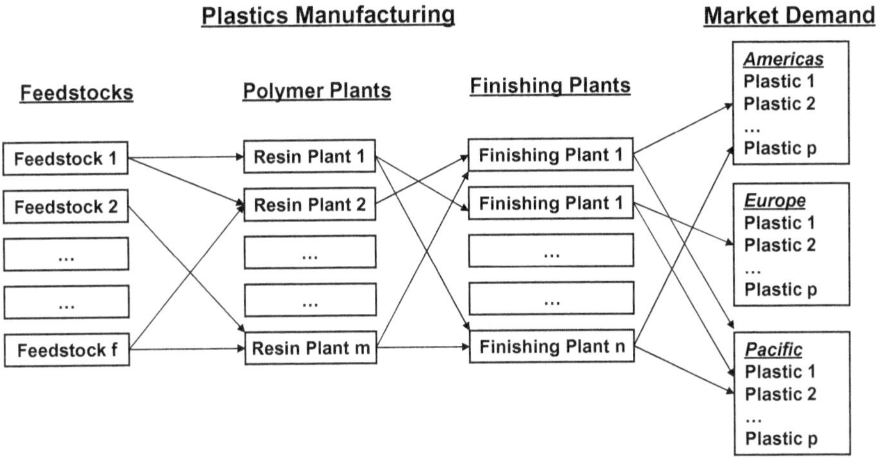

Figure 12.12 Plastics manufacturing supply chain.

The required feedstock materials (e.g., crude oil and natural gas) are shipped to each of the polymer plants for conversion into one of 10 polymer types. Polymers are then combined in a finishing plant and suitable chemical additives and pigments for coloring are added to produce more than 20 types of thermoplastic. The resin is then shipped to industrial customers worldwide for use in their products.

Problem Formulation The goal is to maximize net revenues—that is, total revenues from all resins produced less the manufacturing, raw material, transportation and other costs—over the planning horizon. Constraints include:

- Demand: Production for each product should not exceed (estimated) demand.
- Manufacturing capacity: The polymer and finishing plants each have their own capacities and rates of production and are capable of making only a subset of products.
- Network flow: There are restrictions on the types and amounts of raw materials and finished products that can be transported among sites.

The "decision variables" for which solutions are desired are (1) the amount of each polymer to be produced at each polymer plant, and (2) the amount of each thermoplastic resin to be produced at each finishing plant over a specified time period.

Implementation Both the objective function, expressing the goal of maximizing net revenues, and the constraints can be expressed in a linear form, making this an LP problem. The resulting model involved about 2,000 decision variables and 500 constraints. The Lindo mathematical programming software was used to obtain the optimal solution. Analyses were also conducted to assess the sensitivity of the optimum to plausible deviations from the estimated values of the input variables. The evaluations were repeated quarterly using the latest input information.

12.7.4 More Advanced Mathematical Programming Tools

In many practical situations, the assumptions underlying linear programming are not met. Various extensions have been developed to address more complex problems. These include:

- Nonlinear programming: The function to be optimized or the constraints, or both, are nonlinear.
- Stochastic programming: The random variability in the input information is explicitly recognized (this was addressed in the production planning problem by sensitivity analyses).
- Integer programming: Some or all of the variables to be optimized can take on only integer values (e.g., the number of machines or the number of operators) and a satisfactory approximation is not obtained by ignoring the integer nature of these variables.

More complex problems can involve a combination of the preceding methods.[13]

12.7.5 Further Reading

Detailed descriptions of the tools of OR/MS are provided in texts on the subject (see Hillier and Lieberman 2005, Winston 2004, and Bronson and Naadimuthu 1997).

12.8 MAJOR TAKEAWAYS

- A business process is a series of interrelated activities that convert inputs into outputs to achieve a business goal.
- The performance of its processes is key to the success of any business and there is a frequent need for improvement. This calls for a team effort among all involved.
- Business process improvement typically requires one to:
  - Understand the business background and problem
  - Define the process
  - Gather data and validate the measurement system
  - Quantify process performance
  - Identify potential sources of performance variability
  - Measure the sources of performance variability
  - Improve the process and validate impact
  - Monitor the process

13. As problems become increasing complex, discrete event simulation is an increasingly attractive alternative, or supplement, to mathematical programming.

- Elementary statistical tools often (but not always) suffice in addressing business improvement projects. A broad perspective and statistical thinking are especially important.
- Many concepts discussed in earlier chapters, such as segmentation, control charts, DOE, the differentiation between control and uncontrolled variables, discrete event simulation and FMEAs, also apply to business process improvement.
- OR/MS also provides important tools, such as mathematical programming, for improving business processes.

DISCUSSION QUESTIONS

General

1. In Section 12.2.1, we give some examples of business processes. Select one or more of these and describe how you might proceed in improving the process.
2. The loan decision example (Section 12.3) required obtaining various types of data (e.g., time to gather bank reference information) that had not been routinely gathered historically. What issues might arise when those involved with the process are asked to gather data, which was not previously part of their jobs? How would you address such issues?
3. What issues might arise if the box plots by sales manager and region shown in Figure 12.10 are not adequately explained? How might you minimize their occurrence?
4. Discuss the applicability of robust design to business processes and give some examples.
5. Consider the waiting time in handling phone calls to an airline ticket booking service. Suggest control and noise variables that impact this process and how you might use the control variables to counteract the noise variables in improving service.
6. Some call centers, when unable to give immediate service, provide customers an estimate of expected waiting time. An airline has asked you to assess whether or not to give callers such information. How would you conduct such an evaluation?
7. In the previous problem, how would you develop a system to provide callers estimates of expected waiting times.
8. Describe some common business processes for which the major concern is (a) reducing variability, (b) increasing or decreasing the mean, and (c) reducing both variability and increasing or decreasing the mean.
9. You are the Dean of Faculty at a university. You have had numerous complaints about uneven course grades, especially among instructors of different sections of the same course. How might you gather the needed data and conduct appropriate analyses to help address this concern, both immediately and in the long run?

Technical

1. The discussion in Section 12.2.3 suggests the need to address the "influx of travelers at ticket counters at peak times of day or year or, less predictably, due to flight cancellations." How would you use a statistical (and OR/MS) approach in each of these situations?

2. The statistical evaluations in the loan decision example (Section 12.3) consisted principally of simple graphical displays. Although these seemed to serve the purpose, they have limitations. Discuss these limitations and propose some further analyses to gain deeper understanding.

3. In Section 12.3.6, it was mentioned that "a multiple regression analysis was performed to relate (various) explanatory variables ... to delay time." How would you conduct such an analysis?

4. In assessing life insurance applications, borderline cases are typically turned over to human underwriters. The underwriter's decision of whether to grant an application (and in which risk category to place it) depends on factors such as the applicant's age, medical condition, smoking status, etc. Different analysts differ in their decisions. Propose a study to assess the resulting variability and the underlying causes and how you would use the results to gain greater consistency.

5. It was stated in Section 12.5.5 that in the skip-tracing example "customers that could not be found with one or two databases, might still be located by using additional databases." A footnote added that "this showed up as a lack of interactions in the statistical analysis." Explain this comment.

6. A more sophisticated procedure in the skip tracing example is to tailor the search to specific characteristics of accounts, such as balance owed, past payment history, and reason for nonresponse. How would you factor these into the planning and analysis of the proposed experiment?

7. Say that in the simple LP illustration in Sidebar 12.5, there are five product types, and a further requirement to build no more than three types. How would this change the mathematical formulation? Conjecture some other possible problem constraints and show how you would handle them.

8. A bank is planning to allow its customers to do online transactions via the Internet. How can discrete event simulation facilitate the planning and development of this service? Briefly explain the basic elements and operation of the resulting model.

9. In the preceding example, how might the bank usefully apply OR/MS tools other than discrete event simulation?

10. In the OR/MS production-planning problem in Section 12.7.3, we say that "analyses were also conducted to assess the sensitivity of the optimum to plausible deviations from the estimated values of the input variables." Elaborate on how to do this.

11. A fast-food restaurant wishes to determine the appropriate number of stations and employee staffing. How might you use queuing theory and discrete event simulation to make such assessments? Compare the two approaches.

12. Elaborate on one of the examples of business process problems that can be addressed using OR/MS methods presented in Section 12.7.1 and indicate how you might proceed in obtaining a solution.

CHAPTER 13

FURTHER APPLICATIONS

13.1 ABOUT THIS CHAPTER

There are many areas of application of statistics in business and industry beyond those discussed in earlier chapters. We briefly consider some of these here. We begin with three industries that have traditionally used statistics extensively: the food, beverage, and related industries; the semiconductor industry; and the communications industry. Next, we discuss statistical image analysis – a relatively new field with important medical, security, and other applications. We then provide short descriptions of the role of statistics in various other areas and conclude with a glance into the future.

We do not cover all the bases; but the discussion should demonstrate the use of statistics in addressing diverse problems in business and industry (and beyond). Many of our comments are based upon inputs from those working in the areas discussed.

13.2 FOOD, BEVERAGE, AND RELATED INDUSTRIES[1]

13.2.1 The Setting: From Soup to Soap

Modern conveniences that "enrich" our lives – such as microwave popcorn, frozen gourmet dinners, and instant chicken soup – would not have been possible without scientific advances in chemistry, biochemistry, microbiology, and physics. The food and beverage industry has become a high-technology business. It provides challenging opportunities for the application of statistics in areas ranging from market studies to product development to manufacturing. Similar issues also arise in applications involving many other consumer products, especially the ones that are purchased repeatedly – such as suntan lotion, detergents, light bulbs, bug killer sprays, and paper tissues.

13.2.2 Controlling Product Variability

Reducing product variability is a major goal in the food, beverage, and related industries. These industries, in fact, provide an ultimate example of the need for robust design.

1. This discussion has benefited greatly from inputs provided by Lynne Hare and Mark Vandeven.

There are differences in the types of applications even within the food and beverage industry, especially between raw foods, such as fresh fruits, vegetables, eggs, and raw meats; and processed foods, such as cereals and soft drinks.

Raw Foods Raw foods are often subject to variability in appearance and taste. These differences are often due to variations in nature, especially fluctuations in temperature and rainfall and differences in soil fertility. Such environmental (or noise) variables are difficult or impossible to control. Consumers recognize this and expect – and sometimes even enjoy – some product variability, such as between early season, mid-season and late season fruits. In addition, for most raw foods, the purchaser can examine, and sometimes even sample, the product prior to purchase.

There are, moreover, various control variables that can be set to make raw foods more homogeneous. These range from the type and amount of fertilizer to apply to when to harvest and market. In addition, the product itself might be controlled by, for example, spraying with varying amounts of preservative.

The challenges of growing foods more scientifically led to the development of the statistical design of experiments in agriculture during the early part of the last century (Section 2.2.7). These applications tended to focus on improving product yield (e.g., growing the most vegetables or the biggest tomatoes), rather than on minimizing variability.

Processed Products Consumers are less willing to accept variability in processed products. In buying chocolate chip cookies, canned soups, bottled beer, and so on, they expect a product to look and taste the same from one purchase to the next. We want cheeses from different packages of a particular brand to taste the same, despite the fact that they come from milk from different cows on different farms at different times. Also, cereal manufacturers need combat the tendency of ingredients with different densities (e.g., nuts, raisins, and grains) to separate during packaging. Similar requirements hold for other repetitively purchased products. Tissues, for example, need to have the same soft feel from one package to the next.

Manufacturers strive to achieve product homogeneity for processed products by:

- Working with suppliers to minimize variability.
- Understanding the factors that induce variability during manufacture and addressing these.
- Varying processing conditions to adjust for the variability in raw products. For example, to achieve homogeneity in roasted peanuts, measurements on such factors as sugar content and moisture are obtained on crop samples during harvesting. This information is then used to set processing (e.g., roasting) conditions.
- Blending product grown at different times or in different places. The processing of flour, for example, involves mixing varieties of soft and hard grains, whose properties vary over time. Homogeneity in the final product is attained by adjusting the mixing proportions, based upon product or process measurements.

As for other products, producers need also be concerned with robustness to usage conditions. A dishwasher detergent, for example, needs to be effective in cleaning dishes regardless of water hardness.

13.2.3 Data Issues

In the food, beverage, and related industries, as in other businesses, data are obtained to help determine what features a product needs to have (and what must be done) to capture the highest possible consumer interest. The ultimate judgment for food and beverage products often involves an assessment of taste and appearance; health considerations are also becoming increasingly significant. Other characteristics may be important for nonfood products; the users of a shampoo might, for example, be concerned with aroma.

Sources of Data Foods, beverages, and related products are typically assessed by:

- Consumers (or consumer panels): These generally consist of (potential) purchasers who provide their personal preferences about product characteristics. These are solicited during design and marketing (Section 3.3.1), or they might be on a (possibly prototype) product that has already been made. Consumers describe what they like and dislike about the product, how it compares with competitive products, and gauge their ultimate satisfaction. The results may be used to make improvements and, possibly, to formulate advertising claims (Section 9.8.3).
- Instruments: These involve measurements on such product properties as color, texture, and composition to quantify chemical or physical characteristics.
- Experts (or panels of experts): These typically are trained employees who make sensory assessments of product properties that cannot be measured by instruments (e.g., juiciness, puffiness, and aroma), generally without making value judgments. A typical (and highly envied) example is that of wine tasters. Such evaluations are most often made on final product as a quality check.

A common (simplified) sequence of activities is displayed in Figure 13.1.

Types of Data Measurements obtained by instruments are, by and large, similar to those for other products. In contrast, evaluations by consumers tend to be more subjective and are often expressed on an ordinal (intensity or acceptance) scale, such as from 1 to 5. Such numbers may express preferences running from completely unimportant to highly important or assessments of the product as poor, fair, good, very

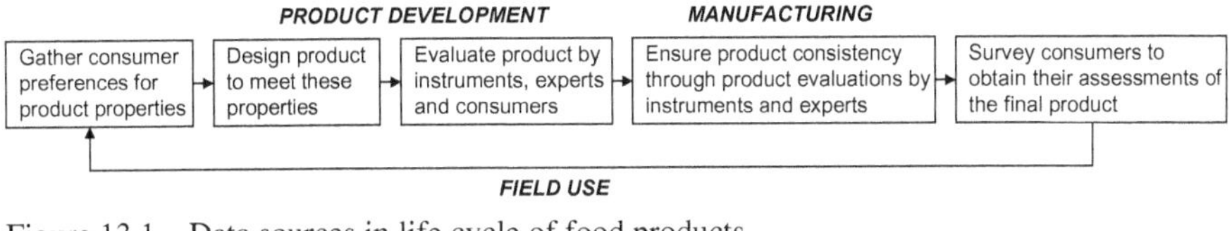

Figure 13.1 Data sources in life cycle of food products.

good, and outstanding. Relative product rankings may be used for assessing competitive products. Sensory assessments by experts, such as an assessment of the degree of dryness of a wine or softness of a tissue, may also be on such a scale.

Data Gathering As in other applications, the data-gathering process is highly important. We generally strive to obtain a random sample of product, and to come as close as possible to a random sample of consumer evaluators.[2]

When many samples are to be evaluated in the same session, protocols that minimize the impact of the sequence of evaluation need to be developed; this is especially important in taste tests.[3]

The resulting sensory and preference data are inherently variable; there are differences in how individuals perceive product attributes, how consumers and testers use scoring scales, and their stamina over multiple samples. Benchmarks for consistently assigning scores need to be established. GRR studies (Section 3.4.2) might be conducted to assess measurement consistency.

Data Analysis As we have seen, the data resulting from consumer preference and expert sensory studies are often ordinal (e.g., intensity scale from 1 to 5) or categorical (e.g., fruity, acidity, and bitter taste). Statistical analysis of such data requires special methods (see Agresti 2002, 2007; Hildebrand, Laing, and Rosenthal 1977). (Also see Lea, Naes, and Rodbotten 1997 for a discussion of specialized methods for the analysis of sensory data.)

13.2.4 Product Formulation: Mixture Problems

> **As a Southern general wrote, mixing mint juleps "is a rite that must not be entrusted to a novice, a statistician, nor a Yankee."**
> —W. Faulkner

Statistically planned experiments are frequently used to help develop the best possible food, beverage or related product, based upon consumer preferences, manufacturing efficiency, cost considerations – or, most likely, a combination (see Hare 2006). These typically call for determining an optimum product formulation, as well as optimum processing conditions.

Processed foods generally involve mixtures of ingredients. For a baked product, for example, we may wish to determine the best blend or mix of ingredients (e.g., yeast, flour, water, shortening, eggs, salt, and spices). This suggests the use of a specialized type of designed experiment, known as a mixture experiment – so-called because it involves a mixture of product ingredients that is constrained to add to 100%. Cornell (2002) describes a mixture experiment to identify the optimal blend

2. In contrast, expert evaluators are generally selected based on their proficiencies.

3. Due to fatigue, etc. it is sometimes inadvisable to have each evaluator assess each product. This leads to the use of randomized incomplete block experimental designs to obtain data for making comparisons.

of three fish species (mullet, sheephead, and croaker) and the best settings of the process factors (duration of deep fat frying, cooking temperature, and time) to make fish patties with consistent texture.

See Cornell (2002) and Smith (2005) for detailed discussions of the planning and analysis of mixture experiments.

13.2.5 Manufacturing

The manufacture of processed foods involves such operations as mixing, storing, drying, condensation, and heat treatment. These operations are similar to those encountered for many continuous processes and considerations similar to those discussed in earlier chapters apply.

13.2.6 Ensuring Food Safety

The Basic Concern Ensuring a safe product is a key concern to the food and beverage industry. Potential hazards can be biological (e.g., a microbe), chemical (e.g., a toxin), or physical, such as glass or metal fragments implanted in the food.[4]

The Traditional Approach to Ensuring Safety Industry and regulators have traditionally depended on spot-checks of manufacturing and random sampling of final product to help ensure food safety. This approach, though necessary, tends to be reactive and often fails to identify problems and their causes speedily.

A Proactive Approach The U.S. Food and Drug Administration and Department of Agriculture are promoting a preventive approach to food safety, known as Hazard Analysis and Critical Control Point (HACCP) similar to the proactive approach for product design, described in Chapter 3, "for use in all segments of the food industry from growing, harvesting, processing, manufacturing, distributing, and merchandising to preparing food for consumption."[5] Issues addressed range from analyzing "potential safety and health hazards" to establishing "recordkeeping that ensures good data are consistently obtained and maintained on potential hazards and their causes." (See Surak 2007 for further information.)

13.2.7 Food Spoilage

Spoilage of a product over time is a further concern of the food and beverage industry, especially for such products as milk, cheese, and meats. There are again strict guidelines and standards to which the industry need adhere. Many foods are labeled with a date by which the product should be sold or used. The determination of such dates raises issues similar to those for pharmaceuticals (Section 10.4.4).

4. The U.S. media recently reported an *E. coli* outbreak, attributed to raw spinach, and peanut butter contaminated with salmonella.

5. Per the HACCP website http://www.cfsan.fda.gov/~comm/haccpov.html.

13.3 SEMICONDUCTOR INDUSTRY[6]

13.3.1 The Setting

Semiconductors, by controlling the flow of electrical signals, are the engine behind much of today's advanced electronic technologies (e.g., computers, cell phones, digital televisions, airplane control systems, etc). They may be in the form of discrete devices, such as individual transistors, resistors, capacitors and diodes. Or they may be connected together in large numbers on a continuous substrate as an integrated circuit (IC).

The basic element is a chip, built in a tiny rectangle on a thin wafer sawed from a cylindrical ingot of extremely pure, crystalline silicon. A chip may have millions of transistors. The dimensions are, moreover, getting continuously smaller – even as the specifications on properties are getting tighter. A single wafer can yield hundreds, thousands, or even tens of thousands of chips.

13.3.2 Business Features

The Manufacturing Process Semiconductor manufacturing takes place in clean rooms and involves complex processes, starting with the fabrication of a seed crystal. This is sliced into wafers that may range in size from a small chocolate chip cookie to a large pizza. The processing of wafers typically requires many steps; several wafers are often worked together in lots. The wafers are eventually diced and fabricated into ICs or discrete devices.

Process Measurements Automated measurements are taken during and after each step, or after several steps, of the fabrication process on all or selected wafers and chips and on the generally expensive tools doing the processing. More intensive measurements might be taken when the process is deemed "unhealthy." Measurements may be used for early-stage rejection (to avoid the cost of further processing of inferior material) or rework or for adaptive control (see Sidebar 13.1). Further measurements, often based on machine sensor trace data, are used for fault detection, and still others for control charting (Section 6.7) and for providing information for improving the process.

SIDEBAR 13.1: INTEGRATING ENGINEERING PROCESS CONTROL (EPC) AND SPC IN THE SEMICONDUCTOR INDUSTRY

The integration of EPC and SPC (Section 8.2) in semiconductor fabrication may involve using the results from one stage of wafer processing to automatically make changes in the process settings in subsequent steps. This can get a drifting process back on target and/or reduce variability, while at the same time allowing the variability to be monitored by control charts. Sachs, Hu, and Ingolfsson (1995) describe an application aimed a keeping the thickness of epitaxy layers on wafers on target.

6. This discussion has benefited greatly from inputs provided by Mike Clayton, Veronica Czitrom, Mary Lanzerotti, Paul Tobias, and Emmanuel Yashchin. Various definitions were adapted from the Semiconductor Industry Association's online glossary.

End-of-Line Measurements A series of automated measurements on electrical properties, such as on current and threshold voltage, are typically taken on each chip at the end of the process. The chip is accepted if all measurements meet specifications (see Sidebar 13.2).

SIDEBAR 13.2: END-OF-LINE YIELD

End-of-line yield is the percentage of chips that meets final product specifications on all measurements. End-of-line yield is often low for new products; yields of less than 50% are not uncommon and may be initially acceptable. Due to the high volume of product and the low cost of inspection, economic analysis may suggest 100% end-of-line inspection to control quality of early product – while striving to improve the process upstream.

Over time, end-of-line yield needs to improve dramatically. The ability of an operation to raise its yield rapidly is a hallmark of its success and involves all stages of production. Mature processes typically and consistently yield over 90% acceptable product.

The accepted chips may be sorted into groups, based upon their gigahertz rating and possibly other properties. It is important that the process yields a high percentage of "fast-running" chips; these command the highest price. If a chip's electrical characteristics do not fully meet prime product requirements, it may, however, still be useable as a less-demanding product.

Complex statistical yield models have been developed. Articles on this subject appear in the *IBM Journal of Research and Development* and in *Semiconductor International*, including a series of papers by Atchison and Ross (1999).

A Data-Intensive Business Quoting Czitrom and Spagon (1997), "the chemistry and physics of these (semiconductor) processes are not well understood. ... Semiconductor manufacturing is thus a measurement intensive business."

Statistical methods harness the data to improve each of the many process steps and tie them together. They have also helped shift industry thinking from the compilation of yield rates to understanding underlying causes and how to address them. Data-mining methods are used to find patterns of defects traceable to a particular process step; statistically designed experiments and simulation are also increasingly employed. These are integrated with whatever physical models are available.

13.3.3 Understanding the Impact of Process Variables upon Performance

A key industry goal is to improve productivity and yield across the board, and to establish optimum process specifications. It is, therefore, critical to gain an understanding of how the many factors associated with processing (e.g., photolithography, etching, stripping, diffusion, ion implantation, and deposition) impact electrical performance and yield.

Such understanding is sometimes gained, especially during early stages of development, through laboratory-scale investigations. Frequently, however, studies are conducted directly on a (pilot) production line with good units sold to customers. A manufacturer might, for example, make a variety of design masks to determine which provides the best results. Experiments may focus on a specific process step or several such steps. Such studies raise various technical challenges (see Sidebar 13.3).

SIDEBAR 13.3: SOME TECHNICAL CHALLENGES

Challenges in studies to understand the impact of process variables upon semiconductor performance include:

- The frequent need to consider a large number of potential impacting variables, especially in experiments involving multiple processing stages. This typically results in the intitial use of screening experiments (such as fractional factorial plans) to efficiently identify a smaller set of critical factors.
- Interactions between variables in different steps of the process – for example, between implant and anneal.
- The multivariate nature of such studies – due to multiple, and often correlated, response variables. (A response variable at one stage might be a process variable in a subsequent stage.)
- The relatively long elapsed time to conduct even a single production run. The fabrication process often spans weeks and even months.
- The constant pressure to upgrade, calibrate, and maintain state-of-the-art metrology instruments capable of measuring nanometer features with accuracy and precision.

(See Baron, Takken, Yashchin, and Lanzerotti 2004 for further discussion and references.)

13.3.4 Monitoring and Controlling Critical Process Variables

Control charts, especially $\overline{X}$ and range charts, are often used to monitor critical parameters and to detect changes throughout the manufacturing process. These help decide when corrective action needs to be taken to ensure the process stays on target and to reduce variability. Understanding the structure of the process is critical to establishing the most effective groupings for this purpose; wafers or sometimes locations on a wafer are often used.

Due to the numerous operations involved in semiconductor fabrication and the many measurements taken, the number of such control charts can be quite large. Therefore, it is useful to develop an automated approach for assessing the results and providing appropriate notifications for out-of-control situations, while striving to minimize the number of "false alarms."[7]

13.3.5 Assessing Product Life

Accelerated life tests (Section 5.7.8), especially at high-voltage stress, are frequently used to assess the life of ICs, as well as that of printed circuit boards, and larger

7. A more advanced approach involves monitoring individual estimated variance components (see Yashchin 1994).

assemblies. Observed failure mechanisms need to be understood and modeled both physically and statistically to ensure they do not have a deleterious impact on usable product life (see Li, Christiansen, Gill, Sullivan, Yashchin, and Filippi 2006; Li, Yashchin, Christiansen, Gill, Filippi, and Sullivan 2006).

13.3.6 Product Burn-In

Newly designed ICs are frequently subjected to burn-in (Section 5.11) prior to shipment to minimize infant mortalities in the field. Such burn-in might be discontinued as the product matures. Statistical concepts help determine how long units should be burned in, based upon cost considerations (see Kim and Kuo 1998).

13.3.7 Sources of Further Information

Czitrom (2003) provides a detailed discussion of statistical approaches and methods dealing with process improvement for the semiconductor industry (see Sidebar 13.4) and Czitrom and Spagon (1997) describe 24 statistical case studies.

SIDEBAR 13.4: STATISTICAL METHODS USED IN SEMICONDUCTOR INDUSTRY

Many of the statistical methods Czitrom (2003) describes have already been mentioned in this and earlier chapters. These include:

- GRR studies to quantify measurement variability and act thereon.
- Multi-vari charts to compare characteristics over different positions on a wafer, different wafers and different manufacturing lots.
- Variance component analysis to quantify the relative contribution to total variability of, say, lots, wafers, positions within a wafer, individual units, and measurement error.

Other methods tend to be unique to the semiconductor industry. For example, wafer maps provide a graphical comparison of a measured performance characteristic by wafer site.

Also, the (NIST/Sematech) online Engineering Statistics Handbook contains many case studies, several from the semiconductor industry.

13.4 COMMUNICATIONS INDUSTRY[8]

13.4.1 Setting the Stage

The communications industry is concerned with the transmission of voice, data, text, sound, and video. This industry has, almost since the advent of the telephone in the

8. This discussion has benefited greatly from inputs provided by Diane Lambert and Scott Vander Wiel.

early twentieth century, played an important role in the development of statistical applications and concepts. As indicated earlier, the study of processes through statistical monitoring can be traced back to Shewhart's work at Bell Labs. Since then, the industry has benefited from the involvement of many key figures in statistics.

This attention is not surprising in light of the massive amounts of data generated every minute of every day. Each communication, such as a telephone call, can produce data on who initiated the communication, who received it, where and when it was placed, how long it lasted, how it was paid for, and so on. In addition, there is extensive systems information, such as on switching performance.

All of this creates formidable challenges in recording and maintaining data and using such data to the advantages of the business.[9]

13.4.2 Usage Assessments

Telephone usage data are explored (or mined) by network engineers and statisticians to identify meaningful patterns and relationships, both short-term and long-term. Such studies help address practical questions such as determining:

- The capabilities needed to allow calls to be successfully placed and completed 99.99% of the time.
- Peak usage times (e.g., 10:00 a.m., U.S. Eastern Standard Time, on Mother's Day) in the United States and the probability of successfully making a connection on the first try at such times.
- A company's penetration into different geographical markets.
- The impact of new technologies, such as Voice Over Internet Protocol (VOIP), on customers' calling habits.

See Raftery, Tanner and Wells (2001) for further discussion and added applications. Also, Lambert and Pinheiro (2001) studied wireless calling records for 96,000 customers, making about 18 million calls in three months. Their work characterized statistically the calling patterns of individual customers with the ultimate goal of identifying unusual events that might signal fraudulent calls – a topic considered directly by Cahill, Chen, Lambert, Pinheiro, and Sun (2002).

13.4.3 Reliability and Service Improvement

The communications industry, like other businesses, strives for high reliability and service. The complexity of communications systems in general, and phone networks in particular, make them subject to hardware failures or possible performance degradation over time resulting in deteriorating message quality. The reasons for failures during operations need to be quickly identified and addressed.

Software issues can, in addition, lead to service breakdowns, potentially involving large geographic areas; software reliability (Section 5.12) is, therefore, of particular interest.

9. References to articles that discuss specific application areas, beyond those cited here, are to be posted on the book's ftp site.

Storms are a further major reason for disruption in service of communications products and are best addressed proactively by making systems more robust to adverse weather. Service providers also strive to predict emergencies and to develop and evaluate the effectiveness of alternative contingency plans.

13.4.4 Business Assessments

Pricing The communications industry is highly competitive. The cost of making calls is very important in gaining market share. Service providers need to know the impact of alternative pricing strategies and may test these out in selected markets. This leads to such statistical issues as deciding which market areas to select, the duration of tests, and the selection, and relevance, of the prior (and possibly subsequent) time periods and market areas with which comparisons are to be made.

Competitive Evaluations Communications service providers – and especially those of wireless products – often wish to make claims about the superiority of their product's performance. Becker, Clark, and Lambert (1998) describe a study that involved hundreds of millions of test calls to quantify reliability for different circuit-based service providers. The study required placing calls on carefully selected test phones between designated cities, say every 15 minutes, using a company's own network and those of competitors and recording fraction of calls blocked.

Logistics Statistical and related concepts are applied to address various logistical issues, such as optimizing cell tower placement.

13.4.5 Internet Opportunities

The Internet has had a major impact on the communications industry. Many companies have expanded their services to become Internet providers to private and commercial customers. Statistical methods are being used to gain a better understanding of Internet traffic patterns to help improve performance.

Also, companies wish to leverage data from Internet business transactions to learn about customer behavior. Online retailers, for example, may personalize their offerings, or their advertising, by tailoring them to individual customer groups, using, perhaps, multivariate statistical analysis methods, such as cluster analyses, to establish such groups.

13.5 STATISTICAL IMAGE ANALYSIS FOR MEDICAL, SECURITY, AND OTHER APPLICATIONS[10]

13.5.1 Introduction

The digital image – simply a matrix of numeric *pixels* (picture elements) represented as brightness on a computer screen or as a printout – is the natural data output format

10. This section was written by Colin McCulloch.

for a wide variety of applications. From digital cameras and flatbed scanners to medical imaging devices, digital images are being produced ubiquitously in staggering volumes. In fact, in many imaging domains so much data are being produced that it has become nearly impossible for humans to review it all. State-of-the-art computed (axial) tomography (CAT) scanners can, for example, generate a thousand images of a person's lungs in just a few seconds. This amount of information can be overwhelming for radiologists looking for lung cancer, especially when they must review scans from dozens of patients each day.

Automatic image analysis holds the promise of sifting through large amounts of data to identify, quantify, and summarize the most important characteristics. Systems are already in use that automatically read addresses on envelopes, detect breast cancer in mammograms, and match crime scene fingerprints to databases of known felons.

Many of these applications require modeling uncertainty in the data and therefore rely, to a significant degree, on statistical concepts and analysis. This section briefly describes two of the most important and challenging areas of statistical image analysis: image restoration and pattern recognition.

13.5.2 Image Restoration

Goal The goal of image restoration is to take an image that demonstrates some defect, such as uneven lighting or being out-of-focus, and alters it to remove the defect. The statistical element of this problem arises from the need to model how an image with a defect can arise when using a particular imaging device such as a digital camera. Although we might be able to model the mechanics of the camera almost deterministically, we are usually uncertain as to exactly how it was used and in what environment. Statistical models can help quantify and address such uncertainty.

Application to Digital Photography One of the advantages of digital photography over conventional film is its ability to correct small "errors" by the photographer. This is accomplished by using one of various easy-to-use and powerful photo-editing software packages.

Many of the image-editing tools in these packages can be considered to be implementations of statistical image analysis. For example, one of the most widely used tools, the UnSharp Mask (USM) filter, addresses the familiar blur seen in out-of-focus photographs caused by the optics of the camera when the focus distance is slightly off the subject of interest. The USM filter recovers sharpness in such pictures.

This filter is based on the notion that an out-of-focus picture can be modeled as a *Gaussian blur* of the desired (but unobserved) in-focus picture. A Gaussian blur is simply a weighted moving average over all the pixels in the image, where the weights are defined by a Gaussian distribution.[11] The model assumes that at each pixel of the in-focus image, the blurred image is made up of the weighted average

11. The term "Gaussian distribution" is widely used in the imaging literature for the normal distribution.

of nearby pixels, weighted more heavily for close-by pixels. To recover sharpness from the out-of-focus image, the USM filter estimates the parts of the image lost due to the camera being out of focus by subtracting a slightly Gaussian blurred version of the out-of-focus image from the original. This difference image, which accentuates the sharp portions of the image (those areas with adjacent structures of large differential brightness), is then added to the original out-of-focus image to increase sharpness and to produce an estimate of the in-focus image.

Simple Example Figure 13.2 gives a stylized example of the Gaussian blur model. In the left panel we have drawn a simple happy face in black on a white background. The middle panel is a Gaussian blur of this image. The goal in image sharpening is to estimate the original in-focus image from this out-of-focus image. The right panel is the result of this estimation process using the USM filter. Like all statistical estimation procedures, it is not a perfect reconstruction. For example, the separation between the left eye and the left side of the face, which is clear in the original image, is less defined in the estimated image. And, all the lines in the estimated image are thicker than in the original. Such artifacts are unavoidable because of the limited information contained in the middle image, which was the input to the filter. However, the contrast in the original image has been largely recovered and, overall, the sharpened estimated image tends to be more pleasing to the eye than the out-of-focus image.

Photography Example Figures 13.3 and 13.4 illustrate this process in a photography setting. The original slightly out-of-focus image, shown in the left panel of Figure 13.3, was taken using a digital camera. The sharpened image is shown in the

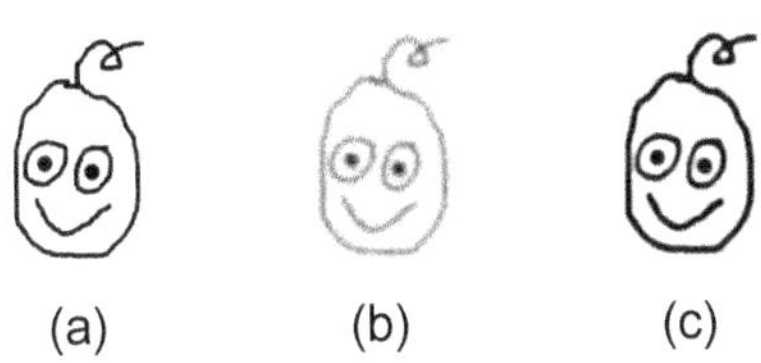

Figure 13.2 A simple example of a Gaussian blur model and the reconstruction of the original image, using a USM filter, showing (a) an in-focus image, (b) its Gaussian blur, and (c) the restored image.

Figure 13.3 Original slightly out-of-focus image (left) and sharpened image (right).

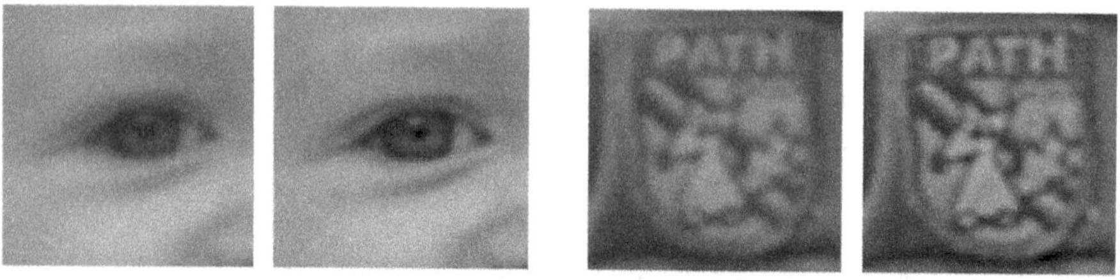

Figure 13.4 Two portions of the image in Figure 13.3, showing the original slightly out-of-focus image (left) and the sharpened image (right).

right panel. It provides greater contrast between adjacent structures with differential brightness as compared to the original image. For example, in the sharpened image, the reflected light in the boy's eyes is more noticeable, and the patch on his coat is better defined. Figure 13.4 highlights these differences by zooming in on the boy's eyes and the patch on his coat.

The USM filter gives the appearance of sharpness to blurry images by increasing contrast between light and dark adjacent structures. It cannot recover detail lost due to a camera being completely out of focus. This is because the only input to the filter is the out-of-focus image; any fine structure not actually present in an image cannot be recovered.

13.5.3 Pattern Recognition

Goal Image pattern recognition is used to determine if a pattern of features is present in an image, and if so, to quantify the location, size, and shape of the pattern. This has numerous important medical applications, such as in detecting tumors. We will, however, discuss an application dealing with photographic face recognition.

There are various components of uncertainty in such problems, including whether or not a particular feature is present in the image, biological variability in the locations of features, and noise in the image data itself.

Face Recognition Automatic face recognition has become an important tool in criminal investigations and in various security settings. Photographs of a suspected criminal are matched with an often large database of pictures of known persons. The resulting system takes a digital image of a suspect's face and analyzes it to identify and characterize various "biometric features" of that person. Typical features are the locations of salient points on the face involving a person's eyes, mouth, nose and chin. The system then scans a database of known persons searching for matches with individuals whose images display similar features.

Figure 13.5 illustrates several biometric features of an image for a particular person (i.e., the centers and corners of his eyes, corners of the mouth, and the tip and underside of the nose, overlaid as white squares.)

At least two parts of this problem rely on statistical analysis. First, we must automatically locate the positions of the biometric features in the given image. Second, we must determine whether the arrangement of these features appears to be consistent with those of any specific image(s) in an available database.

Figure 13.5 Photograph with several biometric feature estimates overlaid.

Locating Biometric Features One method of locating biometric features in an image is to apply a template of features (the white squares in Figure 13.5) to the image and adjust it (via scaling, rotating, and warping) until the best template location is found. The best location is typically defined as the one which deforms the template the least while maximizing some image metric on the pixels underneath each of the template points. This metric could, for instance, be the correlation of the pixels in a small neighborhood around the feature position with those of a gold-standard feature image. Such a gold-standard image could be defined a priori to be an average image estimated from the database or possibly an external image whose biometric features had been defined by extensive physical analysis.

Matching with Database Constructing an effective face recognition system calls for matching the biometric features of an unidentified person, such as the one shown in Figure 13.5, with those of an identified person (or persons) in the database. To do this, natural variability in biometric features of different individuals must be properly modeled. For example, the biometric features representing the inner corners of the left and right eye are almost horizontal to each other for most persons (in properly posed photos), but the distance between them varies appreciably from one person to the next.

To capture such variability, a multivariate statistical model to represent feature locations can be fitted, using the data in the image database. This model is then used to derive metrics of template variability (sometimes called *distance*) of the biometric features for the person under scrutiny from those for each person in the database. Given a new image to be matched, we first locate its biometric features, as outlined above. Then we decide which photos, if any, in the database match this new image. This is done by using the statistical model to calculate the distance from the features detected in the image under scrutiny to each of the image feature sets in the image database. The persons in the database whose images have the smallest distances are determined to be the best matches to the person under scrutiny.

We could also establish a system that flags those matches whose distances are smaller than a specified decision threshold value. This requires trading *sensitivity* against *specificity* in the matching algorithm. High specificity means that if a true match does not exist, then the algorithm will correctly fail to come up with one. High sensitivity means that if there is a true match in the database, the algorithm will likely find it. These competing requirements resemble the balancing act between Type I and Type II errors for a simple statistical hypothesis test.

The preceding method might be enhanced by using a Bayesian approach, as described briefly in Sidebar 13.5.

SIDEBAR 13.5: BAYESIAN METHODS FOR PATTERN RECOGNITION

In using a Bayesian approach for pattern recognition we specify, for each image in the database, a *prior probability* of a match to the image under scrutiny – before considering any information about the image, as gleaned from the database. Thus, if we had added information on subjects in the database (beyond their images) – for instance on age, gender, or date of birth – we would use such information to assign higher prior probabilities to individuals whose characteristics seem more likely to match those of the person under scrutiny. If, on the other hand, we have no such prior information, we might say each image in the database is equally likely to match the image under scrutiny and make each of the prior probabilities equal.

Using a statistical model we would then, for each image in the database, calculate the probability (or *likelihood)* of observing the biometric features of the subject under scrutiny under the assumption that the two subjects are the same. Using Bayes' Rule, the posterior probability that the two images actually match is proportional to the product of the likelihood and the prior probability. This posterior probability is then calculated for each image in the database, and those with the highest posterior probabilities are identified as the most likely matches to the person to be identified.

13.5.4 Further Information

McCulloch, Laading, Wilson, and Johnson (1996) give a detailed face recognition example. See Duda, Hart, and Stork (2000) for a more general discussion of pattern recognition, and Winkler (2006) for other examples of image analysis discussed from a statistical perspective.

13.6 OTHER APPLICATION AREAS

13.6.1 Health Care

Patient Care There is a continuing need worldwide to improve patient care. Statistics plays an important role in such areas as the:

- Comparison of the effectiveness of alternative treatments of a disease and of different approaches to patient care, such as, for example, the length of hospital stays after childbirth.
- Assessment of waiting times in doctors' offices or in hospitals and identification of their likely causes.
- Evaluation and reduction of errors in filling prescriptions, administering medications, performing surgeries, or in diagnoses.

Occasionally, such applications call for conducting a designed experiment. Most of the time, however, experimentation is inappropriate and we must, instead, rely on data from observational studies. As always, planning to get good data from such studies is essential.

Process Improvement Health care organizations, due to their highly process oriented nature, present important opportunities for improvement similar to those for businesses.

The scheduling of doctors in a hospital emergency room provides a typical example. How many doctors are likely to be needed at any given time is predicted from past data on incoming patient flow. Doctors are then scheduled, taking into account their preferences and the need to avoid fatigue. Scheduling outpatient procedures and physical resource planning (e.g., assignment of scanning equipment) are some additional applications.

Further Reading Health care practitioners have come to use quality improvement methodologies, such as PDSA, as well as statistical and operations research methods. Relevant articles, for example, Van Den Heuvel, Does, and Bisgaard (2005), appear in *Six Sigma Forum Magazine* and also *Quality Progress.* At a more technical level, Woodall (2006) presents a detailed review of issues that arise in the application of control charts to health care.

13.6.2 Epidemiological Studies

Epidemiology is the branch of medicine that deals with the study of the causes, distribution, and control of disease in populations. Examples of important issues that are addressed using a statistical approach are:

- Identification of an impending epidemic by, for example, studying occurrence rates in different regions of a country or continent, or the entire world, over time.
- Developing hypotheses concerning likely causes of diseases by relating these to health outcomes. Such evaluations may assess the impact of environmental conditions by geographical region, for example, correlating air quality to cancer incidence in different localities. Or they may study personal habits and be conducted on individuals, such as relating the incidence of cancer to smoking.
- Assessing the impact of different preventative strategies, such as publicity about methods for prevention, on occurrence of a particular disease.

Evaluations such as the preceding rely heavily on observational studies. They also provide important opportunities for using a proactive approach for disease prevention.

(See Kahn and Sempos 1989 for a discussion of statistical methods in epidemiology, and Ahrens and Pigeot 2007 and Kleinbaum, Sullivan, and Barker 2007 for more general discussions.)

13.6.3 National Security

Banks (2005) considers the application of statistics to national defense and homeland security. He describes eight opportunity areas such as:

- Probabilistic risk assessment: This balances cost of investment against risk reduction. Statistics has long been used in this area, dating back to studies of the safety of nuclear reactors.
- Cybersecurity: This aims for early detection of sabotage directed at defense networks, financial services, public utilities, etc. Automated statistical algorithms are used to detect system anomalies quickly and adopt appropriate countermeasures.
- Privacy protection: This includes determining how much noise to add to data to protect personal identity, and other ways of balancing the needs for security and confidentiality.

Banks' main point, however, is that, at the time of writing, the statistics profession in the United States was not yet sufficiently involved in such activities.

13.6.4 Space Exploration

Space exploration presents special challenges since again human life may be at stake. Ensuring high reliability prior to launch, especially for "manned" space vehicles, is critical. Improper analysis of the available data on the susceptibility to failure of O-rings in cold weather led to failure to postpone the fatal 1986 Challenger space shuttle launch (see Dalal, Fowlkes, and Hoadley 1989).

Statistics is also used to assess the risk caused by external factors, such as ice and insulation foam striking a shuttle during launch, for addressing the uncertainty in the physical models that predict where a launched space vehicle will be at a particular point in time, and for many other applications. Also, Rhew and Parker (2007) describe a study in which the design of experiments was used to "efficiently identify and rank the primary contributors to the integrated drag over the vehicle's ascent trajectory" for the launch alert system of the NASA Crew Exploratory Vehicle.

13.6.5 Oil Exploration

Exploration for oil (and for valuable minerals) may provide the ultimate application of decision-making that balances statistical Type I errors (claiming that oil is present when it is not) and Type II errors (failing to detect oil if it is there). Moreover, if it seems likely that there is oil, we need to estimate the quantity and exact location. These assessments typically involve the use of relationships established from past data between physically relevant predictor variables and oil discovery. Such data are of varying relevance and precision; thus, the evaluations might involve using an appropriate weighting scheme and/or Bayesian methods.

(See Agterberg and Bonham-Carter 1999 and Harbaugh, Doveton and Davis 1977 for further discussion.)

13.6.6 Environmental Studies

Environmental concerns are becoming increasingly important to business and industry. Issues may arise as a result of the manufacturing process itself, such as the discharge of undesirable liquids into the ground or river. Or they may be the consequence of the use of products, such as automobiles or locomotives that emit pollutants. Traditionally, statistics was used to address questions such as "does the discharge exceed the 'allowable' limits established by the company, the industry or a regulating authority?" Sometimes, it was also desired to determine whether "hot spots" or highly contaminated areas were present and, if so, to identify their locations. Over time, the focus has shifted to a more proactive approach; that is, to minimize discharges and to find effective pollutant control methods.

Statistical methods are typically used in environmental studies to:

- Model and monitor (using control charts) pollution at a particular location over time.[12]
- Evaluate seasonal patterns and long-term trends in pollution levels using time series analysis.
- Assess the precision of the instruments used to quantify pollution using GRR studies.
- Characterize the geographical distribution of environmental pollution using spatial methods (see Cressie 1993).

Environmental studies often involve analysis of censored data due to the pollutant level being below (left-censored) or sometimes above (right-censored) the capability of the measurement instrument.

(See Gibbons and Coleman 2001, and Gilbert 1987 for further discussion.)

13.6.7 Chemometrics

Chemometrics is a specialized area that deals with the analysis and modeling of chemical data. A common problem is identifying the composition of an unknown material based on the substance's spectra (e.g., reflectance at various wavelengths). Modern instrumentation often generates large amounts of spectral data – a million or more data points per sample are not atypical. Such data are then used to compare

12. A lognormal distribution is often used to represent discharge amounts.

the spectra of the unknown material with those of known materials in order to determine its composition.

Readily available software packages allow chemometric studies to be carried out directly by a laboratory chemist. These typically include multivariate statistical analysis and modeling methods. (See Beebe, Pell, and Seasholtz 1998 for more information.)

13.7 EMERGING AREAS: A GLANCE INTO THE FUTURE

We always overestimate the change that will occur in the next two years and underestimate the change that will occur in the next ten.
—Bill Gates

What are some of the new areas in business and industry in which statistics will be important in the future? We briefly consider this question, carrying forward our discussion in Section 2.4 and elsewhere. (For more detailed assessments, see Lindsay, Kettenring, and Siegmund 2003, 2004; Raftery, Tanner, and Wells 2001; Steinberg 2008; Straf 2003).

13.7.1 Advances in Information Science

Computers in the future may weigh no more than 1.5 tons.
—Popular Mechanics, 1949

The interplay between information science and statistics was discussed briefly in Section 2.4.1 (see also Wegman, Said, Scott, and Solka 2009). Advances in data gathering and in computing continue to lead to new areas of application, as well as new statistical methods.

Data Gathering The new opportunities presented by the continuing advances in data-capturing capabilities and increased access to databases have been highlighted throughout this book. We have stressed the importance of careful early planning of the data-gathering process to get good data for statistical analysis. This will continue to be critical as added capabilities for using data emerge.

Methods currently in use for tracking tagged units, ranging from product inventory to cattle, are, for example, being extended to such diverse applications as product usage assessment, customer behavior evaluation, fraud detection, field monitoring of products and evaluation of Internet traffic. The advent of radio frequency identification devices (RFID), in particular, calls for "an essential data management infrastructure" to effectively utilize the resulting "data avalanche" (quote from RFID Study Group at Pennsylvania State University 2006).

Computations Advances in computer capabilities and in the speed and ease of calculations continue to revolutionize the way statistics is applied in addressing problems in business and industry. The following are just a few examples of such computer-intensive methods:

- Modern Bayesian methods (see Gelman, Carlin, Stern, and Rubin 2003).
- Markov Chain Monte Carlo simulation methods (see Kass, Carlin, Gelman, and Neal 1998).
- Experimentation on large-scale computational models (e.g., CAD/CAM models involving thousands of parameters; see Lindsay, Kettenring, and Siegmund 2003; 2004).
- New clustering and classification methods for data-mining applications (see Hastie, Tibshirani, and Friedman 2003).

In addition, the increasing availability of user-friendly statistical software and readily accessible information about statistics continues to make statistical methods increasingly accessible to nonstatisticians. This will further challenge statisticians to develop tools that are as simple as possible to use – and maximally robust to potential misuse.

13.7.2 Advances in Application Areas

Bioinformatics and Statistical Genetics The rapidly developing field of bioinformatics has been loosely referred to as any use of computers to handle biological information (see Tramontano 2006). Gene mapping, a key application area, has "the ultimate goal ... to identify the genes that play important roles in the inheritance of particular traits and to explain the role of these genes in relation to one another and ... the environment" (see Siegmund and Yakir 2007). This typically requires the assembly and statistical analysis of large sets of DNA and other information. Genetic research experiments, for example, now routinely collect over 300,000 measurements per subject.

Nanotechnology The ability to manipulate material characteristics at the individual molecule level is leading to important changes in the way products are designed, manufactured, and used. The routine clinical use of nanotechnology to deliver drugs to different locations within the body is an active research area. The need to build high-quality and high-reliability products on a nanoscale creates statistical challenges that seem to resemble those faced, early on, by the semiconductor industry in raising yield. (See Dasgupta, Ma, Joseph, Wang, and Wu 2007; Nembhard 2007; Rue 2006 for further discussion.)

Automated Tracking Implanted health-monitoring devices to provide early warning of, say, an impending heart attack hold much promise. As in other applications of remote monitoring, a major contribution of statistics is to help establish action thresholds that provide the desired balance between maximizing detection sensitivity and minimizing the false alarm rate.

Other Areas People all over the world have come to depend on increasingly complex systems (e.g., economy, health, and transportation) that utilize statistics. For example, dynamic travel routing might be extended from using information on current congestion to using predictions of future conditions. These might employ a statistical algorithm that uses data on past traffic (by time of day and day of week), the impact of (forecasted) weather and of special local events (e.g., a football game) to predict traffic volume.

Statistical methods also help address such "hot topics" as climate change and the assessment of world energy reserves.

In addition, the continuing flow of government and industry regulations calls for new programs of disciplined data gathering and statistical analysis. The Sarbanes–Oxley Act, a U.S. federal law enacted in response to various corporate and accounting scandals, provides an example. Thus, Faltin and Faltin (2003) illustrate how a Six Sigma approach, employing publicly available data and statistical tools, could have predicted WorldCom's ultimate bankruptcy.

Statistics will, undoubtedly, continue to make important contributions to addressing new technical and societal challenges involving business and industry as these emerge; in turn, new applications will stimulate further powerful developments in statistics.

13.8 MAJOR TAKEAWAYS

- Applications of statistics in the food, beverage, and related industries are characterized by:
  - The need (especially for processed foods) to minimize the variability that customers experience from one purchase to the next.
  - Data for product assessment are obtained from consumers, instruments and experts and are often ordinal or categorical.
  - The applicability of mixture designs in planned experiments.
  - Manufacturing issues similar to those encountered for many continuous processes.
  - High concern with food safety and spoilage.
- The semiconductor business is characterized by its many process steps and efforts to improve yield. It is heavily data-intensive and provides numerous opportunities for the use of statistics. Applications include:
  - Studies to understand the impact of process variables upon performance
  - Monitoring and controlling critical process variables
  - Product life assessment
  - Product burn-in
- The communications industry, another highly data-intensive business, has traditionally been a heavy user of statistical methods. Typical problems involve usage assessments, reliability and service improvement, and business assessments. The emergence of the Internet has led to new applications.
- Statistical methods play a prominent role in image analysis for medical, security and other applications including:

 - Restoration of images, such as those obtained in digital photography, to remove defects.
 - Pattern recognition in, for example, diagnosing diseases by medical scanning or identifying criminals by matching their pictures with those in an available database.

- Statistics also makes major contributions in many other application areas that often involve business and industry, including health care, epidemiological studies, national security, space exploration, oil exploration, environmental studies and chemometrics (the analysis and modeling of chemical data).
- Technological advances in both information sciences and application areas continue to lead to new applications in bioinformatics and statistical genetics, nanotechnology, automated tracking, and numerous other areas, and to provide the impetus for the development of powerful new statistical methods.

DISCUSSION QUESTIONS

General

1. Why do the food, beverage and related industries "provide an ultimate example of the need for robust design," as stated in Section 13.2.2?
2. Assume that you are the manufacturer of a popular brand of chocolate chip cookies. Propose a program to help ensure product consistency.
3. Assume that you are the manufacturer of a canned fish product. Propose a program to help ensure product safety.
4. How does the "relatively long elapsed time to conduct even a single production run" (Sidebar 13.3) in semiconductor manufacture impact the statistical evaluations?
5. Practices in use in the semiconductor industry, such as end-of-line inspection, accepting relatively low product yield, and burn-in, tend to be contrary to the goals of ensuring *up front* high quality and reliability. Discuss this apparent disconnect and why and when it might be appropriate for the semiconductor industry.
6. A telephone company has asked you to develop a usage assessment system to provide information to improve service. What would you propose?
7. Suggest some applications of statistical image analysis beyond those presented in this chapter.
8. What do you think are the most important contributors to uncertainty in a face recognition system?
9. Suggest some measurable characteristics associated with the correct administration of prescription drugs in a hospital.
10. Applications such as pattern recognition (Section 13.5.3) and oil exploration (Section 13.6.5) call for developing a scheme that strikes the best possible balance between incorrectly identifying a signal when it is not present (Type I error) versus failing to identify a signal when one exists (Type II error). Explain this comment and suggest what might be done to increase the chances of making the right call.

11. Suggest and briefly describe some likely future applications of statistics in business and industry beyond those discussed in this chapter.

Technical

1. We state in Section 13.2.2 that to achieve homogeneity in processing roasted peanuts, "measurements on such factors as sugar content and moisture are obtained on crop samples during harvesting. This information is then used to set processing (e.g., roasting) conditions." Explain this concept and propose a plan for implementation.
2. In the discussion of the need for reducing the variability in the processing of flour in Section 13.2.2, we state that "Homogeneity in the final product is attained by adjusting the mixing proportions, based upon product or process measurements." Propose a system for doing this.
3. Propose a GRR study to help ensure consistency in judgment over time and between tasters for end-of-line wine tasting.
4. Obtain further details about the HACCP system for proactively ensuring food safety (Section 13.2.6) and discuss how statistics fits in.
5. In Section 13.3.6, we state that "Statistical concepts help determine how long units should be burned in, based upon cost considerations." How would you do this?
6. Variance components analysis was suggested in Sidebar 13.4 "to quantify the relative contribution to total variability of, say, lots, wafers, positions within a wafer, individual units and measurement error." This suggests that each of these are random, rather than fixed, effect (analysis of variance) variables. Discuss this assumption, especially with regard to wafer position, and how the analysis is changed if the assumption is not satisfied.
7. Consider a Bayesian interpretation of the face recognition example (Section 13.5.3). Given a defined set of N prior probabilities and N likelihood values from a database of N images, define the posterior probability of a true match. How can you adjust the prior probabilities to change system sensitivity and specificity?
8. What kind of statistical models might you use to model different contributors to uncertainty in a face recognition system?
9. Pattern recognition (Section 13.5.3) and chemometrics (Section 13.6.7) both address problems that involve comparison of an unknown sample against known ones to find a match. Discuss the similarities and differences in the two problems and the statistical methods used to address them.
10. We state in a footnote to Section 13.6.6 that in modeling pollution, "a lognormal distribution is often used to represent discharge amounts." Why might a lognormal distribution be appropriate in this situation?
11. Select one or more of the "Other Applications" described in Section 13.6. Describe statistical evaluations that might likely be performed.

REFERENCES

Abernethy, R.B., J.E. Breneman, C.H. Medlin and G.L. Reinman (1983). *Weibull Analysis Handbook*, Technical Report AFWAL-TR-83-2079, Washington, DC: Air Force Wright Aeronautical Laboratories.

Adjengue, L., S. Yacout and O. Ilk (2007). "Parameter Estimation for Condition Based Maintenance with Correlated Observations," *Quality Engineering*, Volume 19, Number 3, 197–206.

Agresti, A. (2002). *Categorical Data Analysis*, Second Edition, New York, NY: Wiley.

Agresti, A. (2007). *An Introduction to Categorical Data Analysis*, Second Edition, New York, NY: Wiley.

Agterberg, F.P., and G.F. Bonham-Carter (1999). "Logistic Regression and Weights of Evidence in Mineral Exploration," *Proceedings of APCOM'99 Computer Applications in the Mineral Industries 28th International Symposium*, Golden, CO.

Ahrens, W., and I. Pigeot (Editors) (2007). *Handbook of Epidemiology*, New York, NY: Springer.

Akao, Y. (Editor) (1990). *Quality Function Deployment: Integrating Customer Requirements into Product Design*, Portland, OR: Productivity Press.

Alexander, W.P., S.D. Grimshaw, G.R. McQueen and B.A. Slade (2002). "Some Loans are More Equal than Others: Third Party Originations and Defaults in the Subprime Mortgage Industry," *Real Estate Economics*, Volume 30, Number 4, 667–697.

Andersen, B.A. (2007). *Business Process Improvement Toolbox*, Second Edition, Milwaukee, WI: ASQ Quality Press.

Anderson, M.J. (2004). "Achieving Breakthroughs in Non-Manufacturing Processes via Design of Experiments (DOE) Inspirational Examples," Technical Memo, Minneapolis, MN: Stat-Ease, Inc.

Armitage, P., and T. Colton (Editor) (2005). *Encyclopedia of Biostatistics*, Second Edition, New York, NY: Wiley.

Ascher, H., and H. Feingold (1984). *Repairable Systems Reliability*, New York, NY: Marcel Dekker.

Ash, D., and S. Meester (2002). "Best Practices in Reject Inferencing," *Conference on Credit Modeling and Decisioning*, Wharton FIC, University of Pennsylvania, PA.

Atchison, N., and R. Ross (1999). "Integrated Yield Management," Series of articles in *Semiconductor International*.

Automotive Industry Action Group (1990). *Measurement Systems Analysis Reference Manual*, Detroit, MI: Automotive Industry Action Group.

Banks, D. (2005). "Statistics and Homeland Defense: How Are We Meeting the New Challenge?" *Amstat News*, September Issue, 49–51.

The Role of Statistics in Business and Industry, by Gerald J. Hahn & Necip Doganaksoy

Baron, M., A. Takken, E. Yashchin and M. Lanzerotti (2004). "Defects Modeling and Yield Forecasting in Semiconductor Manufacturing," *Proceedings of 2004 Joint Statistical Meetings.*

Barone, S. (2006). "Statistics-Driven Development of OBD Systems: An Overview," *Quality and Reliability Engineering International*, Volume 22, Issue 6, 615–628.

Becker, R.A., L.A. Clark and D. Lambert (1998). "Events Defined by Duration and Severity, With an Application to Network Reliability," (with discussion), *Technometrics*, Volume 40, Number 3, 177–194.

Bedford, T., and R. Cooke (2001). *Probabilistic Risk Analysis: Foundations and Methods*, Cambridge, UK: Cambridge University Press.

Beebe, K.R., R.J. Pell and M.B. Seasholtz (1998). *Chemometrics: A Practical Guide*, New York, NY: Wiley.

Belsley, A.D., E. Kuh and R.E. Welch (1980). *Regression Diagnostics: Identifying Influential Data and Sources of Collinearity*, New York, NY: Wiley.

Benhabib, B. (2003). *Manufacturing: Design, Production, Automation, and Integration*, New York, NY: Marcel-Dekker.

Beran, J. (2003). *Statistics in Musicology*, Boca Raton, FL: Chapman & Hall/CRC.

Berry, M.J.A., and G. Linoff (1997). *Data Mining Techniques for Marketing, Sales, and Customer Support*, New York, NY: Wiley.

Bisgaard, S. (1997). "Designing Experiments for Tolerancing Assembled Products," *Technometrics*, Volume 39, Number 2, 142–152.

Bisgaard, S. (2005). "Innovation, ENBIS and the Importance of Practice in the Development of Statistics," *Quality and Reliability Engineering International*, Volume 21, Issue 5, 429–438.

Bisgaard, S., and J. DeMast (2006). "After Six Sigma – What's Next?" *Quality Progress*, Volume 39, Number 1, 30–36.

Bisgaard, S., and M. Kulahci (2006). "The Application of Principal Component Analysis for Process Monitoring," *Quality Engineering*, Volume 18, Number 1, 95–103.

Bisgaard, S., and M. Kulahci (2007). "Practical Time Series Modeling," *Quality Engineering*, Volume 19, Number 3, 253–262.

Bisgaard, S., and D.M. Steinberg (1997). "The Design and Analysis of 2^{k-p} Prototype Experiments," *Technometrics*, Volume 39, Number 1, 52–62.

Bluhm, C., L. Overbeck and C. Wagner (2002). *An Introduction to Credit Risk Modeling*, Boca Raton, FL: Chapman & Hall/CRC.

Boland, P.J. (1984). "A Bibliographical Glimpse of William Sealy Gosset," *The American Statistician*, Volume 38, Number 3, 179–183.

Borenstein, M. (1994). "The Case for Confidence Intervals in Controlled Clinical Trials." *Controlled Clinical Trials*, Volume 15, Number 5, 411–428.

Box, G.E.P., and N.R. Draper (1969). *Evolutionary Operation – A Statistical Method for Process Improvement*, New York, NY: Wiley.

Box, G.E.P., et al. (2006). *Improving Almost Anything: Ideas and Essays*, New York, NY: Wiley.

Box, G.E.P., J.S. Hunter and W.G. Hunter (2005). *Statistics for Experimenters: Design, Innovation, and Discovery*, Second Edition, New York, NY: Wiley.

Box, G.E.P., and A. Luceño (1997). *Statistical Control: By Monitoring and Feedback Adjustment*, New York, NY: Wiley.

Box, J.F. (1981). "Gosset, Fisher, and the t Distribution," *The American Statistician*, Volume 35, Number 2, 61–66.

Breiman, L., J.H. Friedman, R.A. Olshen and C.J. Stone (1984). *Classification and Regression Trees*, Belmont, CA: Wadsworth.

Breyfogle III, F.W. (2003). *Implementing Six Sigma: Smarter Solutions Using Statistical Methods*, Second Edition, New York, NY: Wiley.

Britz, G.C., D.W. Emerling, L.B. Hare, R.W. Hoerl, S.J. Janis and J.E. Shade (2000). *Improving Performance Through Statistical Thinking*, Milwaukee, WI: ASQ Quality Press.

Bronson, R., and G. Naadimuthu (1997). *Schaum's Outline of Operations Research*, Second Edition, New York, NY: McGraw-Hill.

Burdick, R.K., C.M. Borror and D.C. Montgomery (2005). *Design and Analysis of Gauge R&R Studies*, ASA-SIAM Series on Statistics and Applied Probability, Philadelphia, PA: SIAM.

Burdick, R.K., and F.A. Graybill (1992). *Confidence Intervals on Variance Components*, New York, NY: Marcel Dekker.

Byrne, D., and J. Quinlan (1993). "Robust Function for Attaining High Reliability at Low Cost," *Proceedings of the Annual Reliability & Maintainability Symposium*, 183–191.

Cahill, M., F. Chen, D. Lambert, J. Pinheiro and D. Sun (2002). "Detecting Fraud in the Real World," Chapter 26 in *Handbook of Massive Datasets*, edited by J. Abello, P.M. Pardalos, and M.G. Resende. Dordrecht, The Netherlands: Klewer Academic Publishers.

Campbell, S. (1998). *Statistics You Can't Trust: A Friendly Guide to Clear Thinking About Statistics in Everyday Life*, Parker, CO: Think Twice Publishing.

Carlyle, W.M., D.C. Montgomery and G.C. Runger (2000). "Optimization Problems and Methods in Quality Control and Improvement," (with discussion), *Journal of Quality Technology*, Volume 32, Number 1, 1–31.

Cawse, J.N. (2000). *Experimental Strategies for Combinatorial and High Throughput Materials Development*, Technical Report 2000CRD081, Niskayuna, NY: GE Global Research.

Cawse, J.N. (Editor) (2002). *Experimental Design for Combinatorial and High Throughput Materials Development*, New York, NY: Wiley.

Chow, S.C. (Editor) (2003). *Encyclopedia of Biopharmaceutical Statistics*, Second Edition, New York, NY: Marcel Dekker.

Chowdhury, S. (2002). *Design for Six Sigma: The Revolutionary Process for Achieving Extraordinary Profits*, Chicago, IL: Dearborn Trade.

Churchill, G.A., and D. Iacobucci (2004). *Marketing Research: Methodological Foundations*, Ninth Edition, Cincinnati, OH: South-Western College Publishing.

Coleman, D.E., and D.C. Montgomery (1993). "A Systematic Approach to Planning for a Designed Industrial Experiment," *Technometrics*, Volume 35, Number 1, 1–12.

Condra, L.W. (1993). *Reliability Improvement with Design of Experiments*, New York, NY: Marcel Dekker.

Cornell, J.A. (2002). *Experiments with Mixtures: Designs, Models, and the Analysis of Mixture Data*, Third Edition, New York, NY: Wiley.

Cox, N.D. (1986). *How to Perform Statistical Tolerance Analysis*, ASQC Basic References in Quality Control, Volume 11, Milwaukee, WI: ASQC.

Craig, J.D. (1918). "The Relationship between the Actuary and the Statistician," *Proceedings of the Casualty and Actuarial and Statistical Society*, Volume 5, Number 11, 1–9.

Cressie, N.A.C. (1993). *Statistics for Spatial Data*, New York, NY: Wiley.

Creveling, C.M. (1997). *Tolerance Design: A Handbook for Developing Optimal Specifications*, Reading, MA: Addison-Wesley.

Crow, L.H. (1974). "Reliability Analysis for Complex Repairable Systems," in *Reliability and Biometry*, edited by F. Proschan and R.J. Serfling, pp. 379–410, ASA-SIAM Series on Statistics and Applied Probability, Philadelphia, PA: SIAM.

Czitrom, V. (2003). "Guidelines for Selecting Factors and Factor Levels for an Industrial Designed Experiment," Chapter 1 in *Handbook of Statistics*, Volume 22: Statistics in Industry, edited by R. Khattree and C.R. Rao. Amsterdam, The Netherlands: Elsevier Science B. V.

Czitrom, V., and P.D. Spagon (1997). *Statistical Case Studies for Industrial Process Improvement*, ASA-SIAM Series on Statistics and Applied Probability, Philadelphia, PA: SIAM.

D'Agostino, R.B., and M. Weintraub (1995). "Meta-Analysis: A Method for Synthesizing Research," *Clinical Pharmacology & Therapeutics*, Volume 58, Number 6, 605–616.

Dalal, S.R., E.B. Fowlkes and B. Hoadley (1989). "Risk Analysis of the Space Shuttle: Pre-Challenger Prediction of Failure," *Journal of the American Statistical Association*, Volume 84, Number 408, 945–977.

Dalal, S.R., and C.L. Mallows (1998). "Factor-Covering Designs for Testing Software," *Technometrics*, Volume 40, Number 3, 234–243.

Dasgupta, T., C. Ma, V.R. Joseph, Z.L. Wang and C.F.J. Wu (2007). "Statistical Design and Modeling in Nanostructure Analysis," *Joint Newsletter of the Sections on Physical and Engineering Sciences and Quality and Productivity of the American Statistical Association*, Volume 15, Number 1, 11–13.

Davenport, T.H., and J.E. Short (1990). "The New Industrial Engineering: Information Technology and Business Process Redesign," *Sloan Management Review*, Volume 31, Number 4, 11–27.

Del Castillo, E. (2002). *Statistical Process Adjustment for Quality Control*, New York, NY: Wiley.

Deming, W.E. (1975). "On Probability As a Basis for Action," *The American Statistician*, Volume 29, Number 4, 146–152.

Deming, W.E. (1986). *Out of the Crisis*, Cambridge, MA: MIT Center for Advanced Engineering Study.

Deming, W.E. (1993). *The New Economics for Industry, Government, Education*, Second Edition, Cambridge, MA: The MIT Press.

Derringer, G., and R. Suich (1980). "Simultaneous Optimization of Several Response Variables," *Journal of Quality Technology*, Volume 12, Number 4, 214–219.

Doganaksoy, N. (2000). "Assessment of Impact of Measurement Variability in the Presence of Multiple Sources of Product Variability," *Quality Engineering*, Volume 13, Number 1, 83–89.

Doganaksoy, N., F.W. Faltin and W.T. Tucker (1991). "Identification of Out of Control Quality Characteristics in a Multivariate Manufacturing Environment," *Communications in Statistics – Theory and Methods*, Volume 20, Number 9, 2775–2790.

Doganaksoy, N., and G.J. Hahn (1994). "Moving from Every Lot Inspection to Audit Sampling," *Journal of Quality Technology*, Volume 26, Number 4, 261–273.

Doganaksoy, N., and G.J. Hahn (1996). "Evaluating the Potential Impact of Blending on Product Consistency," *Journal of Quality Technology*, Volume 28, Number 1, 51–60.

Doganaksoy, N., G. Hahn and W.Q. Meeker (2006a). "Assuring Product Reliability and Safety," Chapter 21 in Statistics: A Guide to the Unknown, edited by R. Peck, G. Casella, G.W. Cobb, R. Hoerl and D. Nolan. Fourth Edition, Belmont, CA: Duxbury Press.

Doganaksoy, N., G.J. Hahn and W.Q. Meeker (2006b). "Improving Reliability Through Warranty Data Analysis," *Quality Progress*, Volume 39, Number 11, 63–67.

Doganaksoy, N., and W. Nelson (1998). "A Method to Compare Two Samples of Recurrence Data," *Life Data Analysis*, Volume 4, Number 1, 51–63.

Doganaksoy, N., J. Schmee and M. VanDeven (1996). "Process Monitoring with Multiple Product Grades," *Journal of Quality Technology*, Volume 28, Number 3, 346–355.

Doganaksoy, N., and M. VanDeven (1997). "Process Monitoring with Multiple Products and Production Lines," *Quality Engineering*, Volume 9, Number 4, 689–702.

Duane, J.T. (1964). "Learning Curve Approach to Reliability Monitoring," *IEEE Transactions on Aerospace*, Volume 2, Number 2, 563–566.

Duda, R.O., P.E. Hart and D.G. Stork (2000). *Pattern Classification*, Second Edition, New York, NY: Wiley.

Evans, J.R., and W.M. Lindsay (2007). *Managing for Quality and Performance Excellence*, Seventh Edition, Mason, OH: Thomson/South-Western.

Everitt, B.S. (1987). *Introduction to Optimization Methods and Their Application in Statistics*, London, UK: Chapman and Hall.

Faltin, D.M., and F.W. Faltin (2003). "Toe the Line: No More WorldComs," *Quality Progress*, Volume 36, Number 1, 29–35.

Flanagan-Hyde, P. (2006). "Observational Studies: The Neglected Stepchild in the Family of Data Gathering," *Stats*, Volume 46, Fall 2006, 19–21.

Gaither, C.C., and A.E. Cavazos-Gaither (1996). *Statistically Speaking: A Dictionary of Quotations*, London, UK: Institute of Physics Publishing.

Gelman, A., J.B. Carlin, H.S. Stern and D. Rubin (2003). *Bayesian Data Analysis*, Second Edition, Boca Raton, FL: Chapman & Hall/CRC.

Gelman, A., and D. Nolan (2002). *Teaching Statistics: A Bag of Tricks*, New York, NY: Oxford University Press.

George, M.L. (2003). *Lean Six Sigma for Service*, New York, NY: McGraw-Hill.

George, M.L., D. Rowlands and B. Kastle (2003). *What is Lean Six Sigma?* New York, NY: McGraw-Hill.

George, M.L., D. Rowlands, M. Price and J. Maxey (2004). *The Lean Six Sigma Pocket Toolbook: A Quick Reference Guide to 100 Tools for Improving Quality and Speed*, New York, NY: McGraw-Hill.

Gibbons, R.D., and D.E. Coleman (2001). *Statistical Method for Detection and Quantification of Environmental Contamination*, New York, NY: Wiley.

Gigerenzer, G. (2002). *Calculated Risks: How to Know when Numbers Deceive You*, New York, NY: Simon & Schuster.

Gilbert, R.O. (1987). *Statistical Methods for Environmental Pollution Monitoring*, New York, NY: Wiley.

Graves, S., and T. Menten (1996). "Designing Experiments to Measure and Improve Reliability," Chapter 11 in *Handbook of Reliability Engineering and Management*, edited by W.G. Ireson, C.F. Coombs and R.Y. Moss. Second Edition, New York, NY: McGraw Hill.

Green, P.E., and V. Srivasan (1990). "Conjoint Analysis in Marketing: New Developments with Implications for Research and Practice," *Journal of Marketing*, Volume 54, Number 4, 3–19.

Green, P.E., and J. Wind (1975). "New Ways to Measure Consumers' Judgments," *Harvard Business Review*, Volume 43, Number 4, 107–117.

Group of 30 (1993). *Derivatives: Practices and Principles*, Washington, DC: Group of 30.

Grove, D.M., and T.P. Davis (1992). *Engineering Quality and Experimental Design*, London, UK: Longman.

Groves, R.M., F.J. Fowler, M.P. Couper, J.M. Lepkowski, E. Singer and R. Tourangeau (2004). *Survey Methodology*, New York, NY: Wiley.

Gryna, F.M. (1998). "Market Research and Marketing," Section 18 in *Juran's Quality Handbook*, edited by J.M. Juran and A.B. Godfrey. Fifth Edition, New York, NY: McGraw-Hill.

Gunter, B.H. (1991). "The Use and Abuse of Cpk Revisited," *Quality Progress*, Volume 24, Number 1, 90–94.

Guyton, A.C., and J.E. Hall (2005). *Textbook of Medical Physiology*, Eleventh Edition, Philadelphia, PA: Saunders.

Hahn, G.J. (1977). "Some Things Engineers Should Know About Experimental Design," *Journal of Quality Technology*, Volume 9, Number 1, 13–20.

Hahn, G.J. (1982). "Statistical Assessment of a Process Change," *Journal of Quality Technology*, Volume 14, Number 1, 1–9.

Hahn, G.J. (2002). "Deming and the Proactive Statistician," *The American Statistician*, Volume 56, Number 4, 290–298.

Hahn, G.J., and M.B. Cockrum (1987). "Adapting Control Charts to Meet Practical Needs: A Chemical Processing Application," *Journal of Applied Statistics*, Volume 14, Number 1, 35–52.

Hahn, G.J., and A.F. Dershowitz (1974). "Evolutionary Operation Today – Some Survey Results and Observations," *Applied Statistics*, Volume 23, Number 2, 214–218.

Hahn, G., N. Doganaksoy and R. Hoerl. (2000). "The Evolution of Six Sigma," *Quality Engineering*, Volume 12, Number 3, 317–326.

Hahn, G.J., N. Doganaksoy and C. Stanard (2001). "Statistical Tools for Six Sigma," *Quality Progress*, Volume 34, Number 9, 78–82.

Hahn, G.J., W.J. Hill, R.W. Hoerl and S.A. Zinkgraf (1999). "The Impact of Six Sigma Improvement – A Glimpse into the Future of Statistics," *The American Statistician*, Volume 53, Number 3, 208–215.

Hahn, G.J., and W.Q. Meeker (1991). *Statistical Intervals: A Guide for Practitioners*, New York, NY: Wiley.

Hahn, G.J., and S.S. Shapiro (1967). *Statistical Models in Engineering*, New York, NY: Wiley. Reissued in Wiley Classics Library, 1994.

Haimowitz, I.J., and T.K. Keyes (2002). "Risk Analysis," Chapter 36 in *Handbook of Data Mining and Knowledge Discovery*, edited by W. Klosgen and J.M. Zytkow. New York, NY: Oxford University Press.

Hall, R.W. (1991). *Queuing Methods for Services and Manufacturing*, Englewood Cliffs, NJ: Prentice Hall.

Hamada, M. (1995). "Using Statistically Designed Experiments to Improve Reliability and to Achieve Robust Reliability," *IEEE Transactions on Reliability*, Volume 44, Number 2, 206–215.

Hamada, M., R.J. MacKay and J.B. Whitney (1993). "Continuous Process Improvement with Observational Studies," *Journal of Quality Technology*, Volume 25, Number 2, 77–84.

Hamilton, M. (1959). "The Assessment of Anxiety States by Rating," *British Journal of Medical Psychology*, Volume 32, Number 1, 50–55.

Hand, D.J., and W.E. Henley (1997). "Statistical Classification Methods in Consumer Credit Scoring: A Review," *Journal of the Royal Statistical Society: Series A (Statistics in Society)*, Volume 160, Issue 3, 523–541.

Harbaugh, J.W., J.H. Doveton and J.C. Davis (1977). *Probability Methods in Oil Exploration*, New York, NY: Wiley.

Hare, L.B. (2001). "Chicken Soup for Processes," *Quality Progress*, Volume 34, Number 8, 76–79.

Hare, L.B. (2005). "Linking Statistical Thinking to Six Sigma," *International Journal of Six Sigma and Competitive Advantage*, Volume 1, Number 4, 389–402.

Hare, L.B. (2006). "From Supplier to Customer with Statistical Guidance," *Proceedings of the Biscuit and Cracker Manufacturers' Association's 81 Annual Technical Conference*, Orlando, FL.

Hare, L.B. (2007). "The Ubiquitous Cpk," *Quality Progress*, Volume 40, Number 1, 72–73.

Harrington, E.C. (1965). "The Desirability Function," *Industrial Quality Control*, Volume 21, Number 10, 494–498.

Hastie, T., R. Tibshirani and J.H. Friedman (2003). *The Elements of Statistical Learning*, New York, NY: Springer.

Heckman, J. (1976). "The Common Structure of Statistical Models of Truncation, Sample Selection, and Limited Dependent Variables and a Simple Estimator for Such Models," *Annals of Economic and Social Measurement*, Volume 5, Number 4, 475–492.

Hedges, L.V., and I. Olkin (1985). *Statistical Methods for Meta-Analysis*, New York, NY: Academic Press.

Hildebrand, D.K., J.D. Laing and H. Rosenthal (1977). *Analysis of Ordinal Data*, Newbury Park, CA: Sage.

Hillier, F.S., and G.J. Lieberman (2005). *Introduction to Operations Research*, Eighth Edition, New York, NY: McGraw-Hill.

Hobbs, G.K. (2000). *Accelerated Reliability Engineering: HALT and HASS*, New York, NY: Wiley.

Hoerl, R. (2001). "Six Sigma Black Belts: What Do They Need to Know?" (with discussion), *Journal of Quality Technology*, Volume 33, Number 4, 391–435.

Hoerl, R., and R.D. Snee (2001). *Statistical Thinking: Improving Business Performance*, Pacific Grove, CA: Duxbury Press.

Hosmer, D.W., and S. Lemeshow (2000). *Applied Logistic Regression*, Second Edition, New York, NY: Wiley.

Hoyland, A., and M. Rausand (2004). *System Reliability Theory: Models and Statistics Methods*, Second Edition, New York, NY: Wiley.

Ionnidis, J.P.A., J.C. Cappelleri and J. Lau (1998). "Issues in Comparisons Between Meta-Analysis and Large Trials," *Journal of the American Medical Association*, Volume 279, Number 14, 1089–1093.

Juran, J.M., and A.B. Godfrey (Editors) (1999). *Juran's Quality Handbook*, Fifth Edition, New York, NY: McGraw Hill.

Kahn, H.A., and C.T. Sempos (1989). *Statistical Methods in Epidemiology*, Monographs in Epidemiology and Biostatistics, Volume 12, New York, NY: Oxford University Press.

Kass, R.E., B.P. Carlin, A. Gelman and R.M. Neal (1998). "Markov Chain Monte Carlo in Practice: A Roundtable Discussion," *The American Statistician*, Volume 52, Number 2, 93–100.

Katz, M. (2006). *Study Design and Statistical Analysis: A Practical Guide for Clinicians*, Cambridge, UK: Cambridge University Press.

Kenett, R.S., and D.M. Steinberg (2006). "New Frontiers in the Design of Experiments," *Quality Progress*, Volume 39, Number 8, 61–66.

Kilian, C.S. (1992). *The World of W. Edwards Deming*, Second Edition, Knoxville, TN: SPC Press.

Kim, T., and W. Kuo (1998). "Optimal Burn-in Decision Making," *Quality and Reliability Engineering International*, Volume 14, Number 6, 417–423.

Kirby, K.E., M.G. Leitnaker, C.F. Moore, G.B. Ranney and R.D. Sanders (1990). "Engineering Control Systems and Quality Improvement in Manufacturing," *Quality Progress*, Volume 23, Number 12, 79–82.

Kish, L. (1995). *Survey Sampling*, Second Edition, New York, NY: Wiley.

Kleinbaum, D.G., K. Sullivan and N. Barker (2007). *A Pocket Guide to Epidemiology*, New York, NY: Springer.

Klion, J. (1992). *Practical Electronic Reliability Engineering*, New York, NY: Van Nostrand Reinhold.

Kotz, S., and N.L. Johnson (2002). "Process Capability Indices – A Review, 1992–2000," (with discussion), *Journal of Quality Technology*, Volume 34, Number 1, 2–19.

Lakshminarayanan, M. (2007). "Pharmaceutical Industry Offers Opportunities for Master's Level Statisticians," *Amstat News*, February Issue, 28–29.

Lambert, D., and J. Pinheiro (2001). "Mining a Stream of Transactions for Customer Patterns," Proceedings of the ACM SIGKDD.

Langley, G.J., K.M. Nolan, T.W. Nolan, C.L. Norman and L.P. Provost (1996). *The Improvement Guide: A Practical Approach to Enhancing Organizational Performance*, New York, NY: Jossey-Bass.

Lawless, J.F. (2002). *Statistical Models and Methods for Lifetime Data*, Second Edition, New York, NY: Wiley.

Lea, P., T. Naes and M. Rodbotten (1997). *Analysis of Variance for Sensory Data*, New York, NY: Wiley.

Leake, D.B. (1996). *Case-Based Reasoning: Experiences, Lessons, & Future Directions*, Menlo Park, CA: AAAI Press.

Ledolter, J., and A. Swersey (2007). *Testing 1–2–3: Experimental Design with Applications in Marketing and Service Operations*, Stanford, CA: Stanford University Press.

Levy, D., and S. Brink (2005). *A Change of Heart: How the People of Framingham, Massachusetts, Helped Unravel the Mysteries of Cardiovascular Disease*, New York, NY: Knopf.

Lewis, E.M. (1992). *An Introduction to Credit Scoring*, San Rafael, CA: Fair Isaac and Company.

Li, B., C. Christiansen, J. Gill, T. Sullivan, E. Yashchin and R. Filippi (2006). "Threshold Electromigration Failure Time and its Statistics for Cu Interconnects," *Journal of Applied Physics*, Volume 100, Issue 11, 114516–114525.

Li, B., E. Yashchin, C. Christiansen, J. Gill, R. Filippi and T. Sullivan (2006). "Application of the Three-Parameter Lognormal Distribution in EM Data Analysis," *Microelectronics Reliability*, Volume 46, Number 12, 2049–2055.

Lindsay, B., J. Kettenring and D. Siegmund (2003). "A Report on the Future of Statistics," *Statistical Science*, Volume 19, Number 3, 387–413.

Lindsay, B., J. Kettenring and D. Siegmund (Editors) (2004). *Statistics: Challenges and Opportunities for the Twenty-First Century*, National Science Foundation-American Statistical Association Report.

Lohr, S. (1999). *Sampling: Design and Analysis*, Pacific. Grove, CA: Duxbury Press.

MacGregor, J.F. (1990). "A Different View of the Funnel Experiment," *Journal of Quality Technology*, Volume 22, Number 4, 255–259.

Makuch, W., J. Dodge, J. Ecker, D. Granfors and G. Hahn (1992). "Managing Consumer Credit Delinquency in the US Economy: A Multi-Billion Dollar Management Science Application," *Interfaces*, Volume 22, Number 1, 90–109.

Mann, N.R. (1985). *The Keys to Excellence: The Story of the Deming Philosophy*, Los Angeles, CA: Prestwick Books.

Manos, A. (2006). "Value Stream Mapping – An Introduction," *Quality Progress*, Volume 39, Number 6, 64–69.

Manos, A. (2007). "The Benefits of Kaizen and Kaizen Events," *Quality Progress*, Volume 40, Number 2, 47–48.

Markowitz, H.M. (1952). "Portfolio Selection," *Journal of Finance*, Volume 7, Number 1, 77–91.

Marra, L.J. (1989). "Sharkbite on the SL Submarine Lightwave Cable System: History, Causes, and Resolution," *IEEE Journal of Oceanic Engineering*, Volume 14, Number 3, 230–237.

Mason, R.L., and J.C. Young (2001). *Multivariate Statistical Process Control with Industrial Application*, ASA-SIAM Series on Statistics and Applied Probability, Philadelphia, PA: SIAM.

McCulloch, C.C., J.K. Laading, A.G. Wilson and V.E. Johnson (1996). "A Shape-Based Framework for Automated Image Segmentation," *The American Statistical Association Proceedings of the Section on Bayesian Statistical Science*, 1–6.

Meeker, W.Q., and L.A. Escobar (1998). *Statistical Methods for Reliability Data*, New York, NY: Wiley.

Meeker, W.Q., and M. Hamada (1995). "Statistical Tools for the Rapid Development and Evaluation of High-Reliability Products," *IEEE Transactions on Reliability*, Volume 44, Number 2, 187–198.

Mendenhall, W. (1968). *Introduction to Linear Models and The Design and Analysis of Experiments*, Belmont, CA: Duxbury Press.

Mentch, C.C. (1980). "Manufacturing Process Quality Optimization Studies," *Journal of Quality Technology*, Volume 12, Number 3, 119–129.

Montgomery, D.C. (2004). *Introduction to Statistical Quality Control*, Fifth Edition, New York, NY: Wiley.

Myers, R.H., and D.C. Montgomery (2002). *Response Surface Methodology: Process and Product Optimization Using Designed Experiments*, Second Edition, New York, NY: Wiley.

Myers, R.H., D.C. Montgomery and G.G. Vining (2002). *Generalized Linear Models, with Applications in Engineering and the Sciences*, New York, NY: Wiley.

Nair, V.N. (Editor) (1992). "Taguchi's Parameter Design: A Panel Discussion," *Technometrics*, Volume 34, Number 2, 127–161.

Nelson, L.S. (1988). "Control Charts: Rational Subgroups and Effective Applications," *Journal of Quality Technology*, Volume 20, Number 1, 73–75.

Nelson, W. (1982). *Applied Life Data Analysis*, New York, NY: Wiley.

Nelson, W. (1990). *Accelerated Testing: Statistical Models, Test Plans, and Data Analyses*, New York, NY: Wiley.

Nelson, W. (2003). *Recurrent Events Data Analysis For Product Repairs, Disease Episodes, And Other Applications*, ASA-SIAM Series on Statistics and Applied Probability, Philadelphia, PA: SIAM.

Nembhard, H.B. (2007). "Nanotechnology: A Big Little Frontier for Quality," *Quality Progress*, Volume 40, Number 7, 23–29.

Nunnally, B.K., and J.S. McConnell (2007). *Six Sigma in the Pharmaceutical Industry*, Boca Raton, FL: CRC Press.

O'Connor, P.D.T. (2002). *Practical Reliability Engineering*, Fourth Edition, New York, NY: Wiley.

Ong, M.K. (1999). *Internal Credit Risk Models*, London, UK: Risk Books.

Ott, E.R., E.G. Schilling and D.V. Neubauer (2005). *Process Quality Control: Troubleshooting and Interpretation of Data*, Fourth Edition, Milwaukee, WI: ASQ Quality Press.

Pearl, J. (1988). *Probabilistic Reasoning in Intelligent Systems: Networks of Plausible Inference*, San Francisco, CA: Morgan Kaufmann.

Pearn, W.L., and S. Kotz (2006). *Encyclopedia and Handbook of Process Capability Indices*, Series on Quality, Reliability and Engineering Statistics, Volume 12, Hackensack, NJ: World Scientific.

Pecht, M. (Editor) (1995). *Product Reliability, Maintainability and Supportability Handbook*, Boca Raton, FL: CRC Press.

Peck, R., G. Casella, G. Cobb, R. Hoerl, D, Nolan, R. Starbuck and H. Stern (Editors) (2005). *Statistics: A Guide to the Unknown*, Fourth Edition, Belmont, CA: Duxbury Press.

Peck, R., L.D. Haugh and A. Goodman (Editors) (1998). *Statistical Case Studies: A Collaboration Between Academe and Industry*, ASA-SIAM Series on Statistics and Applied Probability, Philadelphia, PA: SIAM.

Phadke, M.S. (1989). *Quality Engineering Using Robust Design*, Englewood Cliffs, NJ: Prentice-Hall.

Proschan, M.A., and M.A. Waclawiw (2000). "Practical Guidelines for Multiplicity Adjustment in Clinical Trials," *Controlled Clinical Trials*, Volume 21, Number 8, 527–539.

Pugh, S. (1990). *Total Design: Integrated Methods for Successful Product Engineering*, Reading, MA: Addison-Wesley.

Raftery, A.E., M.A. Tanner and M.T. Wells (Editors) (2001). *Statistics in the 21st Century*, Boca Raton, FL: Chapman & Hall/CRC.

Ramaswamy, R. (1996). *Design and Management of Service Processes*, Englewood Cliffs, NJ: Prentice Hall.

Raynes, S., and A. Rutledge (2003). *The Analysis of Structured Securities: Precise Risk Measurement and Capital Allocation*, New York, NY: Oxford University Press.

RFID Study Group At Pennsylvania State University (2006). "Challenges in RFID Enabled Supply Chain Management," *Quality Progress*, Volume 39, Number 11, 23–28.

Rhew, R.D., and P.A. Parker (2007). "A Parametric Geometry Computational Fluid Dynamics (CFD) Study Utilizing Design of Experiments (DOE)," *AIAA 2007-1615, U.S. Air Force Test and Evaluation Days*, Destin, FL.

Rigdon, S.E., and A.P. Basu (2000). *Statistical Methods for the Reliability of Repairable Systems*, New York, NY: Wiley.

Rosenberg, E., and A. Gleit (1994). "Quantitative Methods in Credit Management: A Survey," *Operations Research*, Volume 42, Number 4, 589–613.

Rue, B.L. (2006). "Assuring Product Quality in the Production of Nanoelectronic Components," *Quality Engineering*, Volume 18, Number 4, 477–490.

Ruggeri, F., R.S. Kenett and F.W. Faltin (Editors) (2008). *Encyclopedia of Statistics in Quality and Reliability*, New York, NY: Wiley.

Ryan, T.P. (2000). *Statistical Methods for Quality Improvement*. New York: Wiley.

Sachs, E., A. Hu and A. Ingolfsson (1995). "Run by Run Process Control: Combining SPC and Feedback Control," *IEEE Transactions on Semiconductor Manufacturing*, Volume 8, Number 1, 26–43.

Scheaffer, R.L., W. Mendenhall and R.L. Ott (2005). *Elementary Survey Sampling*, Sixth Edition, Belmont, CA: Duxbury Press.

Schilling, E.G. (1982). *Acceptance Sampling in Quality Control*, New York, NY: Marcel Dekker.

Schilling, E.G. (1998). "Acceptance Sampling," Section 46 in *Juran's Quality Handbook*, edited by J.M. Juran and A.B. Godfrey. Fifth Edition, New York, NY: McGraw-Hill.

Schlager, N. (Editor) (1994). *When Technology Fails: Significant Technological Disasters, Accidents, and Failures of the Twentieth Century*, Detroit, MI: Gale Research.

Senn, S. (1997). *Statistical Issues in Drug Development*, New York, NY: Wiley.

Shewhart, W.A. (1931). *Economic Control of Manufactured Products*, New York, NY: Van Nostrand. Reprinted in 1981.

Shewhart, W.A. (1939). *Statistical Methods from the Viewpoint of Quality Control*, Mineola, NY: Dover Publications. Reprinted, with an introduction by W.E. Deming, in 1986.

Shingo, S. (1986). *Zero Quality Control: Source Inspection and the Poka-Yoke System*, translated by A.P. Dillion, Portland, OR: Productivity Press.

Siegmund, D., and B. Yakir (2007). *The Statistics of Gene Mapping*, New York, NY: Springer.

Smith, W.F. (2005). *Experimental Design for Formulation*, ASA-SIAM Series on Statistics and Applied Probability, Philadelphia, PA: SIAM.

Snee, R. (1983). "Graphical Analysis of Process Variation Studies," *Journal of Quality Technology*, Volume 15, Number 2, 76–88.

Snee, R. (2006). "If You're Not Keeping Score, It's Just Practice," *Quality Progress*, Volume 39, Number 5, 72–74.

Snee, R.D., and R.W. Hoerl (2005). *Six Sigma Beyond the Factory Floor: Deployment Strategies for Financial Services, Health Care, and the Rest of the Real Economy*, Englewood Cliffs, NJ: Prentice Hall.

Snee, R.D., and R.W. Hoerl (2007). "Integrating Lean and Six Sigma – a Holistic Approach," *Six Sigma Forum Magazine*, Volume 6, Number 3, 15–21.

Spendley, W., G.R. Hext and F.R. Himsworth (1962). "Sequential Application of Simplex Designs in Optimization and Evolutionary Operation," *Technometrics*, Volume 4, Number 4, 441–461.

Spirer, H.F., L. Spirer and A.J. Jaffe (1998). *Misused Statistics*, Second Edition, New York, NY: Marcel Dekker.

Standard & Poor's (2004). "Default and Recovery Rates of Corporate Bond Issuers: A Statistical Review of Moody's Ratings Performance, 1920–2003," *Moody's Investor Service, Global Credit Research*, January Issue.

Steinberg, D.M. (Editor) (2008). "The Future of Industrial Statistics," *Technometrics*, Volume 50, Number 2, 103–127.

Steiner, S.H., and R.J. MacKay (2005). *Statistical Engineering: An Algorithm for Reducing Variation in Manufacturing Processes*, Milwaukee, WI: ASQ Quality Press.

Steiner, S.H., and R.J. MacKay (2006). "Statistical Engineering: A Case Study," *Quality Progress*, Volume 39, Number 6, 33–39.

Straf, M.L. (2003). "Statistics: The Next Generation," *Journal of the American Statistical Association*, Volume 98, Number 461, 1–6.

Surak, J.G. (2007). "A Recipe for Safe Food: ISO 22000 and HACCP," *Quality Progress*, Volume 40, Number 10, 21–27.

Taguchi, G. (1986). *Introduction to Quality Engineering*, White Plains, NY: UNIPUB/Kraus International.

Taguchi, G., S. Chowdhuri and W. Yuin (2004). *Taguchi's Quality Engineering Handbook*, New York, NY: Wiley.

Taguchi, G., and D. Clausing (1990). "Robust Quality," *Harvard Business Review*, Volume 68, Number 1, 65–75.

Tague, N.R. (2004). *The Quality Toolbox*, Second Edition, Milwaukee, WI: ASQ Quality Press.

Taleb, N.N. (2007). *The Black Swan: The Impact of the Highly Improbable*, New York, NY: Random House.

Tobias, P.A., and D.C. Trindade (1995). *Applied Reliability*, Second Edition, New York, NY: Van Nostrand Reinhold.

Tramontano, A. (2006). *Introduction to Bioinformatics*, Boca Raton, FL: CRC Press.

Tyagi, R., P. Kalish, K. Akbay and G. Munshaw (2004). "GE Plastics Optimizes the Two-Echelon Global Fulfillment Network at its High Performance Polymers Division," *Interfaces*, Volume 33, Number 5, 359–366.

Ulrich, K.T., and S.D. Eppinger (1995). *Product Design and Development*, Third Edition, New York, NY: McGraw-Hill.

Utts, J.M. (2005). *Seeing through Statistics*, Third Edition, Brooks/Cole, Belmont, CA: Duxbury Press.

Van Den Heuvel, J., R.J.M.M. Does and S. Bisgaard (2005). "Dutch Hospital Implements Six Sigma," *Six Sigma Forum Magazine*, Volume 4, Number 2, 11–14.

Vander Wiel, S.A., W. Tucker, F. Faltin and N. Doganaksoy (1992). "Algorithmic Statistical Process Control: Concepts and an Application," *Technometrics*, Volume 34, Number 3, 286–297.

Waite, L., and M. Gallagher (2000). *The Case for Marriage: Why Married People Are Happier, Healthier, and Better Off Financially*, New York, NY: Doubleday.

Wegman, E., Y.H. Said, D.W. Scott and J.L. Solka (2009). *Wiley Encyclopedia of Computational Statistics*, New York, NY: Wiley (in press).

Weisberg, S. (2005). *Applied Linear Regression*, Third Edition, New York, NY: Wiley.

Wheeler, D.J. (1999). *Understanding Variation: The Key to Managing Chaos*, Second Edition, Knoxville, TN: SPC Press.

Wheeler D.J. (2005). *The Six Sigma Practitioner's Guide to Data Analysis*, Knoxville, TN: SPC Press.

Wheeler, D.J., and D.S. Chambers (1992). *Understanding Statistical Process Control*, Second Edition, Knoxville, TN: SPC Press.

Windsor, S.E. (2003). "Attribute Gage R&R," *Six Sigma Forum Magazine*, Volume 2, Number 4, 23–28.

Winkler, G. (2006). *Image Analysis, Random Fields and Markov Chain Monte Carlo Methods: A Mathematical Introduction*, Second Edition, New York, NY: Springer.

Winston, W.L. (2004). *Operations Research: Applications and Algorithms*, Fourth Edition, Belmont, CA: Thomson Brooks/Cole.

Woodall, W. (2006). "The Use of Control Charts in Health-Care and Public-Health Surveillance," (with discussion), *Journal of Quality Technology*, Volume 38, Number 2, 89–139.

Yashchin, E. (1994). "Monitoring Variance Components," *Technometrics*, Volume 36, Number 4, 379–393.

Zalik, E., J. Walter, G. Kelling and C. James (1996). "RAROC at Bank of America: From Theory To Practice," *Journal of Applied Corporate Finance*, Volume 9, Number 2, 83–93.

Zelen, M. (1959). "Factorial Experiments in Life Testing," *Technometrics*, Volume 1, Number 3, 269–288.

INDEX

The Role of Statistics in Business and Industry, by Gerald J. Hahn & Necip Doganaksoy